FUNDAMENTALS AND APPLICATIONS OF LITHIUM-ION BATTERIES IN ELECTRIC DRIVE VEHICLES

FUNDAMENTALS AND APPLICATIONS OF LITHIUM-ION BATTERIES IN ELECTRIC DRIVE VEHICLES

Jiuchun Jiang
and
Caiping Zhang
Beijing Jiaotong University, China

This edition first published 2015

Registered Office
John Wiley & Sons Singapore Pte. Ltd., 1 Fusionopolis Walk, #07-01 Solaris South Tower, Singapore 138628.

For details of our global editorial offices, for customer services and for information about how to apply for permission to reuse the copyright material in this book please see our website at www.wiley.com.

Library of Congress Cataloging-in-Publication Data

Jiang, Jiuchun.
Fundamentals and applications of lithium-ion batteries in electric drive vehicles / Jiuchun Jiang, Caiping Zhang.
pages cm
Includes bibliographical references and index.
ISBN 978-1-118-41478-1 (cloth)
1. Electric vehicles–Batteries. 2. Lithium ion batteries. I. Zhang, Caiping (Chemist) II. Title.
TL220.J53 2015
629.25′02–dc23

2014040218

Set in 10/12pt Times by SPi Publisher Services, Pondicherry, India
Printed and bound in Singapore by Markono Print Media Pte Ltd

1 2015

Contents

About the Authors

Professor **Jiuchun Jiang** is the Dean of the School of Electrical Engineering at the Beijing Jiaotong University, China. He has more than 17 years research experiences in renewable energy technology, management of advanced batteries and EV infrastructural facilities. He has more than 50 publications and holds 8 patents. His research has contributed to the commercial battery management system (BMS) products. The developed BMS products ranked first in the domestic market in the last three years. He has also designed a number of large scale battery charging stations, such as for the Beijing Olympic Games, the Shanghai World Expo, and the Guangzhou Asian Games. He was the winner of China National Science and Technology Progress Award and Ministry of Education Science and Technology Progress Award.

Caiping Zhang is an associate professor with the National Active Distribution Network Technology Research Center, School of Electrical Engineering, Beijing Jiaotong University. She has more than 10 years research experience in the field of battery modeling and simulation, states estimation, battery charging, and battery control and optimization in electric vehicles. She also does research on the reuse technology of EV used batteries and battery energy storage systems. She has had more than 20 publications in the last five years.

Foreword

Battery management is not a new concept—monitoring and control concepts were proposed as early as the 1960s to improve battery safety. After years of intensive study, it remains a field needing more research. This is not because we did not learn much during the past 50 years, we did. But the subject of study is rapidly changing. The materials and structure of the battery anode, cathode and electrolytes continue to evolve and improve, and the electrochemistry and aging mechanisms also continue to change. The performance and capacity of batteries degrade due to the disordering and deforming of electrode structure, decomposition of the electrolyte, dissolution of metal, dendrite formation, and so on. The relative importance of these mechanisms is battery-chemistry dependent, and the rate of degradation changes significantly with many factors, including operating temperature, charge and discharge rate, and depth of discharge. Finally, these aging mechanisms happen at different timescales, posing challenges to data collection and analysis. The safety incidents of the Boeing Dreamliner battery systems in 2012 remind us that much remains to be done before advanced high energy density battery systems can be used safely and reliably in challenging applications such as aircraft and electric vehicles.

While the interaction among many chemical and physical reactions makes it a challenging task to fully understand battery safety and reliability, model-based battery managing algorithms start to appear, showing excellent potential in engineering applications. This book by Professor Jiuchun Jiang reports his research outcome and contribution made over the last 17 years. Most notable contents included in this book are his work on the lithium-ion battery performance model, methods to estimate lithium-ion battery state of charge, state of energy, and peak power, charging technique, and battery equalization techniques. These functions are critical in the pursuit of safer and more reliable battery systems. After we gain better understanding and confidence, the cost of battery systems will reduce through reduced over-design. All of these are existing barriers for wider adoption of advanced batteries in transportation applications.

I recommend this book not only because of its solid technical content, but also because of the unique role Professor Jiang plays in the development of battery management systems in China. The results reported in this book are based on his extensive experience in designing commercial battery management systems and charging stations used in large demonstration projects held during the Beijing Olympic Games, Shanghai World Exposition, and Guangzhou Asian Games. I think this book is a must-read for anyone who wants to learn more about vehicle lithium-ion battery management technologies developed and used in China.

Huei Peng
Professor, University of Michigan
Director, US-China Clean Energy
Research Center-Clean Vehicle Consortium

Preface

The power battery is the main power source for electric vehicles; its performance has vital influence on the safety, efficiency and economy of electric vehicle operations. Currently power batteries for electric vehicles mainly include lead-acid, nickel cadmium, nickel metal hydride and lithium-ion batteries. For a long time, the lead-acid battery was widely used because of its mature technology, stable performance and low price. However, its disadvantages of low energy density, long charging time, short life, and lead contamination limit its usefulness in electric vehicles. The nickel cadmium battery has been used for its large charge-discharge rate; however, its disadvantages of memory effect and heavy metal contamination cannot be solved. Nickel-metal hydride batteries have been widely applied in hybrid cars for their large charge-discharge rate and they are environmentally friendly. However, their single cell voltage is low and they should not be connected in parallel, restricting their application in electric vehicles. The lithium-ion batteries are widely accepted because of their high voltage platform, high energy density, good cycle performance, and low self-discharge, and are regarded as a good choice for the new generation of power batteries. The lithium-ion battery cathode material can be lithium cobalt oxide, manganese oxide, lithium iron phosphate, nickel manganese cobalt oxide, lithium nickel cobalt aluminum oxide, and so on.

Currently, one of the key factors restricting the development of electric vehicles is that the battery power is not satisfactory; the battery specific energy, specific power, consistency, longevity, and price are not as good as expected. A battery acts as a power system which converts electrical energy and chemical energy. Its operation is very complex because the reactions are related to temperature, accumulated charge-discharge, charge-discharge rate and other factors. The battery management system (BMS) protection mainly ensures that the battery works within reasonable parameters. It detects voltage, current,and temperature of the battery pack and relays this information. It carries out thermal management, balancing control, charge and discharge control, fault diagnosis, and CAN communication. It also estimates the SOC and SOH at the same time.

The BMS needs people who are familiar with both the electrochemical properties of the battery and its electrical applications. It is necessary to write an instruction book since there are not many people with this compound knowledge. This book provides basic theoretical knowledge and practical resource materials to researchers engaged in electric vehicles and lithium-ion battery development and design, and people who work on the battery management system.

In this book we discuss key technologies and research methods for the lithium-ion power battery management system, and the difficulties encountered with it in electric vehicles. The contents include lithium-ion battery performance modeling and simulation; the theory and methods of estimation of the lithium-ion battery state of charge, state of energy and peak power; lithium-ion battery charge and discharge control technology; consistent evaluation and equalization techniques of the battery pack; and battery management system design and application in electric vehicles.

This book focuses on systematically expounding the theoretical connotation and practical application of the lithium-ion battery management systems. Part of the content of the book is directly derived from real vehicle tests. Through comparative analysis of the different system structures the related concepts are made clear and understanding of the battery management system is deepened.

In order to strengthen the understanding, the book makes deep analysis of some important concepts. Using simulation technology combined with schematic diagrams, it gives a vivid description and detailed analysis of the basic concepts, the estimation methods and the battery charge and discharge control principles, therefore the descriptions are intuitive and vivid, readers can have a clear understanding of the principle of battery management system technology and, combined with case analysis, the readers' perceptual knowledge is enhanced.

The contents are summarized as follows:

Chapter 1 is an introduction, which presents the terms, types and characteristics of the power battery, and the functions and key technologies of the battery management system.

Chapter 2 introduces the operating principle, charge and discharge characteristics, model classification and characteristics of the lithium-ion battery, and performance simulation of the equivalent circuit model.

Chapter 3 introduces the definition and estimation methods for battery SOC and SOE.

Chapter 4 introduces the definition and test methods of battery peak power, and the determination of available power for a battery pack.

Chapter 5 introduces lithium-ion battery optimization charging methods, taking charge life and charge time together into account, and expounds battery discharge control technology combined with vehicle operational states, battery SOE and SOC.

Chapter 6 introduces the reasons for inconsistency of a battery pack and battery consistency evaluation parameter indexes, and describes the battery equalization method and strategy.

Chapter 7 introduces the structure of the BMS, the battery parameter collection scheme, logical control and security alarm theory of BMS, and BMS application analysis in electric vehicles.

This book is a group achievement of the faculties and PhD students of the National Active Distribution Network Technology Research Center (NANTEC), Beijing Jiaotong University (BJTU). The book benefits from their hard work in the field of electric vehicle battery management, and tireless efforts to provide the most advanced knowledge and technology over decades. The faculties involved in the preparation are Weige Zhang, Zhanguo Wang, Minming Gong, Bingxiang Sun, Wei Shi, Feng Wen, Jiapeng Wen, Hongyu Guo, and so on. The students involved are Zeyu Ma, Dafen Chen, Xue Li, Fangdan Zheng, Yanru Zhang, and so on. We would like to express our sincere thanks to them all!

Jiuchun Jiang and Caiping Zhang
NANTEC, BJTU
Beijing, China

1

Introduction

1.1 The Development of Batteries in Electric Drive Vehicles

1.1.1 The Goals

Energy and environmental issues have long been challenges facing the world's automotive industry. In recent years, the grim energy and environmental situation around the world has accelerated the strategic transformation of transportation and energy technology, and thus set off a worldwide upsurge of new energy vehicle development. Under the various scenarios depicted in the technology roadmaps of new energy vehicles, hybrid electric vehicles (HEV), battery electric vehicles (BEV), and fuel cell vehicles (FCEV) are generally considered as important development directions for future automotive energy power systems, and have become a high strategic priority of major automobile manufacturers worldwide.

The power battery is an important component of an electric vehicle (EV), directly providing its source of energy. In general, the goals for a powertrain system in EVs are: excellent safety, high specific energy, high specific power, good temperature characteristics, long cycle life, low cost, no maintenance, low self-discharge, good consistency, no environmental pollution, good recoverability, and recyclability. In BEV, the specific energy determines the total driving distance in the pure electric drive mode; the specific power determines the vehicle dynamics, such as the maximum gradeability and the maximum vehicle speed; and the cycle life and the cost of the powertrain system have direct effect on EV manufacture and running costs. For a long time, battery technology has been a bottleneck in the development of EVs; some existing battery technologies have achieved some of these goals, but it is far more challenging to meet all the goals simultaneously [1].

1.1.2 Trends in Development of the Batteries

Power batteries used in EVs basically include nickel-metal hydride and lithium-ion batteries (LIBs). The nickel-metal hydride batteries are widely used in HEVs owing to their high charge-and-discharge rate and environmentally friendly features. However, the application

Fundamentals and Applications of Lithium-ion Batteries in Electric Drive Vehicles, First Edition.
Jiuchun Jiang and Caiping Zhang.

of nickel-metal hydride batteries in EVs remains limited because they have low voltage and are unsuitable for parallel connection. The LIBs, with the advantages of a high voltage performance platform, such as high energy density (theoretical specific capacity reaches 3860 $mAh\,g^{-1}$), environmentally benign features, wide operating temperature range, low self-discharge rate, no memory effect, high efficiency, and long cycle life, have become widely accepted in recent years, and have become one of the most important components for the new generation of EVs.

LIBs can be classified into lithium cobalt oxide, lithium manganate (LMO), lithium iron phosphate (LFP), lithium-polymer, and lithium nickel-manganese-cobalt (NMC) batteries, which are based on positive active materials. The comparisons of various materials are shown in Table 1.1 [2]. Lithium cobalt oxide and nickel acid lithium batteries, developed earlier, have encountered a bottleneck owing to the use of cobalt and nickel, which have high costs and poor consistency. The LMO and LFP batteries have more application opportunities in EVs in recent years, with the progress in technology and enhancement of safety performance; safety no longer being a concern due to the improvement of consistency and elimination of explosion risk. At the Beijing Olympic Games, 50 pure electric buses used LMO batteries as the power system, the Shanghai World Expo and Guangzhou Asian Games, used 60 and 35 units, respectively. A type of 8-ton sanitation truck produced by Foton Motor and a large number of trolleybuses in Beijing also use LMO and LFP batteries as a power source. Furthermore, EVs developed by most automobile manufacturers in China use LFP batteries as the power system, such as the E6 pure electric taxi by BYD, 2008EV, and 5008EV by Hangzhou Zhongtai, "Tongyue" pure electric cars by JAC, Bonbon MINI pure electric cars by Chang-an Automobile, S18 pure electric cars by Chery, and so on. So far, the E6 pure electric taxi by BYD, 2008EV, and 5008EV by Hangzhou Zhongtai, and "Tongyue" pure electric cars by JAC have achieved small-scale mass production and have been put into demonstration operation.

It is noticeable that the LIBs, which have lithium titanate (LTO) as a negative electrode, have attracted wide attention in recent years, because of their wide working temperature range, good ratio characteristics and long cycle life. However, they have been merely experimentally

Table 1.1 Comparisons of different types of LIB.

Category of lithium batteries	Lithium cobalt oxide	Lithium iron phosphate	Lithium manganate	Lithium titanate	Ternary materials	Lithium-polymer
Advantages	Good reversibility, high energy density	Long cycle life, high safety	Rich resources, high safety	Long cycle life, high safety, good rate characteristics	Good cycling performance and good thermal stability	Strong over-charge abilities
Disadvantages	Poor cobalt resource, bad anti-abuse capabilities	Low energy density, poorly conductive	Poor recycling performance in high temperature	Low density, high cost	High cost, complicated manufacturing process	Low density, long cycle life

demonstrated on EVs owing to their low energy density, higher cost, immature bulk production technology, and so on.

1.1.3 Application Issues of LIBs

Although LIBs, with their superior performance, have been widely used in portable devices, they have limited application in EVs, the main reasons being summarized below.

1.1.3.1 Poor Working Environment

1. A large number of large capacity batteries are used through series and parallel connection. In order to reach the corresponding level of voltage, power, and energy, a large number of large-capacity batteries need to be used in EVs through series and parallel connection, which requires high consistency among the battery pack. Additionally, different from an individual battery, grouping management in a battery pack also requires more advanced technology.
2. Large working current and extreme current fluctuation. Figure 1.1 [3] shows the working current, representative cell voltage and the speed of the Beijing Olympic Games EV bus during the acceleration process. It can be seen that the battery current is high (maximum value over 350 A) and changes quickly (the time to change from 300 to 0 A is <0.5 s), which may result in over-discharge and over-heating, as well as the problem of capacity and low energy utilization, and also may cause difficulty for the online estimation of the battery state.
3. Limited space. This may increase the difficulty of the assembly process, heat radiation and cooling ventilation design of battery systems (including batteries, battery management

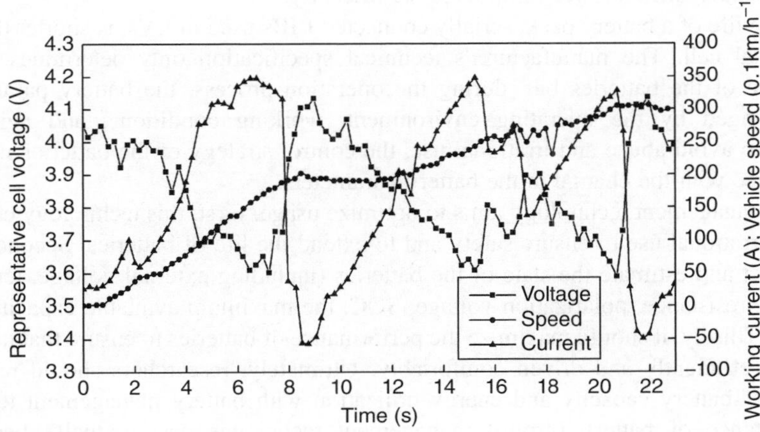

Figure 1.1 Acceleration curve of Olympic EV buses. (Reproduced with permission from Feng Wen, “Study on basic issues of the Li-ion battery pack management technology for Pure Electric Vehicles.”, Beijing Jiaotong University ©2009.)

system (BMS), and protection modules). For example, if the battery works in a high temperature environment for a long time, the decrease in battery capacity will be accelerated, which may even result in thermal runaway and cause safety risks. Further, temperature fluctuation will cause differences between the degradation speed and the self-discharge coefficient, which may lead to accelerated inconsistency of the battery pack, capacity loss, and low energy utilization. Realizing efficient management of the battery pack poses a far more serious challenge in battery research and development.

4. Poor working conditions. Vehicle bumping and shaking requires higher anti-shock and anti-vibration performance; dusty, rainy, and line wear conditions may cause short circuit or other insulation problems.

1.1.3.2 Poor Anti-Abuse Capabilities

The anti-abuse capability of LIBs is insufficient. More specifically, irrational use (such as operation at high or low temperature regularly or for a long time, too high or low state of charge (SOC), over-current, etc.) will substantially shorten the battery life. Such battery abuse may cause battery failure, and even fire, explosion, or other safety problems.

1.1.4 Significance of Battery Management Technology

In order to improve the performance of future LIBs, researchers in the electrochemistry field have conducted further research on LIBs in terms of the electrochemical mechanism, including the effects of temperature [4, 5], voltage, current, and aging on the battery performance [6–8], the influence of over-charge, over-discharge [9], over-current and over-heating [10], and so on. By enhancing the anode and cathode materials, additives, binder, doping and coating, electrolyte formula and technology, the energy density, power density, and safety, the cycle life of individual LIBs has been improved significantly.

The cycle life of a battery pack, serially connected LIBs used in EVs, is shorter than that of an individual cell. The manufacturer's technical specification only determines the initial performance of the batteries but. during the operation process, the battery parameters are always changed by the operating environment, working conditions, and aging status. Therefore, to avoid abuse and irrational use, the control strategy of the batteries needs to be in accordance with the change in the battery parameters.

Battery management technology aims to optimize usage. First, this technology could avoid abuse and irrational use to ensure safety and to extend the life of batteries. Secondly, it may timely detect and estimate the state of the batteries (including external voltage, temperature, current, DC resistance, polarization voltage, SOC, the maximum available capacity, consistency, etc.). Thirdly, it should maximize the performance of batteries to ensure that the vehicles can be run efficiently and driven comfortably. Ultimately, researchers should realize high efficiency of battery capacity and energy utilization with battery management technology. The importance of battery (group) management techniques has gradually been widely recognized by researchers in battery technology. In this book, from the application perspective, basic issues of LIB management technology are discussed in order to provide a theoretical basis and technical support for a secure, efficient and long-life application in EVs.

1.2 Development of Battery Management Technologies

The battery management technologies have developed from no management and simple management to comprehensive management.

1.2.1 No Management

For a long time, lead-acid batteries dominated the market because of their mature process, good anti-abuse capabilities and low price. However, development of the technology of battery management has lagged behind owing to the lack of connection with the market. In single cell applications, SOC estimation and charge–discharge control were based on the cell external voltage (the cell terminal voltage is called the external voltage in order to distinguish it from the terminal voltage of the battery pack). After series connection, a simple expansion was produced on the basis of single cell management technology. Based on the battery pack terminal voltage, SOC estimation and charge–discharge control were realized by researchers.

Practical application results demonstrated that the life of the battery pack in series connection was significantly shorter than that of a cell. By testing the limitation of the life of the battery, it was found that the management pattern was based on the battery terminal voltage which neglects the differences among cells. This situation resulted in some of the batteries in the pack being over-charged or over-discharged, which was the main reason for the reduction in the lifespan of the battery pack. Therefore, battery consistency was examined on a regular basis (such as once a month). In addition, the batteries with lower voltage were separately charged to ensure the battery's consistency, which thereby decreased the probability of over-charge and over-discharge. By periodically (e.g., once every 6 months) fully charging and discharging all cells, the battery pack capacity and states could be determined, which could prevent batteries from working in a fault status for a long time, and, to some extent, could expand the life of the battery pack. This was a rudiment of the BMS, whose functions included fault diagnosis, SOC and capacity estimation, as well as the evaluation of battery pack consistency.

1.2.2 Simple Management

With wide applications of the batteries, the problems of the traditional management approach became apparent, such as non-online detection, low automation, time-consuming periodical maintenance, and serious energy loss. The equipment used to monitor and manage batteries is called the BMS. The basic functions of the BMS are:

1. Online monitoring of battery external parameters, such as voltage, temperature, current, and so on.
2. Battery fault analysis and alarm.
3. Starting the cooling fan when the battery temperature is high.
4. Battery pack SOC estimation.

The BMS effectively reduces manual detection work, and improves automation and security of batteries utilization. However, it has some disadvantages. The BMS replaces the traditional manual operation with automated detection. The traditional manual operation could only discover the problems and raise the alarm, but could neither ensure the consistency of the battery pack, nor provide a guide to battery maintenance. Therefore, the workload and complexity of battery maintenance are not reduced.

Most BMS designers are electrical engineers, so their study focuses on the optimal design of the battery circuit detection, to improve the accuracy, anti-interference and reliability. They regard the batteries as a "black box" due to their insufficient knowledge of the electrochemistry of batteries, and analyze battery status and usage in terms of external characteristics. They consider the battery pack as a "big battery", even though the batteries are serially connected into a group. They have made achievement in management research through a simple expansion based on single cell management technology, and therefore realized state estimate and charge–discharge control on the basis of the terminal voltage of the battery pack.

However, this method cannot ensure the accuracy of the estimation of the battery SOC. The issue that the battery pack has a shorter lifespan than a single cell still exists. This is because the BMS cannot play an effective role in the management and control function, only provide automatic detection of the external characteristics of the batteries and give a fault alarm. Hence, it is just a monitoring system and does not achieve optimal usage and effective management of the batteries.

1.2.3 Comprehensive Management

LIBs, with their excellent performance, have been widely used in portable devices and EVs. The anti-abuse capabilities of LIBs are inefficient. When the simple BMS is applied to LIBs, especially a battery pack in series, safety incidents repeatedly occur, showing that the states estimation and charge–discharge control method, based on the battery external characters, could not ensure the safety and life of the battery pack.

More attention has been paid to battery management technology in recent years, and, with the endeavor of researchers over time, its function can now be defined explicitly:

1. Real-time monitoring of battery states. By measuring external characteristic parameters (such as the external voltage, current, cell temperature, etc.), with the appropriate algorithm, BMS could realize estimation and monitoring of battery internal parameters and states (such as the DC resistance, polarization voltage, maximum available capacity, SOC, etc.)
2. Efficient battery energy utilization. Provide a theoretical basis and data support to battery usage, maintenance and equalization.
3. Prevent over-charge or over-discharge of the battery.
4. Ensure user safety and extend battery life.

In order to achieve the above objectives, researchers focus on battery modeling, SOC estimation, consistency evaluation and equalization. Although battery management technology has developed rapidly, there are some difficulties in the following aspects.

1. Interdisciplinary. Battery management technology involves electrochemistry, electricity, thermology, and so on.

2. Multi-variable coupling. The performances of the battery are affected by mutual coupling components, such as temperature, voltage, current, SOC, working conditions, aging and other factors.
3. Nonlinearity. The battery temperature and degradation are nonlinearly related to battery internal resistance, polarization voltage, discharge capacity and rate characteristics.
4. Universality. The battery performances of various manufacturers differ with regard to self-discharge, temperature performance, capacity, internal resistance, and so on. Therefore, it is important to seek out a refined management method generally applicable to a state estimate algorithm and charge–discharge management.
5. Battery pack consistency. Difficulty remains in accurate states estimation and efficient management resulting from the differences between cells.

1.3 BMS Key Technologies

The development history of the BMS is from an initial independent monitoring system to data interaction and management with the vehicle and charging devices. Eventually, BMS realizes an optimal match with the vehicle system and its normative development procedure. The role of the BMS is thus substantially expanded, its key technologies are:

1. Battery state estimation. Battery state estimation is expanded from just SOC to state of energy (SOE), state of function (SOF), and state of health (SOH). Battery performance is evaluated from the energy, the maximum available charge and discharge power/current, the battery SOH, and other indicators, thus realizing an accurate estimate of the battery state.
2. Battery equalization. With the increase in the number of EVs demonstrations, heavy regular maintenance workload and other issues are becoming increasingly prominent. Battery equalization is becoming the obstacle for the development of EVs. BMS equipped with an equalization function is becoming the standard configuration of a power battery system. The balancing current is designed from tens of milliamperes to several amperes. The equalization pattern includes passive balancing or active balancing or both. The equalization objective is good voltage consistency, and maximum capacity and energy utilization. After the design of thermal management, a consistency evaluation method and systematic resolution of the actual demands of the equalization current, a rational and effective battery equalization system could become a reality.
3. Battery safety management. Battery safety is the basic requirement in the battery systems. A BMS can not only prevent a battery from over-charge, over-discharge, over-heating, and over-current by power control and diagnostic alarm, but also has the functions of high voltage interlock and insulation detection. In addition, from preliminary exploration, humidity sensors and collision sensors are suitable for automotive application.

In this book we will describe and discuss the key technologies and research methods of the lithium-ion power BMS. There are five main parts: LIB performance modeling and simulation; the theory and methods of estimation of the LIB SOC, SOE, SOH, and peak power; LIB charge and discharge control technology; techniques for the consistent evaluation and equalization of the battery pack: and finally BMS design and application in electric drive vehicles. In this book, part of the contents and graphics are taken directly from real vehicle tests. In general, this book describes the relevant concepts and fundamentals in detail through comparative analysis of various systems structures and scenarios.

References

[1] Chen, Q. and Sun, F. (2002) *Modern Electric Vehicles*, Beijing Institute of Technology, Beijing.

[2] Huang, K., Wang, Z., and Liu, S. (2008) *Fundamentals and Key Technologies of Lithium-Ion Batteries*, Chemical Industry Press, Beijing.

[3] Wen, F. (2009) *Study on basic issues of the Li-ion battery pack management technology for pure electric vehicles*. PhD thesis. Beijing Jiaotong University.

[4] Zhang, S.S., Xu, K., and Jow, T.R. (2006) Charge and discharge characteristics of a commercial $LiCoO_2$ based 18650 battery. *Journal of Power Sources*, **160**, 1403–1409.

[5] Xiao, L. (2003) *Studies of the problems related to the development of high performance lithium-ion patteries*. PhD thesis. Wuhan University.

[6] Ramadass, P. (2003) *Capacity fade analysis of commercial Li-ion batteries*. PhD thesis. University of South Carolina.

[7] Zhang, Q. and White, R.E. (2008) Calendar life study of Li-ion pouch cells: Part 2: Simulation. *Journal of Power Sources*, **179**(2), 785–792.

[8] Kwak, G., Park, J., Lee, J. *et al.* (2007) Effects of anode active materials to the storage-capacity fading on commercial lithium-ion batteries. *Journal of Power Sources*, **174**(2), 484–492.

[9] Zhang, S.S., Xu, K., and Jow, T.R. (2006) Study of the charging process of a $LiCoO_2$-based Li-ion battery. *Journal of Power Sources*, **160**, 1349–1354.

[10] Kima, G.-H., Pesaran, A., and Spotnitz, R. (2007) A three-dimensional thermal abuse model for lithium-ion cells. *Journal of Power Sources*, **170**, 476–489.

2

Performance Modeling of Lithium-ion Batteries

The battery model describes the mathematical relationship between voltage, power, current, state of charge (SOC), temperature, and other factors which impact performance during working. It not only shows the basis of state estimation, performance analysis, scientific evaluation and use, but also works as a bridge from external characteristics to the internal state. So the battery model has been of wide interest to researchers.

2.1 Reaction Mechanism of Lithium-ion Batteries

A lithium-ion battery is a high-energy battery in which Li^+ embeds into and escapes from positive and negative materials when charging and discharging. As illustrated in Figure 2.1, from left to right, a battery consists of a cathode current collector, negative electrode active materials, electrolyte, a separator, positive electrode active materials, and an anode current collector. Positive electrode materials of lithium-ion batteries are intercalation compounds of lithium-ion, commonly $LiCoO_2$, $LiNiO_2$, $LiMn_2O_4$, $LiFePO_4$ and $LiNixCo_{1-2x}Mn_xO_2$, and so on. Negative electrode materials are commonly Li_xC_6, TiS_2, V_2O_5, and so on. The electrolyte is an organic solvent in which the lithium salts, such as $LiPF_6$, $LiBF_4$, $LiClO_4$, $LiAsF_6$, and so on, are soluble. The solvents are mainly ethylene carbonate (EC), propylene carbonate (PC), dimethyl carbonate (DMC), chlorine methyl carbonate (ClMC), and so on. The main role of the separator in a battery is to isolate the positive and negative electrodes, while allowing the transport of ions. Recently, a microporous membrane of polyethylene (PE) or polypropylene (PP) has been used commercially as a separator.

Li ions deintercalate from the cathode compound and intercalate into the lattice of the anode during the charging process. The cathode has high potential and poor lithium state, while the anode has low potential and rich lithium state. When discharging, the Li^+ escapes from the anode and embeds into the cathode, producing a rich lithium state at the cathode. So the charging and the discharging process of batteries is also a deintercalation and intercalation process of lithium back and forth between the two electrodes, hence the name "rocking chair batteries". To keep the charging balance, during the charging and discharging process, the same number of

Fundamentals and Applications of Lithium-ion Batteries in Electric Drive Vehicles, First Edition.
Jiuchun Jiang and Caiping Zhang.

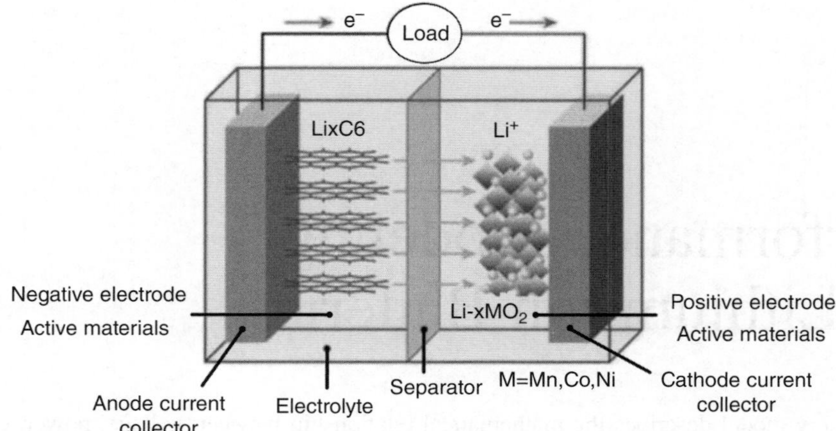

Figure 2.1 Schematic of the discharging process of lithium-ion batteries.

electrons move with the Li^+ between the cathode and anode through the external circuit. Thus a redox reaction occurs between the cathode and the anode.

Considering lithium manganese oxide (LMO) batteries as an example, during charging the Li^+ escapes from the $LiMn_2O_4$ at the cathode, under the electromotive force, the Li^+ passes through the electrolyte and embeds into the carbon interlayer of the graphite. Thus the lithium and carbon interlayer are combined internally. When discharging, the Li^+ escapes from the carbon interlayer of the anode, through an opposite process under the electromotive force, and embeds into the anode $LiMn_2O_4$. The reactions of the batteries are: Reaction in the anode:

$$LiMn_2O_4 \underset{\text{discharge}}{\overset{\text{charge}}{\Leftrightarrow}} Li_{1-x}Mn_2O_4 + xLi^+ + xe^- \tag{2.1}$$

Reaction in the cathode:

$$C + xLi^+ + xe^- \underset{\text{discharge}}{\overset{\text{charge}}{\Leftrightarrow}} Li_xC \tag{2.2}$$

Overall reaction:

$$LiMn_2O_4 + C \underset{\text{discharge}}{\overset{\text{charge}}{\Leftrightarrow}} Li_{1-x}Mn_2O_4 + Li_xC \tag{2.3}$$

120 pure electric buses were demonstrated in the Shanghai World Expo, using two types of LIBs, LMO and lithium iron phosphate batteries as their power sources. Their basic performance parameters are shown in Table 2.1.

Table 2.1 Comparison of the performance parameters of lithium manganese oxide and lithium iron phosphate batteries.

Indicators	Lithium manganese oxide battery	Lithium iron phosphate battery
Nominal cell voltage (V)	3.7	3.2
Nominal capacity (Ah)	360	255
Battery pack voltage (V)	312–437	527–595
Maximum discharge current (A)	360	300
Weight (kg)	~1670	~2050
Working temperature (°C)	−10 to 40	−10 to 40
Safety	Reliable	Reliable

2.2 Testing the Characteristics of Lithium-ion Batteries

2.2.1 Rate Discharge Characteristics

A battery module of 16 lithium-ion cells with nominal capacity of 100 Ah is considered. The relationship between the voltage and the discharged capacity of the battery module under different discharging current at room temperature is shown in Figure 2.2.

Figure 2.2b is a partial enlarged drawing of Figure 2.2a. The discharging capacities are 93.43, 94.43, 94.55, 95.24, and 95.96 Ah, respectively, at the points M1, M2, M3, M4, and M5 with constant current regime 200 A(2 C), 150 A(1.5 C), 100 A(1 C), 50 A(0.5 C), and 33 A(1/3 C), respectively. The open-circuit voltages after keeping in the open-circuit state for 1 h are 54.85, 54.15, 53.44, 52.83, and 52.48 V, respectively. It is seen that the open-circuit voltages increase when the discharging current increases. The decrease in the capacity is not apparent as the discharging current increases. The discharging capacity with the current of 200 A only decreases by 2.6% compared to the discharging capacity with the current of 33 A. The above phenomenon, on the one hand, demonstrates that LMO batteries could keep a high discharge efficiency at the high discharging rate, showing good rate discharging performance. On the other hand, the battery temperature increases rapidly when discharging at high current. The viscosity of the electrolyte is then reduced so that diffusion of the active material to the reaction zone is speeded up, decreasing the concentration polarization and activation polarization of the battery. Hence, the discharge efficiency is improved and the discharge capacity increases due to sufficient active material reaction.

As shown in Figure 2.2a, the working voltage of the battery is relatively stable when the SOC ranges from 20 to 80% (denoted by area B). Homogeneous electrochemical reaction happens inside the battery in this region, which means that the various substances involved in the chemical reaction are in the same phase. The discharge efficiency is high, since most of the chemical energy can be converted into electricity. Because of severe cell polarization and internal resistance, the battery voltage changes rapidly and the discharge efficiency is remarkably decreased when the SOC of a battery increases from 0 to 20% (area A). As shown in Figures 2.3 and 2.4, the internal resistance and polarization resistance of the battery significantly increase when its SOC is within the ranges (0–20%) and (80–100%). The terminal voltage falls rapidly, especially at the end of the discharge. It is suggested that the polarization is serious at the end of the discharge and the discharge efficiency is low. Deep discharge would affect battery cycle life. Hence, deep discharging needs to be avoided to make the battery work in the high efficiency region and to extend the battery life [1].

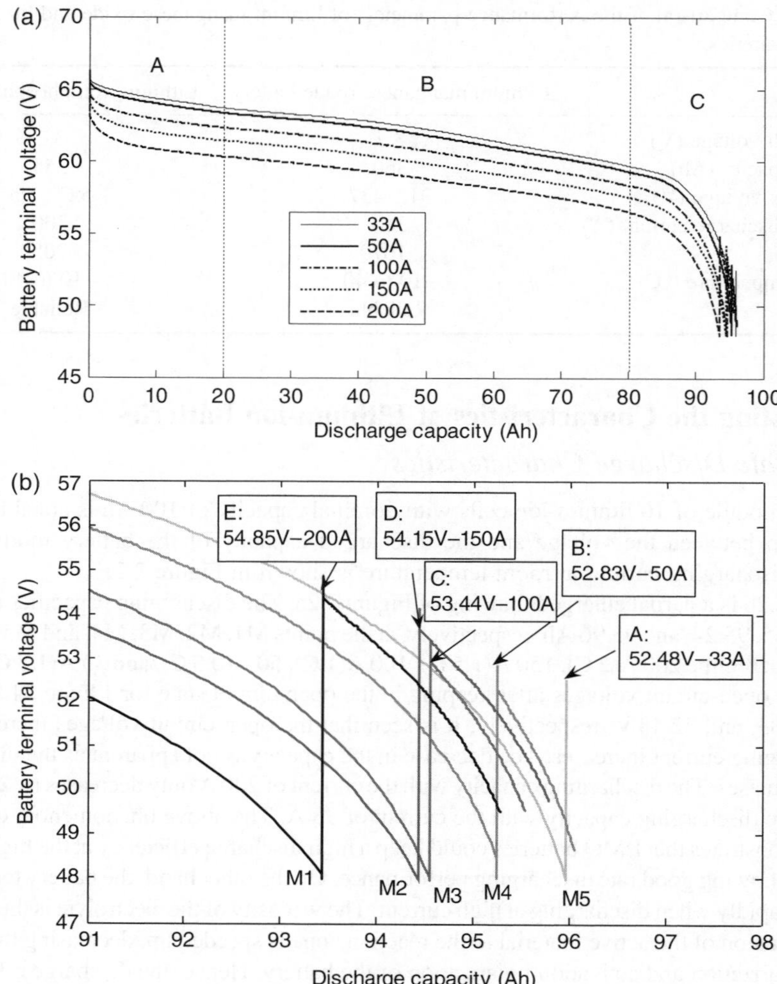

Figure 2.2 Relationship between battery voltage and discharging capacity at various currents of the lithium ion batteries, (b) is a partial enlarged drawing of (a). (Reproduced with permission from Caiping Zhang, "State of Charge Estimation and Peak Power Capability Predict of Lithium-Ion Batteries for Electric Transmission Vehicles", Beijing Institute of Technology, ©2010.)

2.2.2 *Charge and Discharge Characteristics Under Operating Conditions*

Batteries in hybrid electric vehicles are always in the frequent charging–discharging state, while pure electric vehicles have charging conditions under the regenerative braking system. Therefore, the capability of dynamic charging and discharging is an important indicator for the evaluation of battery performance, which lays the basis for the formulation of battery charging and discharging management strategies. The DST (dynamic stress test) cycle conditions test

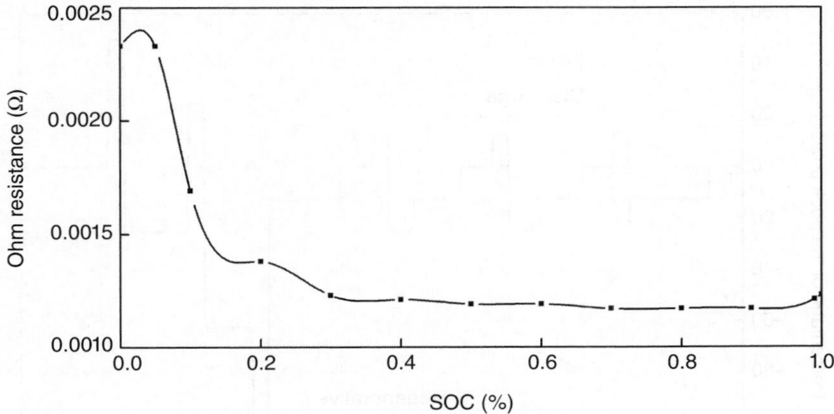

Figure 2.3 Ohmic resistance changes with SOC.

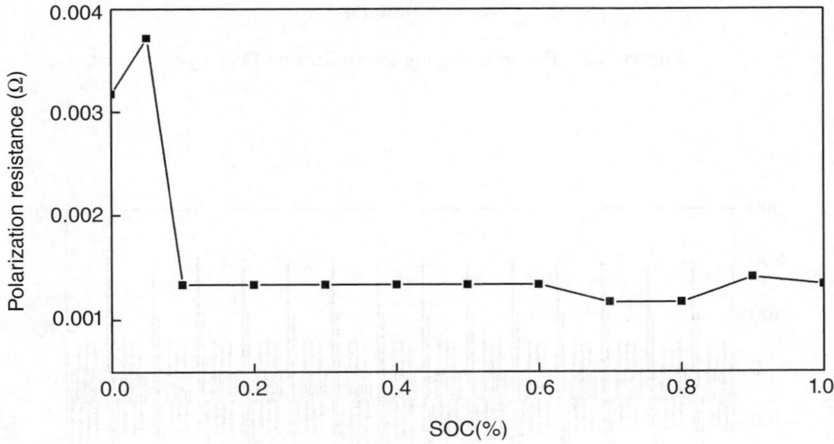

Figure 2.4 Polarization resistance changes with SOC.

method in the "USABC Battery Test Manual" is used to analyze the performance of lithium-ion batteries under working conditions. This test method is as follows: the batteries are charged according to the given charging mechanism, left at open-circuit state for 4 h after fully charged, then tested according to DST cycle conditions. If the voltage of the battery module reaches the minimum restriction, which means the single battery voltage decreases to less than 3.0 V, the discharging of the test process would be completed. The relationship between the charging and discharging power of batteries and time under the DST cycle condition is shown in Figure 2.5. The curves of the change in the current and voltage of batteries in the overall DST cycle process with time are shown in Figure 2.6.

As shown in Table 2.2, the net discharging capacity of a battery under the DST cycle conditions decreases more significantly than the discharging capacity under constant current, shown in Figure 2.6. The available net discharging capacity of the same battery is different

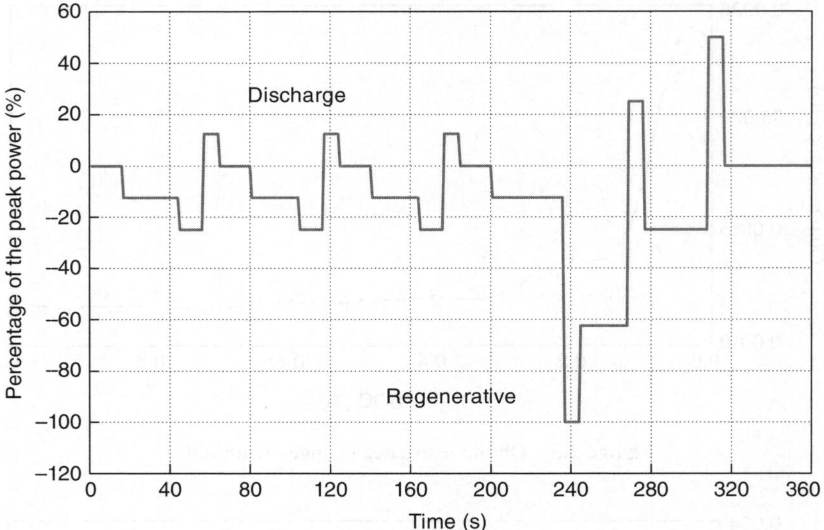

Figure 2.5 Power changing curve for one DST cycle.

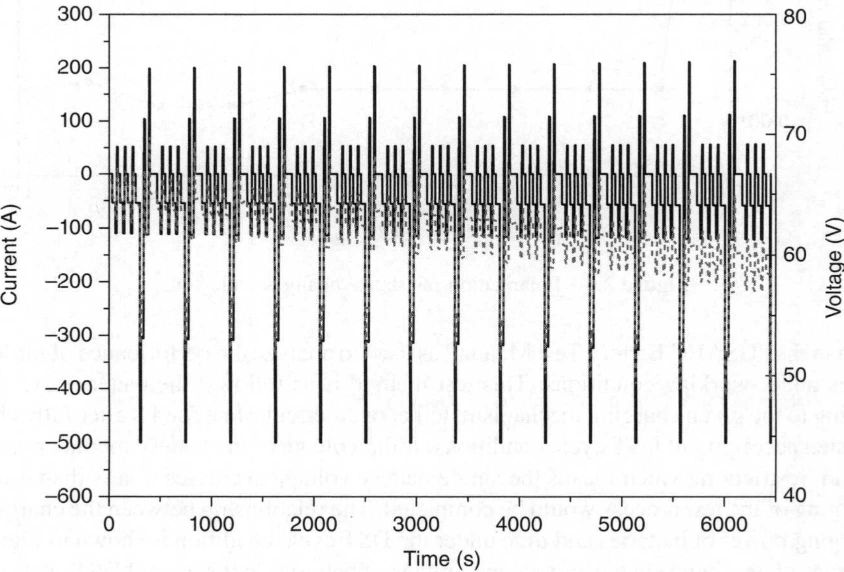

Figure 2.6 The battery voltage and current as a function of time in the overall DST driving cycles. (Reproduced with permission from Caiping Zhang, "State of Charge Estimation and Peak Power Capability Predict of Lithium-Ion Batteries for Electric Transmission Vehicles", Beijing Institute of Technology, ©2010.)

Table 2.2 DST cycle test results of batteries.

Cycle time (s)	Discharging capacity (Ah)	Charging capacity (Ah)	Net discharging capacity (Ah)	Discharging energy (Wh)	Charging energy (Wh)	Net discharging energy (Wh)
6449	103.57	15.03	88.54	5950.89	985.66	4965.23

Discharging capacity: The sum of the capacities of the battery discharged under DST cycle conditions.
Charging capacity: The sum of the capacities of the battery charged under DST cycle conditions.
Net discharging capacity: discharging capacity – charging capacity.

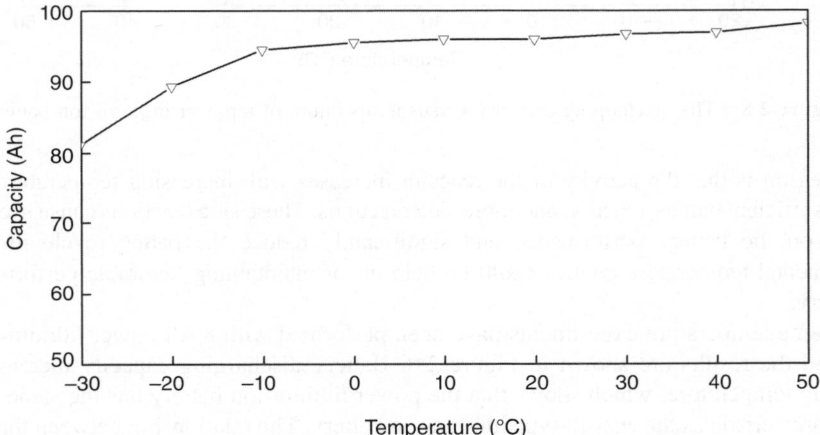

Figure 2.7 The discharging capacity versus temperature of an energy-type lithium-ion battery.

between the dynamic discharging and the constant current discharging. When the battery reaches the discharging cutoff condition, the state of the battery current is limited in the battery discharging mechanism. Thus the influence of battery dynamic charging and discharging efficiency should be considered when estimating battery state [2].

2.2.3 *Impact of Temperature on Capacity*

100 Ah energy type lithium manganese battery modules are tested at the C/3 constant current and with temperature variation of 10 °C between −30 and 50 °C. The capacity–temperature curve is shown in Figure 2.7. In the constant-current discharging mode, it is seen that the discharging capacity declines markedly with decreasing environmental temperature. Battery discharging capacity decreases by about 20% in the working environment with a temperature of −30 °C. This is because of the serious polarization of the battery at low temperature, the active materials cannot be sufficiently utilized. It shows a low efficiency when the discharging voltage declines. It is seen in Figure 2.7 that battery discharging capacity increases with increasing temperature.

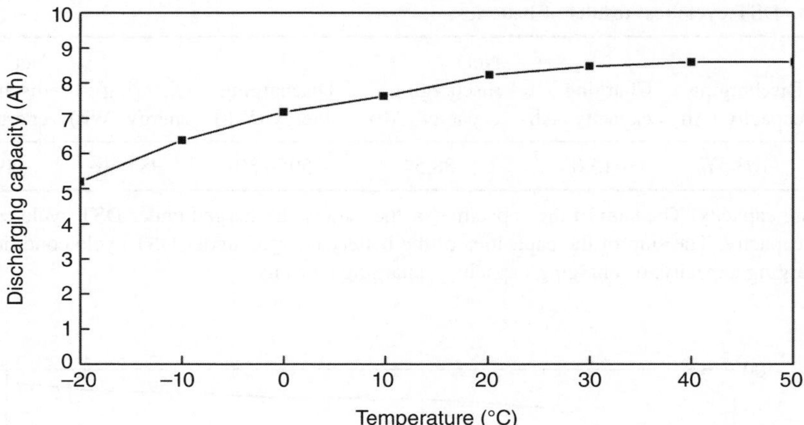

Figure 2.8 The discharging capacity versus temperature of a power lithium- ion battery.

The reason is that the activity of the reactant increases with increasing temperature, which leads to sufficient battery reaction and more side reactions. These side reactions cause irreversible impacts on the battery performance and significantly reduce the battery cycle life. Thus, environmental temperature control could be helpful for maintaining favorable performance of the battery.

The same temperature experiments have been performed with 8 Ah power lithium-ion batteries and the results are shown in Figure 2.8. Battery discharging capacity increases with increasing temperature, which shows that the power lithium-ion battery has the same temperature characteristic as the energy-type lithium-ion battery. The relationship between the battery capacity and temperature is a nonlinear function. In order to improve the estimation accuracy of SOC, the temperature should be considered in the calculation because the SOC is an evaluation standard of the remaining battery capacity.

The battery internal resistance can be tested and analyzed in both frequency and time domains. Electrochemical impedance spectroscopy (EIS) is an electrochemical measurement method applied to the electrochemical system in which batteries are given a disturbance signal of small amplitude sine wave potential (or current). The impedance and the phase angle can be obtained by changing the sine wave frequency in the frequency domain. An EIS plot is shown in Figure 2.9, in which the frequency is decreasing from left to right. It is seen from Figure 2.9 that the EIS plot is an approximate semicircle in the first quadrant and an approximate straight line in the fourth quadrant.

The EIS at high frequency being an approximate straight line in the fourth quadrant may result from inductance. The voltage response lagged the current of the tested battery systems, exhibiting the property of inductance. The presence of this inductance is not caused by the induced current inside the battery, but is due to the physical nature of the electrodes, such as the porosity, surface unevenness, and so on. It is also a result of a viscous system.

The impedance at the axis point (when Image = 0) is not equal to zero between the high frequency band and the middle frequency band. This represents the transport of lithium ions and electrons through the electrolyte, the porous membrane, the connection, the active material particles which are related to the ohmic resistance R_{Ω}.

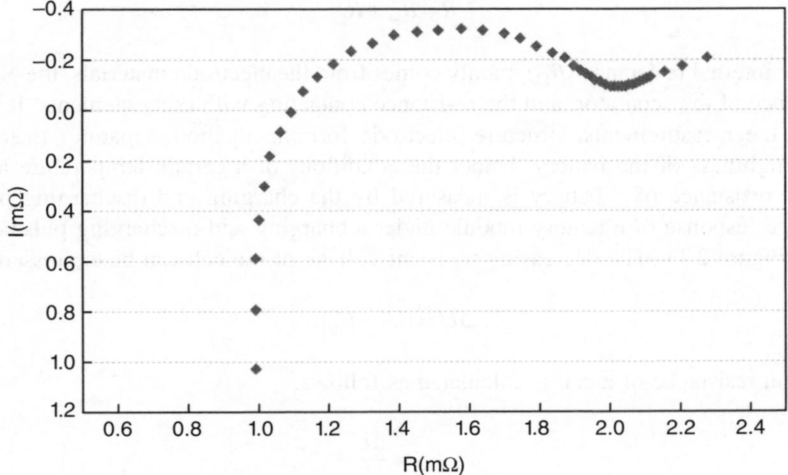

Figure 2.9 EIS of a lithium-ion battery.

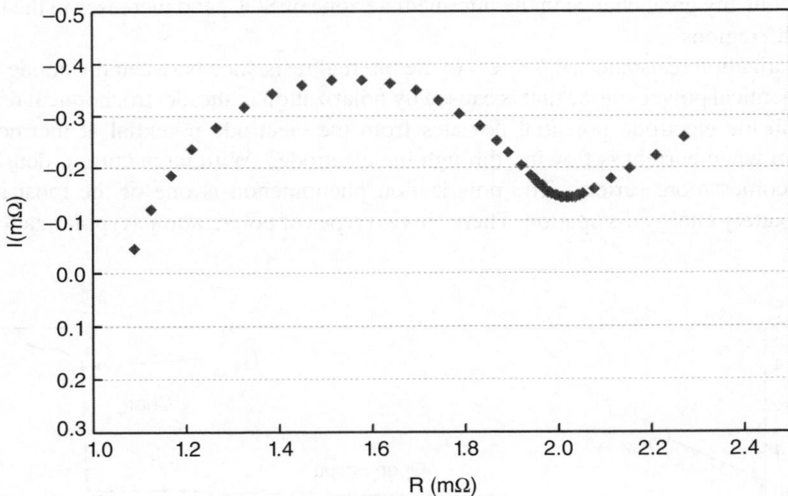

Figure 2.10 Part EIS of a lithium-ion battery.

The Nyquist diagram at the middle frequency and low frequency band is shown in Figure 2.10. The EIS of batteries is made up of the ohmic resistance R_{Ω}, a semicircular band of an R_{ct}/C_{dl} parallel circuit, and a diagonal which reflects the lithium-ion's solid state diffusion process at the low frequency band.

The internal resistance of a battery (R) includes the ohmic internal resistance (R_{Ω}) and the polarization resistance (R_p).

$$R = R_\Omega + R_p \tag{2.4}$$

The ohmic internal resistance (R_Ω) mainly comes from the electrode materials, the electrolyte, the resistance of the separator, and the resistance contacting with other elements. It is closely related to the measurements, structure, electrode forming method, separator materials and assembly tightness of the battery. Under the conditions of a certain temperature and SOC, the ohmic resistance of a battery is measured by the charging and discharging pulse test. The voltage response of a battery module under a charging and discharging pulse current is shown in Figure 2.11. The decreasing transient voltage of the cell can be expressed as:

$$\Delta U = |U_1 - U_0| \tag{2.5}$$

The internal resistance of a cell is calculated as follows:

$$R_O = \frac{\Delta U}{|I|} \tag{2.6}$$

Figure 2.12 shows the charging and discharging internal resistance of a 100 Ah lithium-ion battery under different SOC at room temperature. It is obvious that the resistance value is quite stable without any great change in the intermediate zone of SOC, and increases in the lower and higher SOC regions.

The polarization resistance (R_p) refers to the internal resistance between the anode and cathode of a chemical power source that is caused by polarization in the electrochemical reaction. In polarization the electrode potential deviates from the electrode potential at thermodynamic equilibrium when current is flowing through the electrodes. With large current density polarization becomes more serious. The polarization phenomenon is one of the most important causes of battery energy dissipation. There are two types of polarization: (i) polarization caused

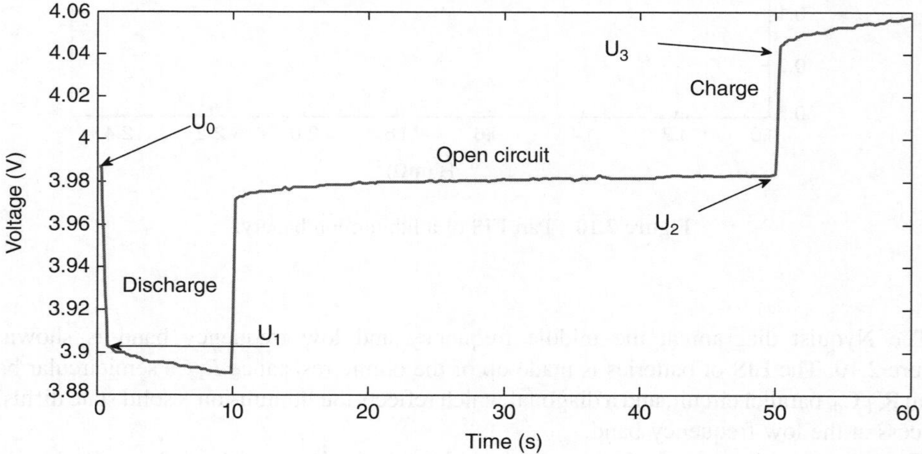

Figure 2.11 Voltage response of a LMO battery at pulse charge and discharge current.

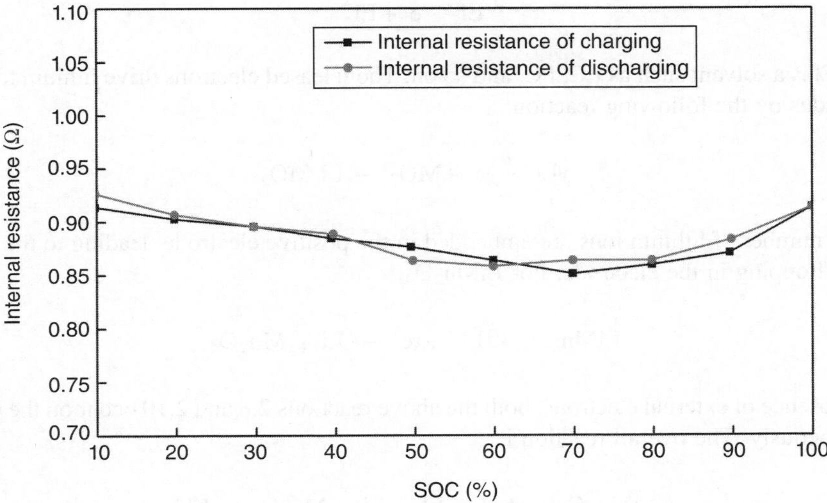

Figure 2.12 Changes of internal resistance as SOC varies when a battery is being charged and discharged.

by the battery resistance, known as ohmic polarization; and (ii) polarization caused by slowing of the ion transport process at the interface layer between the electrode and the electrolyte, known as activation polarization. The polarization resistance is closely related to the nature of the active material, the structure of the electrode, the manufacturing process of the battery, and the working conditions such as current and temperature. Enhancing electrochemical polarization and concentration polarization increases the polarization resistance, and may even cause cathode passivation. When the battery is being discharged by a large current, reduction in temperature has adverse effects on the electrochemical polarization and ion diffusion. So the internal resistance of the battery increases in conditions of low temperature and low humidity. In addition the polarization resistance increases with increase in the logarithm of the current density.

2.2.4 *Self-Discharge*

Although connected in open circuit, the capacity of a chemical power source still naturally attenuates and this phenomenon is called self-discharge. Within a certain time, the ratio of the capacity of a battery after self-discharging to that before self-discharging is called the charging retention capability. The faster the self-discharge, the worse the charging retention ability. The self-discharge rate or capacity retention rate is commonly used to measure the speed of self-discharge of a battery. The self-discharge rate is expressed as the percentage reduction of capacity in a certain time, usually days or months.

The decomposition reaction of the electrolyte and the initial intercalation reaction of lithium mainly lead to self-discharge in a fully charged lithium-ion battery. The oxidation reaction of the electrolyte on the anode is:

$$\mathrm{El} \rightarrow \mathrm{e}^- + \mathrm{El}^+ \tag{2.7}$$

where El is a solvent such as EC, PC, and so on. The released electrons drive lithium to embed into oxides by the following reaction:

$$y\mathrm{Li}^+ + y\mathrm{e}^- + \mathrm{MO}_2 \rightarrow \mathrm{Li}_y\mathrm{MO}_2 \tag{2.8}$$

A large number of lithium ions are embedded in the positive electrode, leading to the state of charge dropping in the electrode. For $LiMn_2O_4$:

$$\mathrm{LiMn_2O_4} + x\mathrm{Li}^+ + x\mathrm{e}^- \rightarrow \mathrm{Li}_{1+x}\mathrm{Mn_2O_4} \tag{2.9}$$

In the absence of external electrons, both the above reactions 2.8 and 2.10 occur on the cathode simultaneously. The overall reaction is:

$$\mathrm{LiMn_2O_4} + x\mathrm{Li}^+ + x\mathrm{El} \rightarrow \mathrm{Li}_{1+x}\mathrm{Mn_2O_4} + x\mathrm{El}^+ \tag{2.10}$$

The self-discharge of a battery is mainly determined by the electrode material, the manufacturing process, the storage conditions, and other factors. The main factors impacting on the self-discharge rate are the storage temperature of the battery, the humidity conditions, and so on. Increasing temperatures may improve the activity of the anode and cathode materials inside a battery, and accelerate the speed of conduction of ions in the electrolyte, reduce the strength of the separator and other auxiliary materials, and improve the reaction rate of self-discharge. If the temperature is too high, it will seriously damage the chemical balance within the battery and result in the occurrence of irreversible reactions which damage the battery. In a low temperature and low humidity environment, the self-discharge rate of the battery is low and the environment is a benefit to battery storage. However, too low temperature may cause irreversibility of the electrode material and the overall performance of the battery will be greatly reduced.

The requirements for storage of a lithium-ion battery are:

- Should be stored in a dry, clean and well-ventilated warehouse with a temperature of 5–40 °C
- Should not be subjected to direct sunlight
- Should be at least 2 m away from any heat source
- Should not be inverted or laid on its side, and should avoid mechanical shock and stress.

2.3 Battery Modeling Method

The battery performance model can be used to estimate the SOC and provide the battery group model with electric vehicle performance simulation. The accuracy of the battery performance model directly affects the availability of electric vehicle simulation results and the estimation accuracy of the battery charging state. Common battery models in the electric vehicle are the equivalent circuit model, the simplified electrochemical model, and the neural network model.

2.3.1 Equivalent Circuit Model

The equivalent circuit model can be used to simulate the dynamic characteristics of the battery. It is made up of circuit elements such as resistors, capacitors, a constant voltage source, and so on. It can be used for various working conditions of the battery, and the state-space equations of the model can be deduced to facilitate analysis and application. Thus this model is widely used in various types of electric vehicle modeling simulations and battery management systems.

The Rint model in Figure 2.13, designed by the Idaho National Laboratory, uses an ideal voltage source to describe the open-circuit voltage of the battery. The battery's internal resistance R and the open-circuit voltage are functions of the SOC and temperature, and the internal resistance value changes when charging under the same SOC [3].

In Figure 2.14, the Thevenin model, which is the most typical circuit model, considers the characteristics of the battery as capacitive and resistive. The model uses an ideal voltage source U_{ocv} to describe the open-circuit voltage of the battery, and the resistance R_{Ω} is the ohmic resistance of the battery, while the capacitor and resistor are connected in parallel in order to describe the battery's over-potential [4].

The RC model in Figure 2.15, which consists of two capacitances and three resistances, is designed by the famous battery manufacturer SAFT. The large capacitance C_{B} describes the energy storage capacity; the small capacitance C_{C} describes surface effects of the battery electrodes; the resistance R_{T} is referred to as the terminal resistance; the resistance R_{E} is referred to

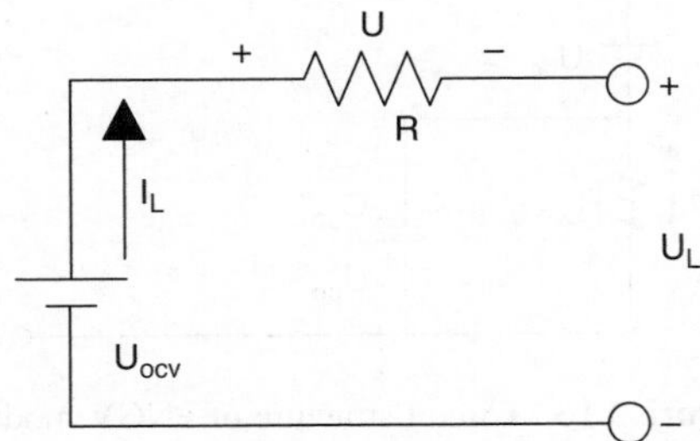

Figure 2.13 Circuit structure of Rint model.

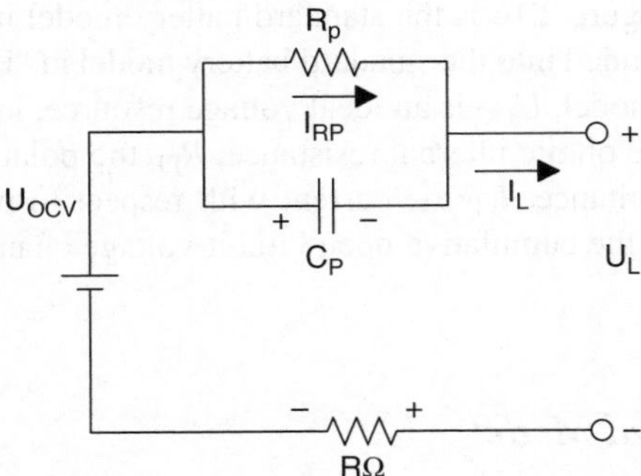

Figure 2.14 Circuit structure of Thevenin model.

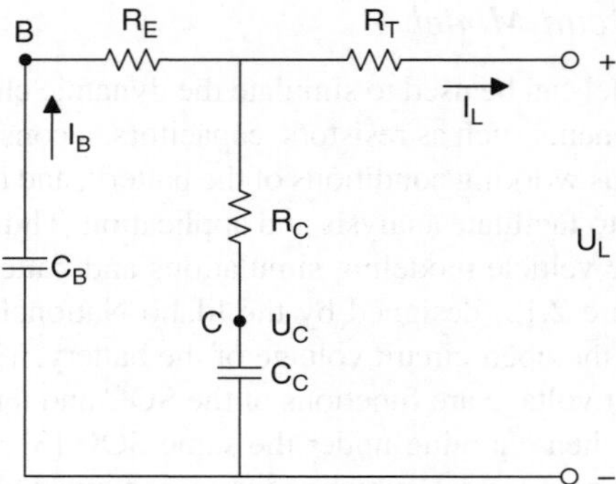

Figure 2.15 Circuit structure of RC model.

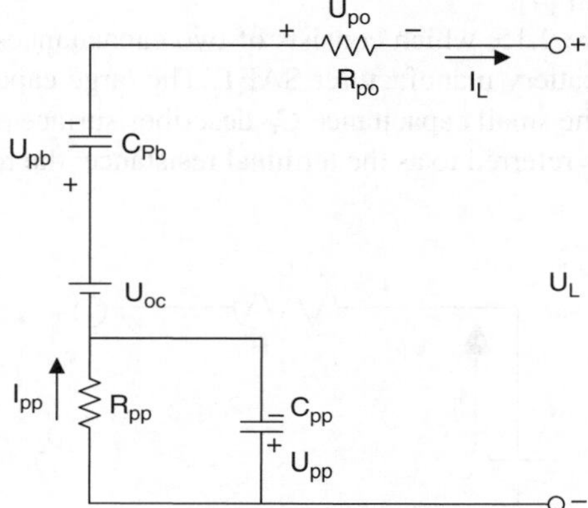

Figure 2.16 Circuit structure of PNGV model.

as the cutoff resistance; and the resistance R_C is referred to as the capacitive resistance. In this model, the cathode of the battery is defined as the zero potential point.

The PNGV model in Figure 2.16 is the standard battery model in the "PNGV Battery Test Manual" in 2001, and extended into the standard battery model in "Freedom CAR Battery Test Manual" in 2003. In this model, U_{OC} is an ideal voltage resource, indicating open-circuit voltage of a battery, R_{PO} is the ohmic internal resistance, R_{PP} the polarization internal resistance, C_{PP} the polarization capacitance, I_{PP} the current with respect to polarization resistance; the capacitance C_{Pb} describes the cumulative open-circuit voltage change with respect to loading time [5].

2.3.2 *Electrochemical Model*

The electrochemical model uses mathematical methods to describe the internal reaction process of a battery, based on electrochemical theory. The Peukert formula is the most typical battery model, seen in Equation 2.12, which expresses that the available charging of a battery decreases with increasing discharging current.

$$I^n T_i = \text{Constant} \tag{2.11}$$

where I is the discharging current, n the constant of a battery, and T_i is the discharging time under current I.

The Shepherd model was proposed in 1965 and is expressed as Equation 2.12. Electrochemical activities are described via the voltage and current of a battery.

$$E_t = E_o - R_i I - K_i\left(\frac{1}{1-f}\right) \tag{2.12}$$

where E_t is the terminal voltage of a battery, E_o the open-circuit voltage of a fully charged battery, R_i the ohmic internal resistance, K_i the polarization resistance, I the transient current and f the net discharging capacity calculated according to the Ah integration method. The Shepherd model is commonly used to analyze hybrid cars, and to calculate battery voltage and SOC together with the Peukert equation under different powers.

The Shepherd model is applicable to a small constant current battery, and is able to find the turning point where the terminal voltage begins to decline rapidly. Such a critical state occurs infrequently in the actual working process of an electric car battery. Unnewehr and Nasar simplified the Shepherd model into Equations 2.13–2.15.

$$E_t = E_o - R_i I - K_i f \tag{2.13}$$

$$E_{oc} = E_o - K_i f \tag{2.14}$$

$$R = R_o - K_R f \tag{2.15}$$

where E_{oc} is the open-circuit voltage, R_o the total internal resistance of a fully charged battery, K_R the experimental constant, and R the battery equivalent resistance.

It is developed into the Nerst model and the Nerst expanded model based on the Unnewehr model, respectively expressed in Equations 2.16 and 2.17.

$$E_t = E_i - R_i I + K_i \ln(f) \tag{2.16}$$

$$E_t = E_o - R_i I + K_i \ln(f) + K_j \ln(1-f) \tag{2.17}$$

Based on the Shepherd model, the Unnewehr model and the Nerst expanded model, Dr. Gregory L. Plett from Colorado University developed a group formulation of the above three electrochemical models, seen in Equation 2.18.

$$U_L = K_0 - R I_L - \frac{K_1}{\text{SOC}} - K_2\text{SOC} + K_3\ln(\text{SOC}) + K_4\ln(1-\text{SOC}) \tag{2.18}$$

where U_L is the battery loading voltage, I_L the current, R the battery internal resistance, and K_0, K_1, K_2, K_3, and K_4 are the mean model coefficients, respectively.

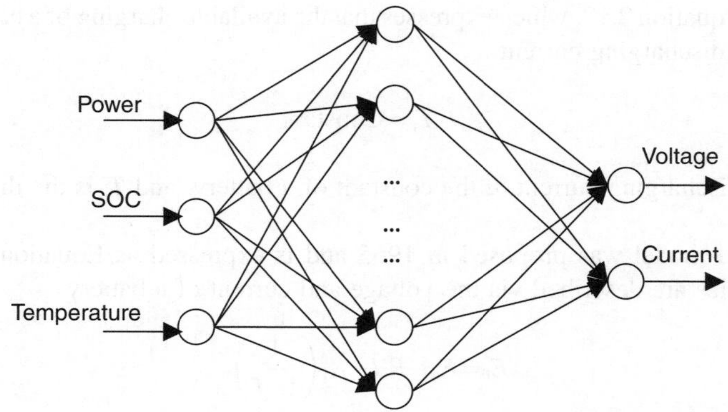

Figure 2.17 Structure of neural network model with power to be input.

2.3.3 Neural Network Model

A battery is a highly nonlinear system, and the basic characteristics of the neural network are nonlinear, with a parallel structure and learning ability. The neural network model could respond to external excitation and is suitable for battery modeling. The BP neural network structure with error back propagation is commonly used for modeling and simulation of battery performance (Figure 2.17).

The paper published in 1998 from the US Sandia International Laboratory is the earliest literature on the neural network modeling of battery performance. ADVISOR has used a neural network model since 1999, designed by Prof. Mahajan from the University of Colorado. It is a double-layer neural network and its inputs are requisite power and SOC, the outputs are the voltage and current of a battery [6].

The selection and quantity of neural network input variables affect the accuracy of the model and the amount of computation required. Error in the neural network method is mainly dependent on the training data and training methods, and all test data which can be used to train the model and optimize the performance of the model. The data-trained neural network model can only be used within the scope of the original training data. The battery working voltage, current, and power are impacted by vehicle operating conditions, which are unpredictable, while it is difficult for a neural network model to simulate every working condition. Thus, there are many limitations in electric vehicle applications.

2.4 Simulation and Comparison of Equivalent Circuit Models

The equivalent circuit models mainly include the Rint model, Thevenin model, and PNGV model. The Rint model comprises a voltage source which represents the battery potential and an internal resistance which is connected to the voltage source in series, and wherein the electrical potential is a function of the battery SOC, while the resistance is a function of temperature, SOC, and the cycle life. The Thevenin model is the Rint model with the addition of a parallel RC network in series, which describes the dynamic characteristics of the battery. The PNGV model is the Thevenin model with the addition of a capacitance C_{pb} in series which describes the battery open-circuit voltage changing with respect to the current loads accumulatively. Here, the effectiveness of these three typical equivalent circuit models and applicability with respect to different working conditions of lithium-ion batteries are analyzed, when these three models are applied to LIB simulation. This establishes a theoretical foundation for the future improvement of the equivalent circuit model.

2.4.1 Model Parameters Identification Principle

The equivalent circuit model parameters are typically estimated by calculating the data obtained from the battery identification experiment via the least squares method. The least squares method is used to solve the problem of seeking a reliable value from a set of measured values and is widely used in system identification and parameter estimation. Its basic theory is measuring a set of data under equal accuracy in pairs, then finding a fitting curve. Minimizing the sum of the squares of the differences between the value of each point on the fitting curve and the measured values gives the best fitting curve.

If the linear model is:

$$y(t) = a_1x_1(t) + a_2x_2(t) + \cdots + a_nx_n(t) \tag{2.19}$$

Herein, $y(t_1), y(t_2), \ldots, y(t_n)$ are observed data, the independent variables $x(t_1), x(t_2), \ldots, x(t_n)$ are known, while $a_1, a_2, \ldots, a_n$ are unknown parameters to be estimated, this formula could be expressed as:

$$\begin{bmatrix} y(t_1) \\ y(t_2) \\ \cdots \\ y(t_m) \end{bmatrix} = \begin{bmatrix} x_1(t_1) & x_2(t_1) & \ldots & x_n(t_1) \\ x_1(t_2) & x_2(t_2) & \ldots & x_n(t_2) \\ \cdots & \cdots & \cdots & \cdots \\ x_1(t_m) & x_2(t_m) & \ldots & x_n(t_m) \end{bmatrix} \times \begin{bmatrix} a_1 \\ a_2 \\ \cdots \\ a_n \end{bmatrix} \tag{2.20}$$

or

$$Y = Xa.$$

Using the least squares method it is found that $a = (X^{\mathrm{T}}X)^{-1}X^{\mathrm{T}}Y$, when $\sum_{i=1}^{m}\left(y(t_i) - \sum_{j=1}^{n} a_j x_j(t_i)\right)^2$ is a minimum.

2.4.2 Implementation Steps of Parameter Identification

To obtain battery model parameters, according to the HPPC (hybrid pulse power characterization) test method in the "Plug-in Hybrid Electric Vehicle Battery Test Manual", 16 manganese lithium batteries with a rated capacity of 100 Ah are connected in series into a battery pack, which is taken as a study object to conduct parameter identification experiments. The testing process is as follows: at room temperature, according to the manufacturer's recommended charging system, the battery module is fully charged, and allowed to stand for 2 h. Then it is discharged by 10%

SOC under a current of 0.3 C (30 A), and allowed to stand for 1 h to recover electrochemical balance and thermal balance, before composite pulse charging and discharging. Hereafter, the HPPC test is implemented: the battery is discharged for 10 s under a current of 1 C, subsequently the battery is allowed to stand for 40 s, and then it is charged for 10 s under a current of 0.75 C. During this process, the voltage and current are recorded every 0.1 s. The HPPC cycle is illustrated in Figure 2.18. The HPPC test should be implemented at intervals of 10% SOC. It is necessary to complete the change of SOC test points of the battery pack by discharging under a constant current of 0.3 C. It is kept at open-circuit state for 1 h between two HPPC cycles. Eventually the test is finished when the pack is fully discharged. The composite pulse experiments include the charging and discharging process of the battery as well as the intermediate continuance. The experiment can manifest complex internal physical and chemical reaction of the battery through external characteristics. The same parameter identification tests are implemented under different SOC and it is more reasonable to take model parameters as a function of SOC.

Based on the output equation of these three battery models, the HPPC test data of each SOC point are discretized respectively to derive Equations 2.21–2.23. Then the model parameters are identified by using multiple linear regression methods. It is known for Equations 2.22 and 2.23 that, when using the least squares method to identify parameters, the selection of the time constant τ is the key to obtaining accurate parameters. Here, the best τ is determined by the coefficient of determination r^2. The closer the r^2 to 1, the more accurate the estimation, and the better the τ is. According to the above theory, the coefficient of each point under these three battery models is expressed in Tables 2.3–2.5.

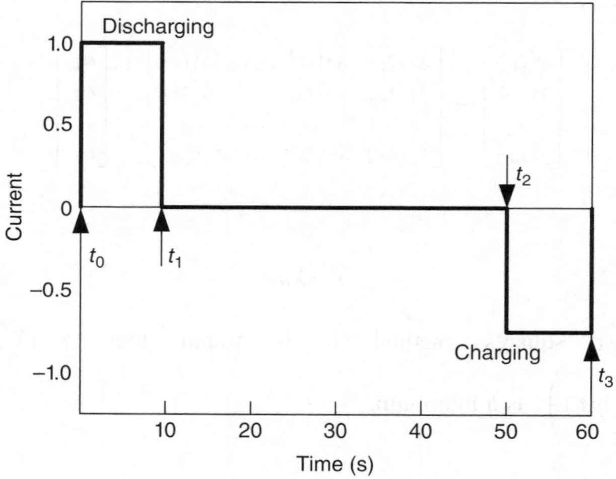

Figure 2.18 HPPC cycling process.

Table 2.3 Rint model parameters for different SOC (at room temperature).

SOC	0.1	0.2	0.3	0.4	0.5	0.6	0.7	0.8	0.9
V_{oc} (V)	59.91	60.891	61.66	62.515	63.153	63.677	64.077	64.63	65.082
R_o (Ω)	0.025	0.0249	0.0249	0.0249	0.0247	0.0246	0.0245	0.0246	0.02469

Table 2.4 Thevenin model parameters for different SOC (at room temperature).

SOC	0.1	0.2	0.3	0.4	0.5	0.6	0.7	0.8	0.9
τ (s)	45	44	38	41	36	36	37	33	28
V_{oc} (V)	60.088	61.06	61.822	62.682	63.312	63.819	64.235	64.79	65.2317
C_p (F)	4563.89	4741.38	4594.92	4617.1	4472.05	4979.25	4567.9	4145.73	3076.92
R_p (mΩ)	9.86	9.28	8.27	8.88	8.05	7.23	8.1	7.96	7.15
R_b (mΩ)	24.8	24.66	24.54	24.53	24.26	24.21	24.05	24.03	23.95

Table 2.5 PNGV model parameters for different SOC (at room temperature).

SOC	0.1	0.2	0.3	0.4	0.5	0.6	0.7	0.8	0.9
τ (s)	17	14	17	15	17	14	20	17	16.5
V_{oc} (V)	60.104	61.081	61.837	62.702	63.329	63.84	64.245	64.8084	65.25
C_{pb} (F)	11 601.19	10 989.36	14 187.56	11 701.29	14 415.29	13 659.21	18 025.46	15 203.78	18 567.93
C_{pp} (F)	5 555.56	5 468.75	5 214.72	5 338.08	5 044.51	5 405.41	5 115.09	4 521.28	4 064.04
R_{pp} (mΩ)	3.06	2.56	3.26	2.81	3.37	2.59	3.91	3.76	4.06
R_{pb} (mΩ)	24.74	24.62	24.52	24.52	24.26	24.2	24.03	24.04	23.95

Rint model:

$$U_{L,i} = -R_o I_{L,i} + U_{ocv} \tag{2.21}$$

Thevenin model:

$$\begin{cases} U_{L,i} = -R_b I_{L,i} - R_p I_{p,i} + U_{oc} \\ I_{p,i} = \left\{1 - \frac{[1-\exp(-\Delta t/\tau)]}{(\Delta t/\tau)}\right\} \times I_{L,i} + \\ \left\{\frac{[1-\exp(-\Delta t/\tau)]}{(\Delta t/\tau)} - \exp(-\Delta t/\tau)\right\} \times I_{L,i\text{-}1} + \exp(-\Delta t/\tau) \times I_{p,i\text{-}1} \end{cases} \tag{2.22}$$

PNGV model:

$$\begin{cases} U_{L,i} = -(1/C_{pb}) \times \left(\sum I_L \Delta t\right)_i - R_o I_{L,i} - R_{pp} I_{p,i} \\ \left(\sum I_L \Delta t\right)_{\mathrm{i}} = \left(\sum I_L \Delta t\right)_{i-1} + (I_{L,i} + I_{L,i-1}) \times (t_i - t_{i-1})/2 \\ I_{p,i} = \left\{1 - \frac{[1-\exp(-\Delta t/\tau)]}{(\Delta t/\tau)}\right\} \times I_{L,i} + \\ \left\{\frac{[1-\exp(-\Delta t/\tau)]}{(\Delta t/\tau)} - \exp(-\Delta t/\tau)\right\} \times I_{L,i-1} + \exp(-\Delta t/\tau) \times I_{p,i-1} \end{cases} \tag{2.23}$$

2.4.3 Comparison of Simulation of Three Equivalent Circuit Models

After obtaining equivalent circuit model parameters at each SOC point, the Rint, Thevenin, and PNGV simulation models are created by using Matlab/Simulink. The output voltage accuracy of the three circuit models' simulation is analyzed and compared under self-defined composite pulse and DST cycling conditions. Herein, the self-defined composite pulse condition is based on working current limits and the duration of the peak current when LIBs are applied to an electric vehicle, which is illustrated in Figure 2.19. Its maximum discharging current is 400 A (4 C), while the maximum charging current is 300 A (3 C) in low SOC. The self-defined composite pulse includes the charging and discharging process of LIBs from low to high current and the standing process, which fully reflects the LIB output characteristics under different working condition inputs.

The comparison of voltage response and test results of the three circuit models under self-defined composite pulse conditions is illustrated in Figure 2.20. It is seen that when being charged and discharged in small current (<1 C), the three circuit models could simulate the actual output voltage of the batteries, with errors all less than 0.25 V, while the errors increase in large current and the maximum error of the PNGV model reaches 2.41 V. The Rint model has the larger errors when being charged, and could not reflect the polarization characteristics

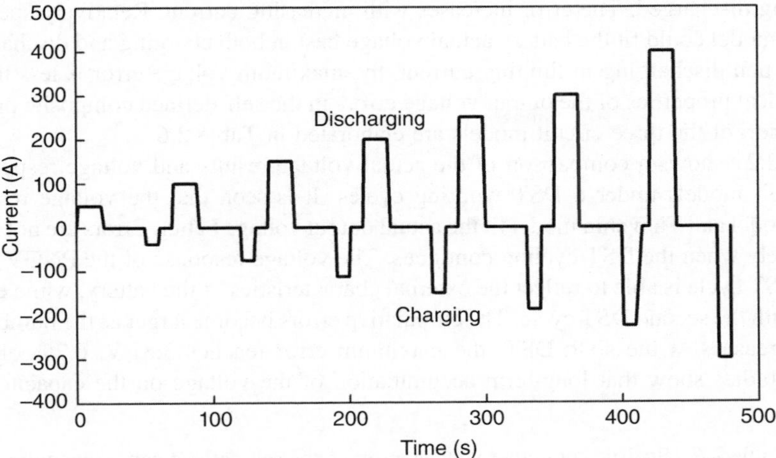

Figure 2.19 Scheme of self-defined composite pulse cycling condition. (Reproduced with permission from Caiping Zhang, "State of Charge Estimation and Peak Power Capability Predict of Lithium-Ion Batteries for Electric Transmission Vehicles", Beijing Institute of Technology, ©2010.)

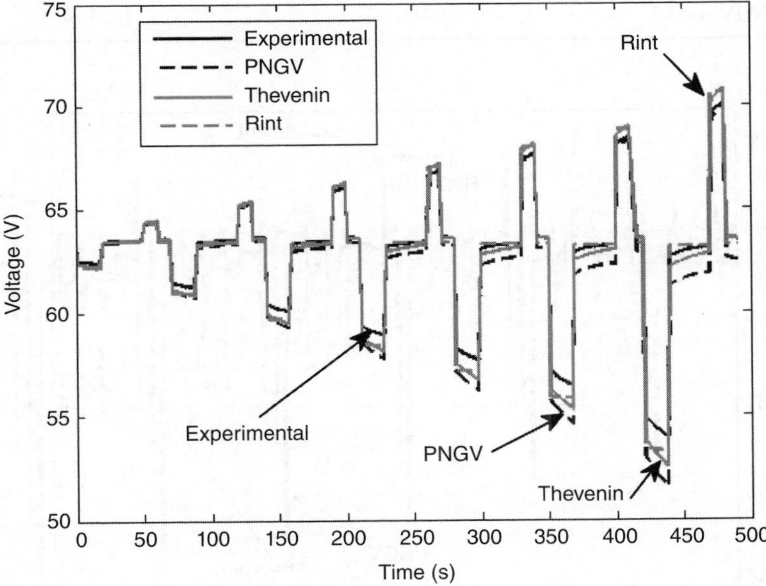

Figure 2.20 Comparison of voltage response and test results of three circuit models under self-defined composite pulse conditions.

during the dynamic process. The Thevenin and PNGV models consider the capacitance of the battery, so both can reflect the polarization characteristics. Herein the PNGV model could accurately simulate actual output voltage when being charged, while the voltage errors are large when being discharged. The error increases with increasing current. Relatively speaking, the Thevenin model could fit the battery actual voltage best in both charging and discharging conditions. When discharging in limiting current, the maximum voltage error is less than 1.5 V. The statistical properties of the output voltage errors in the self-defined composite pulse working condition of the three circuit models are elaborated in Table 2.6.

Figure 2.21 shows a comparison of the actual voltage results and voltage response of the three circuit models under 6 DST working cycles. It is seen that the voltage responses of the Rint model and Thevenin model fit the actual output voltage better. Errors are not increasing cumulatively when the DST cycling continues. The voltage response of the PNGV model for the first DST cycle is able to reflect the external characteristics of the battery, while error accumulates with the second DST cycle. The cumulative errors become larger as the number of DST cycles increases. At the sixth DST, the maximum error reaches 3.87 V, 6.7% of the rated voltage. Studies show that long-term accumulation of the voltage on the capacitor C_{pb} can

Table 2.6 Statistics of output voltage errors in the self-defined composite pulse working condition of three circuit models.

Model	Max voltage error (V)	Average voltage error (V)
Rint model	1.65	0.18
Thevenin model	1.44	0.24
PNGV model	2.41	0.63

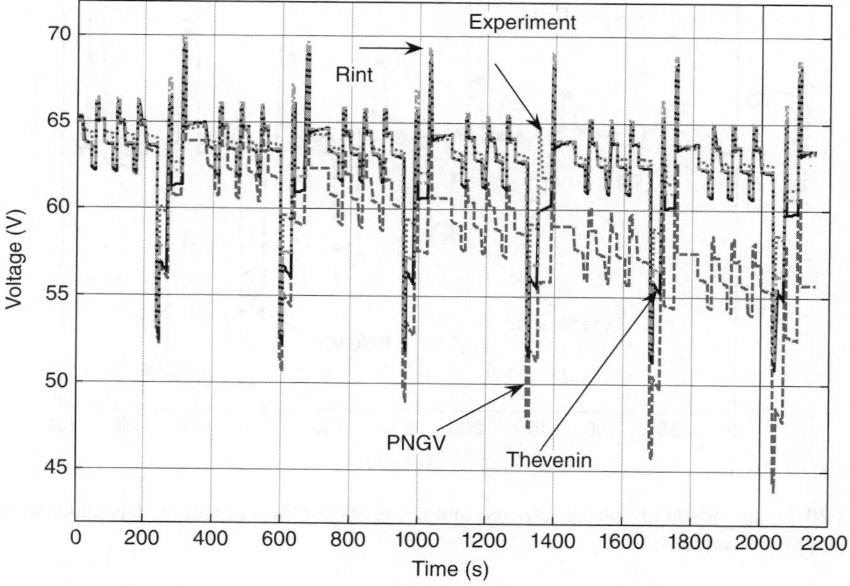

Figure 2.21 Comparison of voltage response and test results of three circuit models under 6 DST working cycles (SOC_0 = 0.8). (Reproduced with permission from Caiping Zhang, "State of Charge Estimation and Peak Power Capability Predict of Lithium-Ion Batteries for Electric Transmission Vehicles", Beijing Institute of Technology, ©2010.)

Table 2.7 Statistics of output voltage errors in 6 DST working cycles for three circuit models.

Model	Maximum voltage error (V)	Average voltage error (V)
Rint model	1.12	0.18
Thevenin model	1.09	0.24
PNGV model	3.87	1.85

increase the error of the PNGV model. It is seen that the changed capacitance C_{pb}, which indicates that the battery's open-circuit voltage changes with the accumulative current loads, should not be used for the simulation of LIB. Statistics of the output voltage errors in the 6 DST cycling of the three circuit models are shown in Table 2.7. Based on the above discussion, the results of the comparison of the simulation for LIBs with the three typical equivalent circuit models are summarized as follows:

1. When the Rint model is applied to the simulation of LIB, the characteristics of the voltage errors are favorable. However, it could not reflect the dynamic polarization characteristics of the LIB and its accuracy is limited.
2. The PNGV model could match different working conditions (charging and discharging) excellently for a short time; however, in a long-time simulation cumulative voltage error caused by the capacitance C_{pb} increases. So this model is not suitable for LIB simulation.
3. Compared with the PNGV model, the Thevenin model could reflect battery polarization. Cumulative errors do not occur in long-time simulation, and the simulation voltage has better following characteristics of the battery actual working voltage. It could also reflect the actual working characteristics of the LIB, which is suitable for battery simulation. The Thevenin model is relatively simple compared to others, if needed an equivalent circuit model could be created by improving the Thevenin model.

Besides SOC and current, the impact of temperature on model parameters should be considered in the battery model. The voltage comparison curve at low or high temperature and at room temperature is shown in Figure 2.22. The discharging voltage of the battery is lower than that at room temperature, about 10 V when discharging current at 4 °C, and the charging voltage is higher than that at room temperature. The above results are due to DC internal resistance, because the ohmic resistance increases with decreasing temperature. In addition, the conductivity of the electrolyte in the battery decreases at low temperature, causing increased polarization resistance. Battery polarization is lower at high temperature than at room temperature, because the conductivity of the battery increases at high temperature, causing a decrease in the polarization resistance. Thus in actual application the internal resistances at different temperatures are stored in battery models. Temperature is compensated by model parameters in order to improve the model's output accuracy.

2.5 Battery Modeling Method Based on a Battery Discharging Curve

Figure 2.23 shows the voltage curve of an 8 Ah cell when it is discharged with a constant current of 80 A. The overall discharging curve is composed of three areas, the exponential, the plateau, and the cutoff areas, according to the variation of the cell's discharging voltage curve.

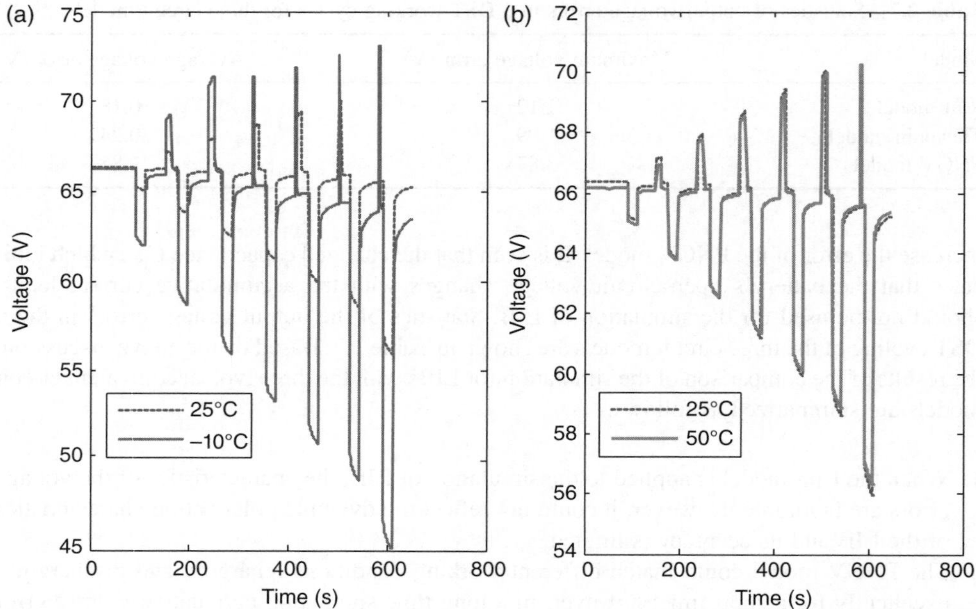

Figure 2.22 Comparison of voltage responses under composite pulse working condition between (a) low temperature (–10 °C) and room temperature (25 °C) and (b) high temperature (50 °C) and room temperature. (Reproduced with permission from Caiping Zhang, "State of Charge Estimation and Peak Power Capability Predict of Lithium-Ion Batteries for Electric Transmission Vehicles", Beijing Institute of Technology, ©2010.)

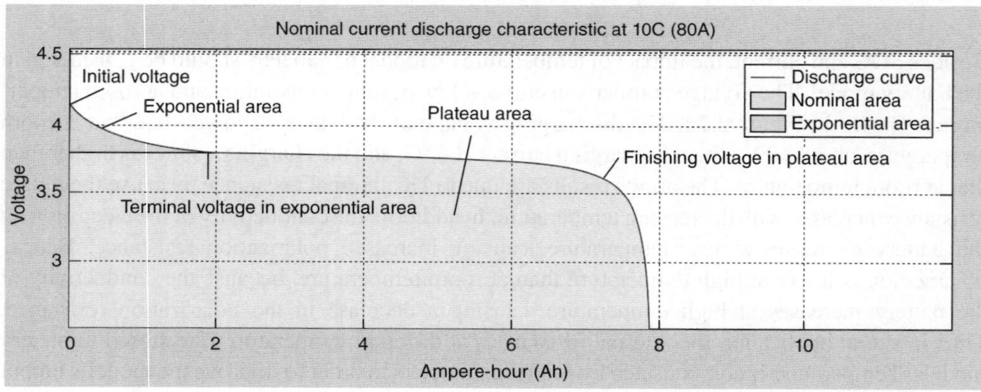

Figure 2.23 Discharging curve of LIB.

The exponential area refers to the area where the voltage falls at an exponential rate during the discharging process, and occurs mainly in the initial stage of discharging. The cutoff point of this area varies depending on the materials of the battery system, the plateau area is the area in which the voltage falls slowly during the discharging process. In this area, the rate of chemical reaction is relatively stable, and the flow of ions and electrons inside the battery reaches equilibrium substantially, so the voltage of the battery falls slowly. In the cutoff area, the voltage of the battery falls rapidly and the voltage of polarization increases rapidly. After a short time of discharging the voltage of the battery drops to a cutoff voltage. The total discharging capacity and cutoff voltage can be measured in this area. Thus, we can consider three points on the curve: V_{full} is the initial voltage when discharging at the fully charged state, excluding the open-circuit voltage after fully charged; V_{exp} is the terminal voltage in the exponential area; V_{nom} is the finishing voltage in the plateau area during the discharging process. Besides these three key points, the following are needed in order to build a model: the type of battery, the design capacity, the initial SOC, the maximum capacity, the discharging current, the internal resistance at room temperature, the discharging capacity at the end point of the plateau area, the discharging capacity in the exponential area, the battery's response time, and so on.

Figure 2.24 shows the equivalent circuit model of a cell, similar to the traditional resistance–capacitance model; the battery's external voltage is mainly constituted by the open-circuit voltage, the polarization voltage and the ohmic voltage drop. The model primarily connects a controllable voltage source and a resistance in series, in order to form the internal circuit of the battery. The controllable voltage source is influenced by the battery SOC, the current magnitude of the charging and discharging, the current direction and the current duration time (i.e., the current fluctuation frequency) at the same time. Different from the traditional equivalent circuit, the model will filter the actual current to get a filtered current i^*, and then multiply it by the polarization resistance (the polarization resistance is associated with the battery SOC) to get the polarization voltage of the battery. This model can be used for modeling lead-acid batteries, nickel metal hydride batteries, and lithium ion batteries. Compared with other circuit models, it is universal, and the modeling process is simpler and more convenient.

For all the secondary batteries, the model gives the uniform expression of external voltage of a cell:

$$V_{batt} = E_0 - K\frac{Q}{Q-it}it - Ri + A\exp(-Bit) - K\frac{Q}{Q-it}i^* \tag{2.24}$$

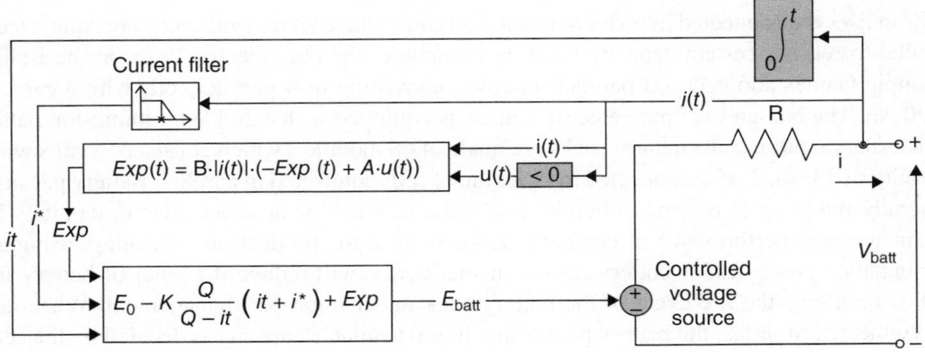

Figure 2.24 The secondary battery equivalent circuit model.

For a LIB, the state equations of the battery external voltage in the charging and discharging conditions are specifically expressed as follows:
Charge:

$$f_{\mathrm{Cha}}(it, i^*, i) = E_0 - K\frac{Q}{Q-it}it + A\exp(-Bit) \tag{2.25}$$

Discharge:

$$f_{\mathrm{Dis}}(it, i^*, i) = E_0 - K\frac{Q}{it+0.1Q}it + A\exp(-Bit) \tag{2.26}$$

The definitions of the symbols and units in the equations are given below:

$f_{\mathrm{Cha}}, f_{\mathrm{Dis}}$: Represent the battery's external voltage (V) in the charging and discharging state, respectively. They are related to the SOC, current and the duration of the current
E_0: Nominal voltage of a battery (V), which is directly related to the material system of a battery
exp(*S*): Dynamic voltage in the exponential area (V), which indicates the voltage change trend in the exponential area
Sel(*s*): Battery charging and discharging status factor; Sel(s) = 0, when discharging, while Sel (s) = 1 when charging
K: Polarization factor (mΩ Ah^{-1}), indicating the change in polarization resistance as the discharging depth changes
i^*: The actual current is filtered by a low-pass filter (A), and the filter's cutoff is frequently related to the polarization recovery time
i: The actual current of a cell (A)
it: Algebraic sum of charging and discharging capacities of a battery from fully charged to the current state (Ah), i.e. the depth of discharge (DOD), corresponding to the battery (SOC)
Q: The maximum actually available capacity of a battery (Ah), general power type battery refers to the discharging capacity at 1 C rate
A: Voltage factor in the exponential area (V)
B: Capacity factor in the exponential area (Ah^{-1})

2.6 Battery Pack Modeling

Cells in EVs are connected in series to reach a certain voltage level, while cells are connected in parallel to reach a certain capacity level. For example, the pure electric buses at the Beijing Olympic Games adopted 104 parallel modules, consisting of 4 parallel cells with a capacity of 90 Ah. The Nissan Leaf pure electric vehicle is equipped with a 24 kWh lithium-ion battery with ternary materials. Its battery packs are made of 48 modules, which consist of 4 cells with a capacity of 33 Ah, 2 cells connected in series and 2 cells connected in parallel. Battery packs are generally made up of dozens to hundreds of cells in series or in series after in parallel. The parameters and performance of cells are inconsistent in the production, screening, using and maintenance processes. A battery pack's inconsistency will reduce the level of battery use, and even affects the life cycle of the battery pack under improper management. When cells are connected in series, the battery pack's maximum available capacity is decided by the cells' capacity and SOC, and the voltage inconsistency may lead to overcharging or discharging in some cells. When cells are connected in parallel, because of the differences in branch internal resistance, the SOC and capacity can cause current imbalance in the parallel branch. Therefore the performance of the battery pack is not a simple superposition of the performance of cells. It is necessaryto model the battery pack to accurately simulate its charging and discharging characteristics, to provide information on its parameters for the battery management system, and to improve its performance and extend its life cycle.

2.6.1 Battery Pack Modeling

Battery parameters, such as ohmic resistance, open-circuit voltage, polarization resistance and polarization capacity, change when the battery is in a different SOC. Inconsistent SOC of cells causes inconsistency of their parameters in the series battery pack. The equivalent circuit model of the battery pack needs to consider the difference in the respective cell's parameters. After several cycles, the SOC of the cells will change due to different capacity values and environmental conditions, which will increase the difficulty of battery pack modeling.

Jonghoon Kim from the Seoul National University screened out batteries with approximately identical battery capacity and internal resistance, and then connected them in series or/and parallel to pack [6]. Eventually he obtained the equivalent circuit of a series and parallel battery pack by comparing parameters between packs and cells. The battery pack equivalent circuit obtained by Jonghoon Kim is built on the strict basis of battery screening experiments. There are some differences in the actual use of the capacity, resistance and other parameters of the cells. Therefore the accuracy of the pack's equivalent circuit obtained by experimental means needs to be verified.

Researchers from the University of Hawaii adopted statistical knowledge to build pack simulation models based on accurate establishment of a cell simulation model with consideration of the differences in the parameters of the cells of the pack. The accurate simulation of the model is dependent on the accuracy of identification of the cell parameters [7].

Most battery pack modeling researchers took the battery pack as a cell to implement the identification of experimental parameters. The battery equivalent circuit model obtained through this method has some limitations, and it ignores the inconsistencies and over-charge/over-discharge on some of the cells.

2.6.2 Simulation of Battery Pack Model

The simulation model of the battery pack, built in the Matlab/Simulink, includes a SOC calculation module, a cell parameters module for a look-up table in the charging and discharging process, and a parallel topology equation solving module. The SOC calculation module uses an Ah integral equation to calculate the cell's SOC in the charging and discharging process. The cell parameters module creates a table for the cell's parameters, which are identified by experiment. A parallel topology equation solver module is used to solve the equation of state.

The equivalent circuit model of the battery pack is simulated according to Figure 2.25, in which the identification of charging and discharging parameters of the cell are based on experiment. The first-order equivalent circuit model of the battery pack in Figure 2.26 is obtained from circuit knowledge and the results of simulation and experiment [8]. In the equivalent module, the OCV is the sum of the cells' OCV_n, R_Ω is the sum of the cells' $R_{\Omega n}$, R_p is the sum of the cells' R_{pn}, and C_p is approximately the sum of the cells' C_{pn} approximately. The flow chart for the simulation model of the simplified series battery pack is shown in Figure 2.27.

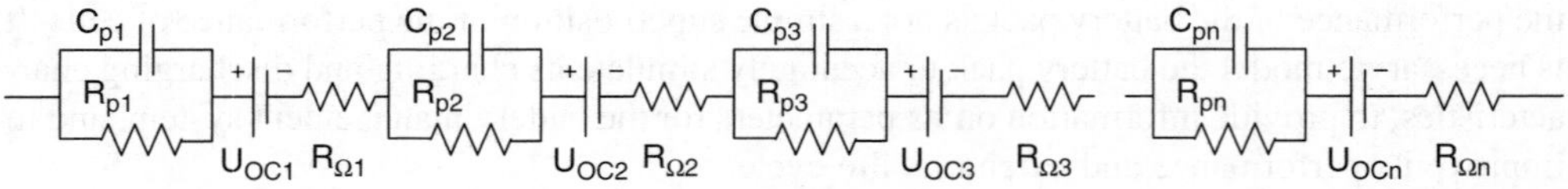

Cp1
Rp1
UOC1
RΩ1
Cp2
Rp2
UOC2
RΩ2
Cp3
Rp3
UOC3
RΩ3
Cpn
Rpn
UOCn
RΩn

Figure 2.25 Thevenin model of series battery pack.

iCp
Cp(series)
iRp
Rp(series)
i
UL(series)
UOCV(series)
RΩ(series)

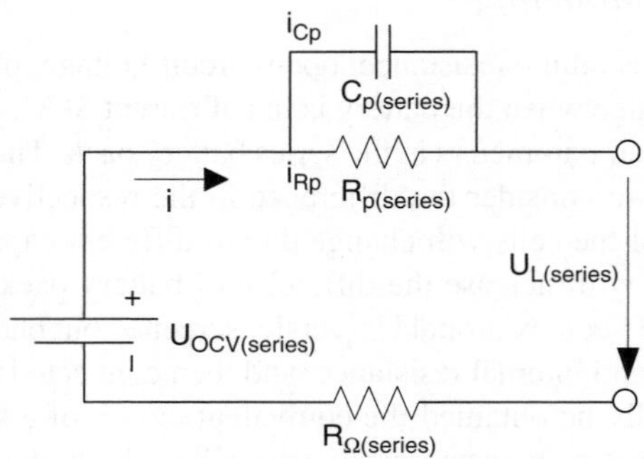

Figure 2.26 Thevenin model of simplified series battery pack.

The simulated models of serial and parallel battery packs are established by differential equations of the equivalent circuit, as shown in Figure 2.28. The S-function in Matlab is used to solve the state equation of the battery circuit model, and then the current on each parallel branch is calculated. The battery branch parameter calculation module uses Ah integration to calculate the SOC of a cell in the charging and discharging processes, and then the module updates the sum of the open-circuit voltage of the series battery pack by using the new SOC of the cells. The ohmic voltage drops and polarization of a series battery pack are received via statistics. Eventually the series battery packs' parameters are returned to the parallel branches current calculation module [9, 10].

The results of different distributions of ohm resistances, SOC and cell capacity are processed in a rating experiment via an experiment platform for battery packs. In the experiment, two batteries with a capacity of 60 Ah are set as 1#A, 2#A, and the battery with a capacity of 20 Ah is set as 3#B.

In the simulation and analysis of the difference of ohmic resistance, the difference between the external voltage V_o and the open-circuit voltage V_{ocv} is expressed as V_Δ, which reflects the difference in the internal voltage of the cell. 1#A and 2#A are connected in parallel to analyze the impact of the difference in ohmic resistances when SOC and capacity are consistent. The current changing on parallel branches during charging process at a current rate of 0.5 C is shown in Figure 2.29. Unbalanced current is quite obvious at the end of charging, and the cell 1#A has reached a charging rate of 0.8 C. The ohmic resistance of 1#A is a little smaller than that of 2#A, which causes the current difference of initial charging, while the unbalanced current with 4#A exists in the voltage plateau during the charging process. When the voltage passes the voltage plateau, the cumulative SOC difference causes the current of 1#A to increase significantly. The difference in SOC between the two cells at the current crossing points in Figure 2.29 is about 7%. Related literature states with regard to the changing of R_p with SOC that "polarization is serious in the low and high SOC, while it changes slowly in the

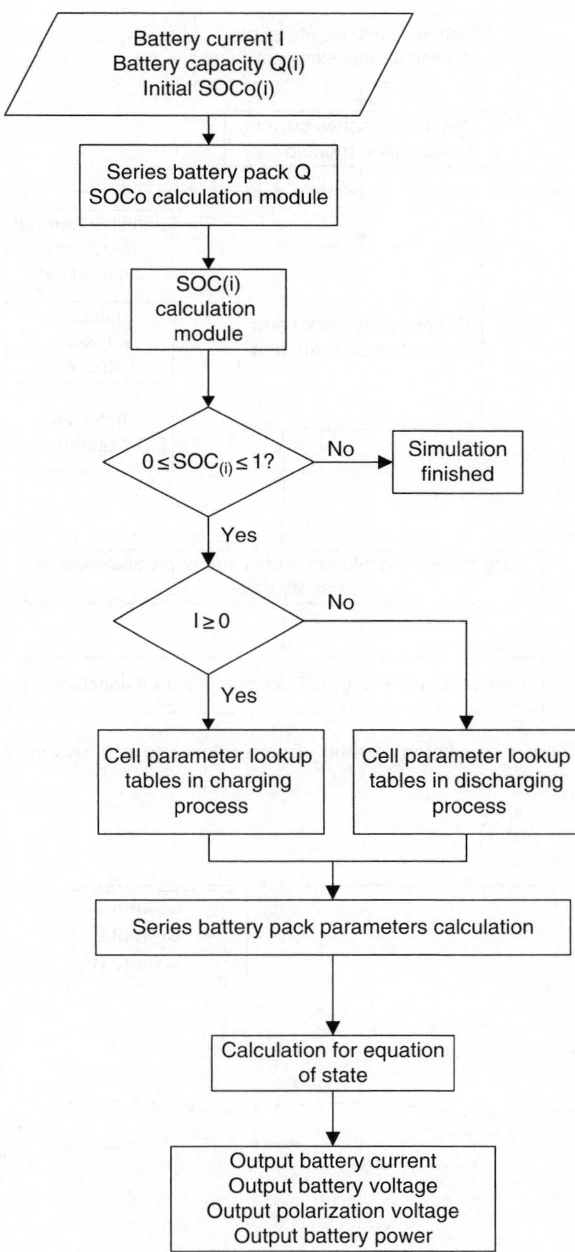

Figure 2.27 Flow chart of simulation model of simplified series battery pack.

voltage plateau". Therefore the difference in polarization voltage U_p, which leads to a large unbalanced current, is large at the end of charging due to the different SOC in the experiment.

Analysis shows that, when the ohm resistances on the branches are considered as an influence factor for the series and parallel simulated model, the current imbalance at the voltage

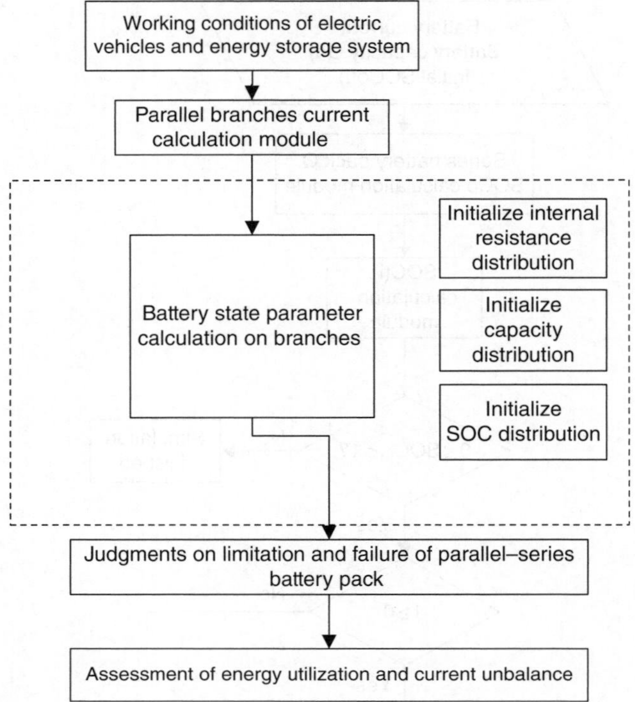

Figure 2.28 Computer simulation method of performance of batteries in series and parallel.

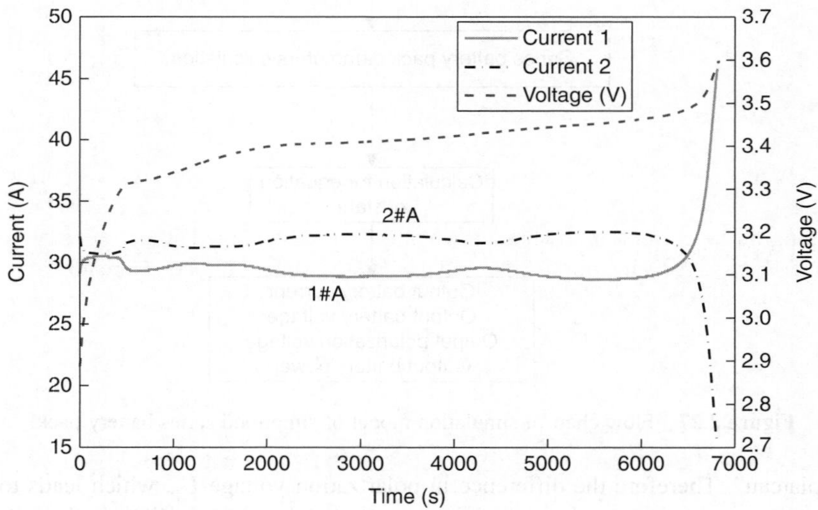

Figure 2.29 Current imbalance due to difference of ohm resistance.

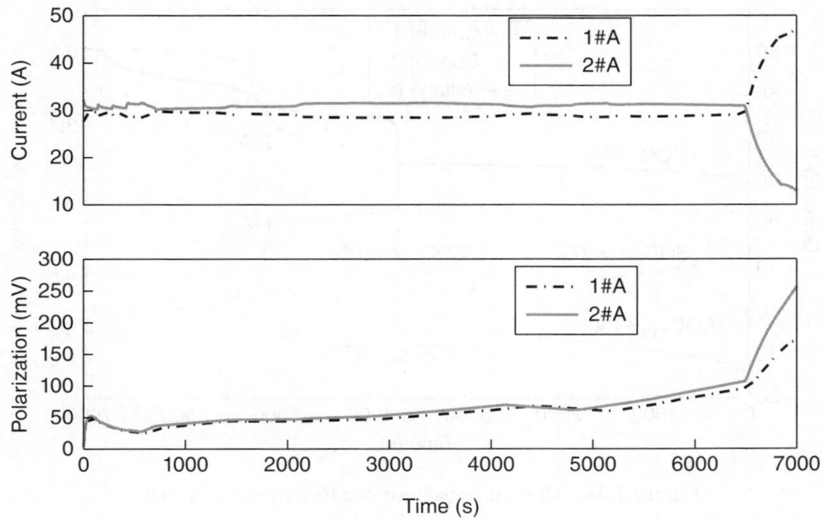

Figure 2.30 Simulation results of difference of ohm resistance.

plateau, which is decided by the difference V_Δ, is caused by the initial difference of the internal ohm resistance, while the current imbalance at the end of charging is caused by the polarization difference. Apparently, the accuracy of simulation on an unbalanced current is dependent on the accuracy of V_Δ, which is comprised of ohm voltage drops V_Ω and polarization voltage u_p. The simulation results for the difference of internal ohm resistance are shown in Figure 2.30. It shows that the accuracy of current changing and imbalance is high, and the average unbalanced current is 4 A initially and at the voltage plateau, and current imbalance is in line with experimental results at the end of charging. In addition, the polarization of the $LiFePO_4$ battery is serious at the end of charging and discharging, and the change is different from the described exponential trend of the circuit model, which makes the trend of current changing different after the current crossing point in the experimental curve.

Simulation and analysis of the difference in SOC are shown below: first 1#A is charged to a SOC of 100%, meanwhile 2#A is discharged to a SOC of 10%. Then 1#A and 2#A are connected in parallel to analyze the impact of the difference in SOC. Figure 2.31 shows the initial current imbalance when 1#A and 2#A are connected in parallel with different initial SOC, and the changing of the current on the branch when the pack is charged at a current rate of 0.5 C. When differences exist in the initial SOC and the battery pack is unloaded, the maximum rate of imbalance current reaches 0.3 C. Under the extreme conditions, the initial transient current is very large when fully discharged or a charged battery pack is connected in parallel. Parallel cells with different SOC are charged at a rate of 0.5 C, the unbalanced current is obvious at the end of charging, and 2#A reaches the charging rate of 1 C. From Figure 2.31, it is found that the battery 1#A enters a current dropping process earlier and drops to 0 A, which causes battery 1#A to be over-charged. Conversely, the current of battery 2#A does not drop to the limit current when the charging finishes, which causes the SOC of 2#A to be less than 100%.

Analysis shows that the over-current, over-charge, current imbalance and time for current balance could be predicted accurately through simulation, when the SOC on the branches is taken into account in the battery pack simulation. The simulation results of the difference of SOC are

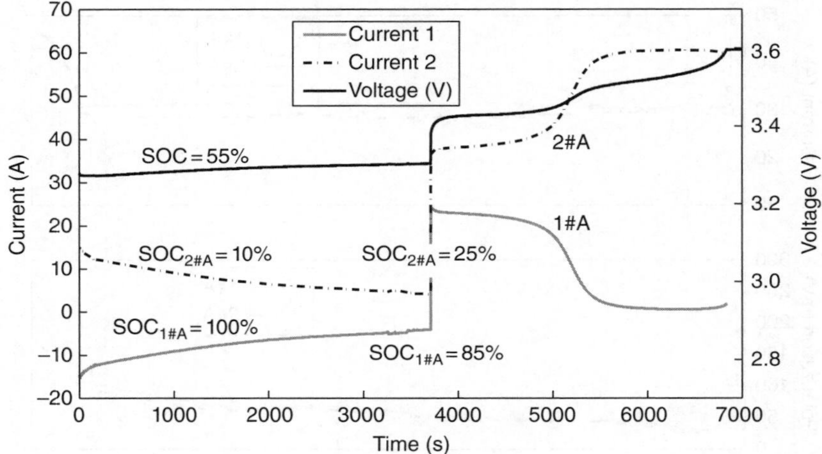

Figure 2.31 Current imbalance due to difference of SOC.

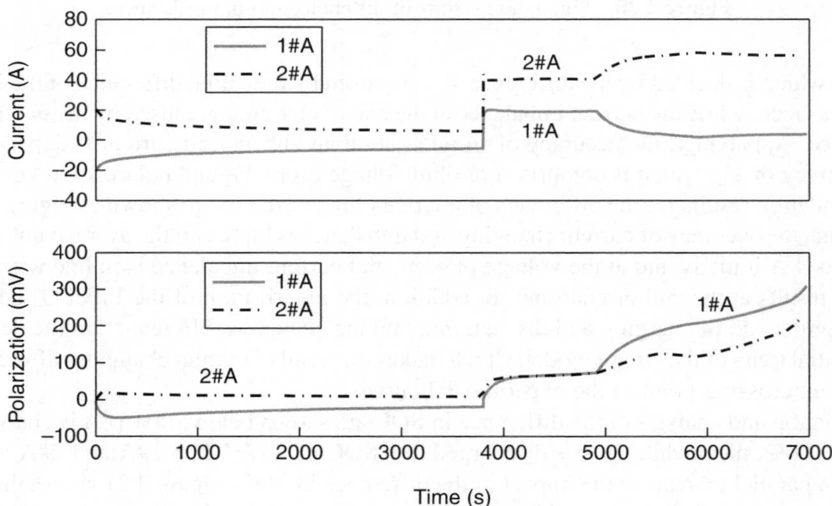

Figure 2.32 Simulation results of difference of SOC. (Reproduced with permission from Weige Zhang, Weishi, Jiuchun Jiang, Jun Jiang, "Numerical Simulation Technique of Series-Parallel Power Lithium Ion Battery", Power System Technology (in Chinese)., vol 36, no.10, 70–74, 2012.)

shown in Figure 2.32. They demonstrate that the time for current balance is in accordance with experimental results; besides, 1#A begins to be charged from a SOC of 85%, and the SOC reaches approximately 100% after 5000 s, while the current drops to about 0 A and the polarization voltage increases abruptly. Because the polarization voltage of 2#A increases due to the parallel connection, the SOC of 2#A reaches 96% when the parallel voltage reaches 3.6 V.

For the simulation and analysis of capacity difference in the case of battery recycling, and taking battery capacity difference as a key factor in the current imbalance, 1#A and 3#B, of the same type but with different specifications, are connected in parallel. Figure 2.33 shows the

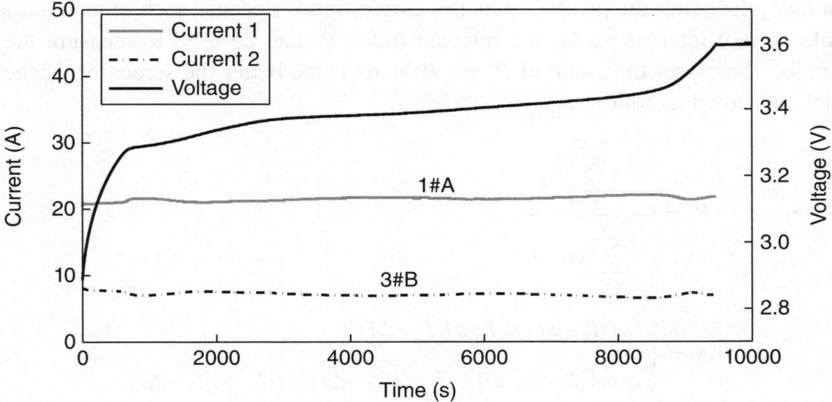

Figure 2.33 Current imbalance due to capacity difference.

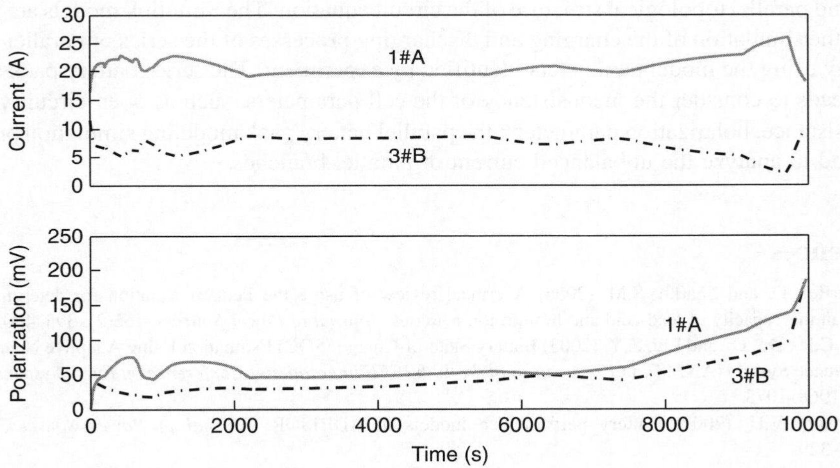

Figure 2.34 Simulation results of capacity difference.

changing of branch current when parallel cells are charged at a rate of 0.35 C. The overall charging process is quite stable. The spontaneous balanced phenomenon is formed by the V_{ocv} of 3#B and 1#A due to the great capacity difference between the two cells, which shows that reasonable methods of screening in the pack can effectively improve the capacity utilization of the battery recycling, and thereby reduce the total cost of the batteries.

Simulation results of the capacity difference are shown in Figure 2.34. From the simulation results, whether the battery packs with capacity difference are connected in parallel or not, the deviation of the branch current obtained from the designed average current is assessed and confirmed. The deviation degree can be represented by the sum of the residuals squared, S, to judge the effect, that is:

$$S=\sum_{t=1}^{N}(I_t-\bar{I})^2 \tag{2.27}$$

The smaller S, the better the parallel branches current ratio. Because each branch has a certain acceptable current interval $\bar{I} \pm \Delta I$, the relevant index R^2 can be used to measure the branch current ratio. The closer the value of R^2 (or R) is to 1, the better the screening of the battery in parallel, as shown as follows:

$$R^2 = 1 - \frac{\sum_{t=1}^{N} A_t^2}{\sum_{t=1}^{N} (I_t - \bar{I})^2}, \qquad (2.28)$$

$$A_t = \begin{cases} 0, (|\bar{I}| - \Delta I) \le |I_t| \le (|\bar{I}| + \Delta I) \\ \min\{|I_t - \bar{I} \pm \Delta I|\}, |I_t| \le (|\bar{I}| - \Delta I) \cup (|I_t| \ge (|\bar{I}| + \Delta I) \end{cases}$$

The battery pack modeling analyzes series and parallel battery equivalent circuit models based on the basic electrical equivalent circuit, and the differential equations are obtained from the series and parallel topological structure of the circuit equation. The Simulink models are built to realize the simulation of the charging and discharging processes of the series or parallel battery packs by using the model parameters identified by experiment. The series battery packs' modeling needs to consider the inconsistency of the cell parameters, such as open-circuit voltage, ohm resistance, polarization parameters; the parallel battery pack modeling simulation provides a method to analyze the unbalanced current of parallel branches.

References

[1] Doerffel, D. and Sharkh, S.M. (2006) A critical review of using the Peukert equation for determining the remaining capacity of lead-acid and lithium-ion batteries. *Journal of Power Sources*, **155**(2), 395–400.

[2] Cai, C.H., Du, D., and Liu, Z.Y. (2003) Battery State-of-Charge (SOC) Estimation Using Adaptive Neuro-Fuzzy Inference System (ANFIS). in *Proceedings of the 12th IEEE International Conference on Fuzzy Systems*., IEEE, pp. 1068–1073.

[3] Johnson, V.H. (2002) Battery performance models in ADBISOR. *Journal of Power Sources*, **110**(8), 321–329.

[4] *Battery Application Manual* Gates Energy Products, Inc., Gainesville, FL, 1989.

[5] Idaho National Engineering & Environmental Laboratory (INEEL) (2001) PNGV Battery Test Manual, Revision 3, DOE/ID-10597, 2001, PDF version.

[6] Gerard, O., Patillon, J.N., and D'Alche-Buc, F. (1997) Neural Network Adaptive Modeling of Battery Discharge Behavior, in *Artificial Neural Networks – ICANN '97, Proceedings of the 7th International Conference, Lausanne, Switzerland, October 1997*, Springer, pp. 1095–1100.

[7] Dubarry, M. and Vuillaume, N. (2009) BY Liaw From single cell model to battery pack simulation for Li-ion batteries. *Journal of Power Sources*, **186**, 500–507.

[8] Kim, J., Shin, J., Jeon, C., and Cho, B. (2011) High accuracy state-of-charge estimation of li-ion battery pack based on screening process. in *Applied Power Electronics Conference and Exposition APEC 2011*, IEEE, pp. 1984–1991.

[9] Wei S., Jiuchun J., Weige Z., *et al.* (2012) Research on charging and discharging characteristics of parallel connection for LiFePO4 Li-ion batteries. *Chinese High Technology Letters*, **22**(2), 205–210 (in Chinese).

[10] Weige Z., Wei S., Jiuchun J., *et al.* Numerical simulation technique of series–parallel power lithium-ion battery. *Power System Technology*, **36**(10), 70–75 (in Chinese).

3

Battery State Estimation

The state estimate of a battery includes the estimate of state of charge (SOC) and state of energy (SOE). SOC in electric vehicles serves a similar function to that of the petrol gauge in the traditional oil-fueled automotive. For hybrid vehicles, an accurate estimate of SOC is the precondition for the optimization of the vehicle control. Effectively estimating SOC and involving it in the vehicle control play an important role in collecting the battery information, optimizing the vehicle operation and driving experience, enhancing the utilization rate of battery capacity and energy, preventing the battery from over-charging and over-discharging, and ensuring the safety and service life during its use. SOE analyzes the remaining capacity from the viewpoint of energy, and, in terms of usage, it can build a simple (linear) relation between the remaining capacity and the time of use or running distance of an electric vehicle. In actual use, the main function of the battery is to store and release energy. The running distance and endurance time of the electric vehicle are directly related to the quantity of energy released by the battery. Hence, describing the SOC and the state of the battery from the viewpoint of energy is of more practical significance.

3.1 Definition of SOC

3.1.1 The Maximum Available Capacity

The discharging capacity is closely related to the current, ambient temperature, and the running conditions. In different running conditions, the difference in battery capacity depends on the fact that the current, the charge or the discharge method, as well as the ohm pressure drop U_R and the polarized voltage U_P, change with environmental variations (at low temperature, high current and continuously discharging state, U_R and U_P are larger). Even when discharging stops, the external voltage U_O equals the discharge cutoff voltage. However, the SOC is different and it leads to the difference in discharge capacity, because U_{OCV}, where $U_O = U_{OCV} - U_R - U_P$, is different. Therefore, the open-circuit voltage (OCV) deviates because the ohm pressure drop and the polarized voltage are the original reason for the difference in

Fundamentals and Applications of Lithium-ion Batteries in Electric Drive Vehicles, First Edition.
Jiuchun Jiang and Caiping Zhang.

the discharge, which eventually results in the coupling between the battery's SOC and the running conditions.

When the discharge current $I \rightarrow 0$, U_R, and U_P can be ignored, thus $U_O \equiv U_{OCV}$. When discharging stops, $U_O = U_{OCV} = U_{Dch\_end}$. At this time, the battery SOC is consistent, and then the discharge capacity will reach the maximum and tends to be stable. Examples are as follows:

Discharge the nominal capacity 90 Ah Li-MnO_2 battery which is fully charged by 1/3, 1/2, and 3/4 C to 3.3 V, respectively. The discharge capacities are 89, 88, and 87 Ah. The discharge curves are shown in Figure 3.1, while the curves of the restoration of the voltage with resting time are shown in Figure 3.2. The voltage, which is recorded after sufficient resting time,

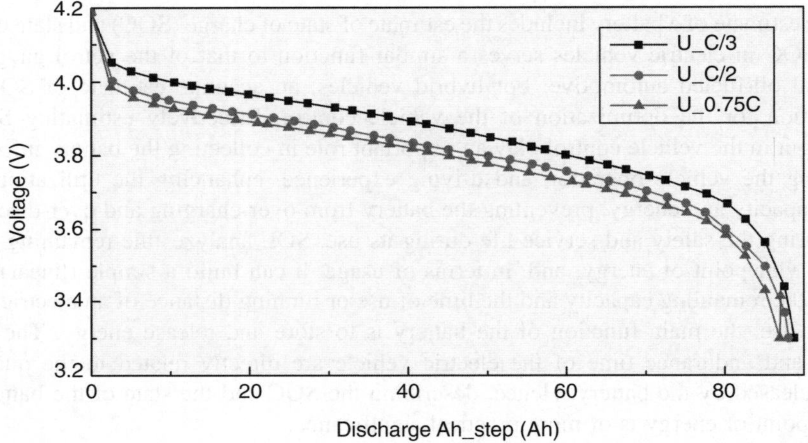

Figure 3.1 Discharge curves with different discharge currents.

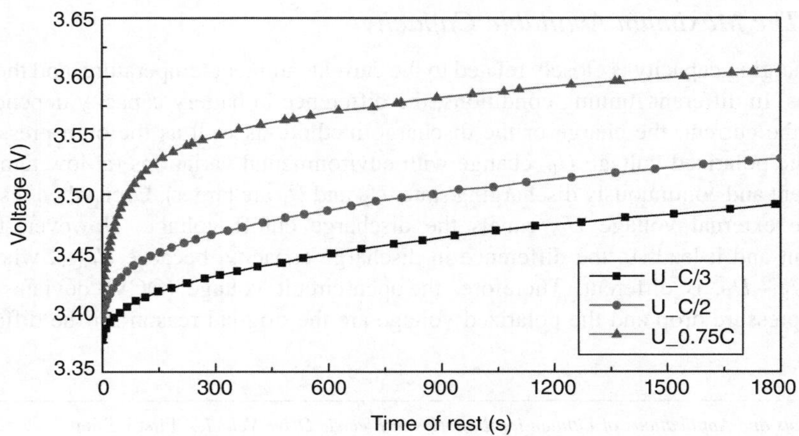

Figure 3.2 Restoring curves of voltage with different currents.

increases with the discharge current. This shows that under the condition of high discharge current, the maximum available capacity of the battery is not reduced, but instead, has not been fully released.

The total discharge capacity is the same when lowering the discharge current after the high current discharge. Taking the example of discharge testing with different current: the battery is first sufficiently charged and then the discharging procedures are as shown in Table 3.1. The curves of the three discharge processes are shown in Figure 3.3.

When the first stage finishes, the discharge capacity is 62, 64, and 65 Ah, respectively. But when lowering the discharge current gradually to the end of step 6, the total capacity of the three discharges are 67 Ah. Therefore, it is seen that by lowering the discharge current after discharging with different current to the cutoff voltage, the battery can continue to discharge, and the total discharge capacity is the same.

Table 3.1 Procedures to lower the discharge current after discharging at different rates.

First	Second	Third
Step1: discharge at 30 A to the end voltage 3.3 V	Step 1: discharge at 60 A to the end voltage 3.3 V	Step 1: discharge at 90 A to the end voltage 3.3 V
Step 2: discharge at 20 A to the end voltage 3.3 V	Step 2: discharge at 30 A to the end voltage 3.3 V	Step 2: discharge at 60 A to the end voltage 3.3 V
Step 3: discharge at 10 A to the end voltage 3.3 V	Step 3: discharge at 20 A to the end voltage 3.3 V	Step 3: discharge at 30 A to the end voltage 3.3 V
Step 4: discharge at 3 A to the end voltage 3.3 V	Step 4: discharge at 10 A to the end voltage 3.3 V	Step 4: discharge at 20 A to the end voltage 3.3 V
	Step 5: discharge at 3 A to the end voltage 3.3 V	Step 5: discharge at 10 A to the end voltage 3.3 V
		Step 6: discharge at 3 A to the end voltage 3.3 V

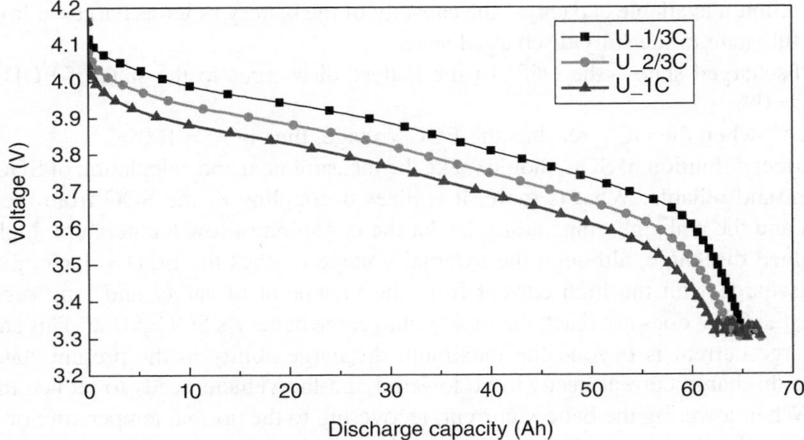

Figure 3.3 Lowering the discharge current after discharging with different rate.

Theoretically, the principle of the conservation of the number of recyclable lithium ions claims that the number of recyclable lithium ions is the same in the same or adjacent discharging process. What changes in the process is the ambient temperature, the running condition and the consequent ohm pressure drop and polarized voltage. So when the external voltage reaches the end of discharge (EOD) voltage, the intercalation/deintercalation amount of lithium ion is different, which leads to the difference in the discharge capacity. However, these ions still exist in the negative electrode of the battery, that is, the number of recyclable lithium ions does not decrease. So when improving the electric ambient, these ions can still be used and the battery can continue to be discharged, which leads to the total discharge capacity conservation.

Based on the above, the concept of the maximum available capacity of a battery is reached and defined as the total discharge capacity in the process of discharging the sufficiently charged battery to the EOD voltage with an ample small current. It is different from the actual capacity in that the state of the maximum available capacity requires the OCV to equal the cutoff voltage when the discharge is finished. Under different current conditions, related to the temperature and discharge model, though the discharge capacity is different, the maximum available capacity is the same. In this theory, once the cutoff charge or discharge voltage is determined, then the maximum available capacity of the battery is fixed. It shows that the number of recyclable lithium ions only relates to the design capacity and aging degree of the battery.

3.1.2 Definition of Single Cell SOC

The SOC refers to the remaining capacity (Q_{rem}) as a percentage of the maximum available capacity:

$$\mathrm{SOC} = \frac{Q_{\mathrm{rem}}}{Q_{\mathrm{max}}} \times 100\% \tag{3.1}$$

Remaining capacity – the capacity of the battery as it discharges from the present state to the fully discharged state.

The maximum available capacity – the capacity of the battery as it discharges at low current from the full state to the fully discharged state.

Fully discharged state – the OCV of the battery discharges to the state of EOD voltage, thus SOC = 0%.

Full SOC–when the OCV reaches the EOC voltage, thus SOC = 100%.

The correct definition of SOC should make the measurement and calculation of SOC simple, convenient and reliable. What is more, it realizes decoupling of the SOC from the running condition and the ambient temperature. Under the condition of low temperature, high current and sustained discharge, although the external voltage reaches the EOD voltage, the battery cannot discharge with the high current from the viewpoint of safety and long service life. And since the OCV does not reach the EOD voltage, the battery's SOC $\neq 0$ %. This shows that the discharge current is beyond the maximum discharge ability in the present state, which means the discharge current needs to be lowered and the vehicle needs to reduce the power to drive. When lowering the battery current, recovering to the normal temperature or standing for a little while, the ohm pressure drops and the polarization voltage decreases to a certain extent and the remaining energy of the battery can be released.

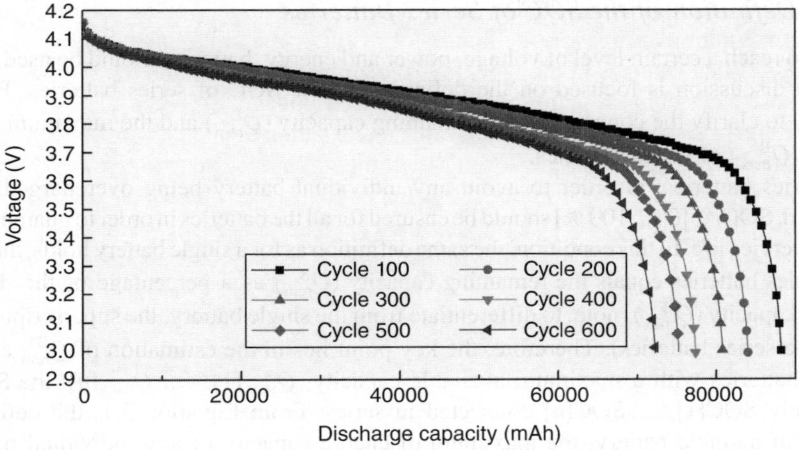

Figure 3.4 The discharge curve after different numbers of cycles.

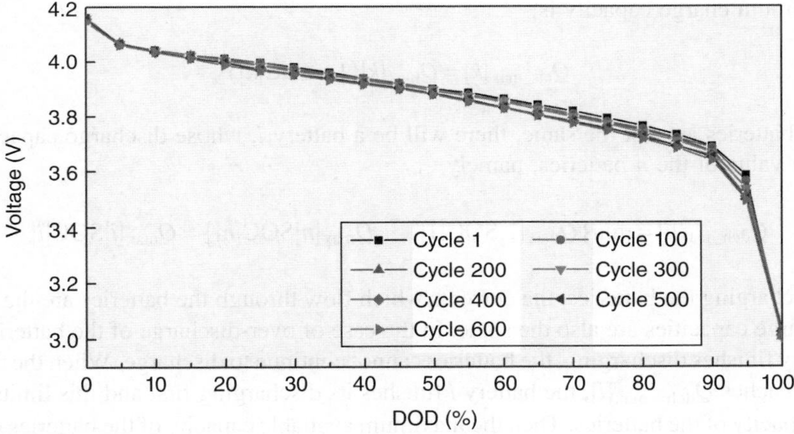

Figure 3.5 The discharge curve of the battery at different aging stages.

To consider the effect of aging, the maximum available capacity of different aging stages should be reviewed. Figure 3.4 shows the discharge curves obtained after different numbers of cycles. It is seen that with increasing number of cycles, the battery ages gradually and the discharge capacity decreases.

Adopting the maximum available capacity to revise the battery's SOC at different aging stages, the changes in discharge voltage with the SOC are shown in Figure 3.5. It can be seen that the discharge curves tend to be consistent after revising. Therefore, revising the maximum available capacity of the battery during aging plays an important role in enhancing the accuracy of SOC estimation.

3.1.3 Definition of the SOC of Series Batteries

In order to reach a certain level of voltage, power and energy, batteries should be used in series, so herein discussion is focused on the definition of the SOC of series batteries. First, it is necessary to clarify the concepts of the remaining capacity ($Q_{\text{rem}}^{\text{B}}$) and the maximum available capacity ($Q_{\text{max}}^{\text{B}}$) of series batteries.

For series batteries, in order to avoid any individual battery being overcharged or over-discharged, $\text{SOC}_i \in [0\,\%, 100\,\%]$ should be ensured for all the batteries in order to guarantee safety and long service life. In this condition, the same definition as for a single battery holds, the (SOC^{B}) of the series batteries equals the remaining capacity ($Q_{\text{rem}}^{\text{B}}$) as a percentage of the maximum available capacity ($Q_{\text{max}}^{\text{B}}$) (note, to differentiate from the single battery, the superscript B is used to indicate series batteries). Therefore, the key point lies in the estimation of $Q_{\text{rem}}^{\text{B}}$ and $Q_{\text{max}}^{\text{B}}$.

For n batteries with a maximum available capacity, $Q_{\text{max}}[1]$, …, $Q_{\text{max}}[n]$, the SOCs are respectively SOC[1], …, SOC[n] connected in series. From Equation 3.1, the definition of the SOC of a single battery, the maximum discharge capacity of any individual battery (k) in the series $Q_{\text{dch\_max}}[k]$ is:

$$Q_{\text{dch\_max}}[k] = Q_{\text{rem}}[k] = Q_{\text{max}}[k]\text{SOC}[k] \tag{3.2}$$

The maximum charge capacity is:

$$Q_{\text{ch\_max}}[k] = Q_{\text{max}}[k](1-\text{SOC}[k]) \tag{3.3}$$

As these batteries are not the same, there will be a battery i, whose discharge capacity is the minimum value of the n batteries, namely,

$$Q_{\text{dch\_max}}[i] = \min\{Q_{\text{max}}[1]\text{SOC}[1], \ldots, Q_{\text{max}}[n]\text{SOC}[n]\} = Q_{\text{max}}[i]\text{SOC}[i] \tag{3.4}$$

When discharging the batteries, the currents which flow through the batteries are the same, so the discharge capacities are also the same. In the case of over-discharge of the batteries, when any battery finishes discharging, the batteries cannot continue to discharge. When the discharge capacity reaches $Q_{\text{dch\_max}}[i]$, the battery i finishes its discharging first and this limits the discharge capacity of the batteries. Then the maximum available capacity of the batteries $Q_{\text{dch\_max}}^{\text{B}}$, namely, the remaining capacity $Q_{\text{rem}}^{\text{B}}$ equals the maximum discharge capacity of battery i:

$$Q_{\text{rem}}^{\text{B}} = Q_{\text{dch\_max}}^{\text{B}} = Q_{\text{dch\_max}}[i] = \min\{Q_{\text{max}}[1]\text{SOC}[1], \ldots, Q_{\text{max}}[n]\text{SOC}[n]\} \tag{3.5}$$

Similarly, there must be a battery j whose charge capacity will be the minimum:

$$\begin{aligned} Q_{\text{ch\_max}}[j] &= \min\{Q_{\text{max}}[1](1-\text{SOC}[1]), \ldots, Q_{\text{max}}[n](1-\text{SOC}[n])\} \\ &= Q_{\text{max}}[j](1-\text{SOC}[j]) \end{aligned} \tag{3.6}$$

When charging the series batteries, the charge capacities are the same because the currents through every battery are the same. In the case of over-charging the batteries, when any battery finishes charging, the batteries cannot continue to charge. When the charge capacity reaches $Q_{ch\_max}[j]$, the battery j finishes its charging first and this limits the charge capacity of the batteries. Then the maximum available capacity of the batteries $Q^B_{ch\_max}$ equals the maximum charge capacity:

$$\begin{aligned} Q^B_{ch\_max} &= Q_{ch\_max}[j] = Q_{max}[j](1-\mathrm{SOC}[j]) \\ &= \min\{Q_{max}[1](1-\mathrm{SOC}[1]),\ldots,Q_{max}[n](1-\mathrm{SOC}[n])\} \end{aligned} \tag{3.7}$$

Therefore, the maximum available capacity of the series batteries is:

$$\begin{aligned} Q^B_{max} &= Q^B_{dch\_max} + Q^B_{ch\_max} = Q_{dch\_max}[i] + Q_{ch\_max}[j] \\ &= \min\{Q_{max}[1]\mathrm{SOC}[1],\ldots,Q_{max}[n]\mathrm{SOC}[n]\} + \\ &\quad \min\{Q_{max}[1](1-\mathrm{SOC}[1]),\ldots,Q_{max}[n](1-\mathrm{SOC}[n])\} \end{aligned} \tag{3.8}$$

From the above, it is seen that the maximum available capacity (Q^B_{max}) of the batteries is closely related to the maximum available capacity Q^B_{max} and the SOC of a single battery.

When $i=j$, Q^B_{max} equals $Q_{max}[i]$ of the battery whose maximum available capacity is the smallest in the group the available capacity is the smallest. This shows why Q^B_{max} of the series batteries is reduced.

Based on the analyses above, the SOC of the series batteries can be defined as:

$$\begin{aligned} \mathrm{SOC}^B &= \frac{Q^B_{rem}}{Q^B_{max}} \times 100\% \\ &= \frac{\min(Q_{max}[1]\mathrm{SOC}[1],\ldots,Q_{max}[n]\mathrm{SOC}[n])}{\min(Q_{max}[1]SOC[1],\ldots,Q_{max}[n]\mathrm{SOC}[n]) + \min(Q_{max}[1](1-\mathrm{SOC}[1]),\ldots,Q_{max}[n](1-\mathrm{SOC}[n]))} \times 100\% \\ &= \frac{Q_{max}[i]\mathrm{SOC}[i]}{Q_{max}[i]\mathrm{SOC}[i] + Q_{max}[j](1-\mathrm{SOC}[j])} \times 100\% \end{aligned} \tag{3.9}$$

1. When $i=j$, the formula can be simplified as:

$$\mathrm{SOC}^B = \frac{Q_{max}[i]\mathrm{SOC}[i]}{Q_{max}[i]} \times 100\% = \mathrm{SOC}[i] \tag{3.10}$$

   Then the estimation of SOC^B for series batteries can be described as the estimation of SOC of a single battery.
2. When $i \neq j$, if the battery pack is discharged till the electric quantity is $Q_{max}[i] \times \mathrm{SOC}[i]$, then the battery i discharges completely and $\mathrm{SOC}^B = 0\ \%$. On the contrary, if the battery pack is charged till the electric quantity is $Q_{max}[j] \times (1 - \mathrm{SOC}[j])$, then the battery j charges completely, and the remaining capacity will be $Q_{max}[i] \times \mathrm{SOC}[i] + Q_{max}[j] \times (1 - \mathrm{SOC}[j])$ and $\mathrm{SOC}^B = 100\ \%$.

By adopting the definition of SOC of the series batteries the definition and estimation of batteries can be solved effectively under inconsistent conditions and become more universal. However, this is only possible if $SOC^B \in [0\ \%, 100\ \%]$, then $SOC \in [0\ \%, 100\ \%]$ must hold for all the single batteries in order to avoid over-charge and over-discharge, which offers data support and enhances the safety of the batteries.

Examples to verify this are shown below:

Example 1
Take two batteries with the same maximum available capacity 100 Ah, but different SOC: SOC[1] = 100 %, SOC[2] = 0 % in series. Then Equation 3.5 shows that $Q^B_{rem} = \min(100 \times 100\%, 100 \times 0\%) = 0$ Ah, and $Q^B_{max} = 0$ Ah can be reached by applying Equation 3.8. The batteries cannot be charged or discharged, so the definition is scientific. It shows that the maximum available capacity and remaining energy of the series batteries is closely related to the SOC of a single battery.

Example 2
Take two batteries with the same SOC of 100% and the maximum available capacities of 50 and 100 Ah in series. Then $Q^B_{rem} = \min(50 \times 100\%, 100 \times 100\%) = 50$ Ah, $Q^B_{max} = 50$ Ah. This shows that the maximum available capacity is closely related to the capacity of every single battery.

Example 3
Take two batteries with the parameters SOC[1] = 100 %, $Q_{max}[1] = 50$ Ah and SOC[2] = 25 %, $Q_{max}[2] = 100$ Ah in series. Then $Q^B_{rem} = \min(50 \times 100\%, 100 \times 25\%) = 25$ Ah, $Q^B_{max} = 25$ Ah. This shows that the capacity of the series batteries may be smaller than the capacity of the smallest single battery,which is closely related to the SOC of a single battery in the pack.

3.2 Discussion on the Estimation of the SOC of a Battery

3.2.1 Load Voltage Detection

During the discharging process of a battery, when the load current is stable, the change in load voltage with SOC is similar to that of the OCV with SOC, so we can use the load voltage to estimate the SOC. Load voltage curves of a lithium-ion battery at different discharge rates are shown in Figure 3.6.

The advantage of load voltage detection is that when discharging with constant current, real-time estimation of the SOC of batteries is realized effectively. However, in practical application, the drastically fluctuating voltage of batteries makes load voltage detection difficult. To solve this problem, it is necessary to store a huge amount of voltage data and build a mathematical model of the dynamic load voltage with SOC, current and temperature. Load voltage detection is not commonly used in real vehicles, but instead is often used to judge the cut-off of batteries' charge or discharge.

3.2.2 Electromotive Force Method

The electromotive force of a battery represents the electric field force of the battery and describes the inner driving force when the battery is outputting electric energy. The electromotive force of a battery can be obtained in theory from the thermodynamic parameters of the

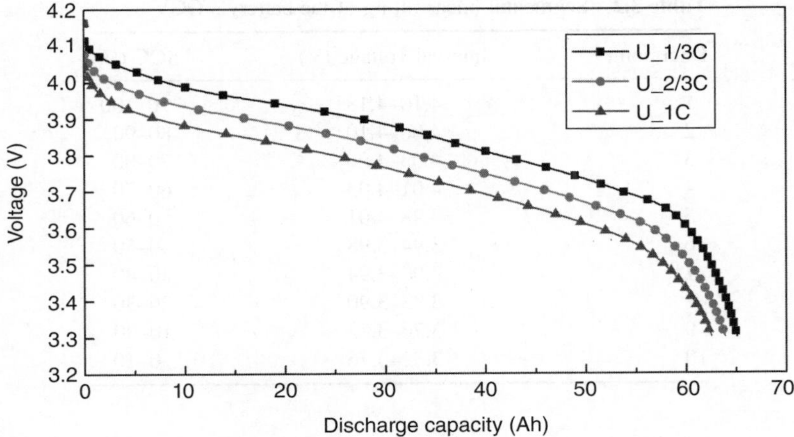

Figure 3.6 Load voltage curves of a lithium-ion battery at different rates.

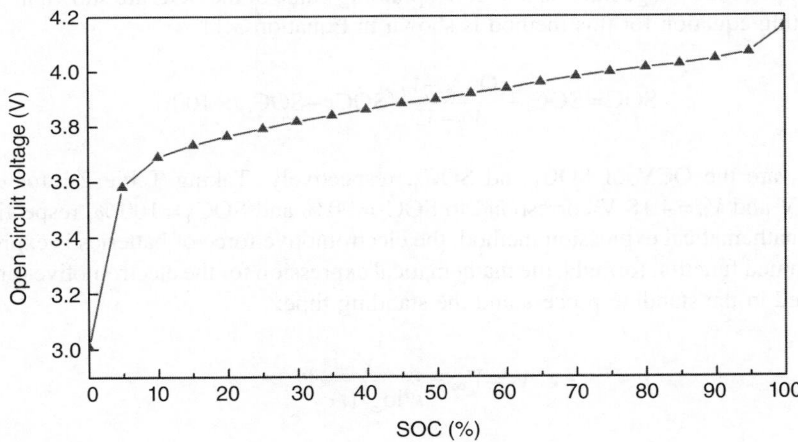

Figure 3.7 Load voltage curves of lithium-ion battery at different rates.

battery and the Nernst equation. In practice the OCV of the battery is often used to approximate the electromotive force. The OCV of a lithium-ion battery, especially a lithium manganese oxide battery, has an approximately linear relationship with SOC and does not change with battery running conditions. Thus, the relation can be used to estimate the SOC. The changes that the OCV of a lithium manganese oxide battery undergoes with the SOC are shown in Figure 3.7. However, the acquisition of the OCV requires that the batteries stand for a long time and are stable, especially under conditions of low temperature and high rate, when the recovery of the battery needs a few hours or even more to keep it stable. There are three methods of estimation of SOC based on the OCV using: look-up tables, segmental linear functions and a mathematical expression.

The look-up table method involves putting the battery parameters into tables linking the parameters to the SOC. The estimation accuracy depends on the size of the data file.

Table 3.2 Segmental linear fitting of the battery's OCV.

Data point	Interval voltage (V)	SOC (%)
1	4.10–4.18	90–100
2	4.06–4.10	80–90
3	4.03–4.06	70–80
4	4.01–4.03	60–70
5	3.98–4.01	50–60
6	3.94–3.98	40–50
7	3.90–3.94	30–40
8	3.83–3.90	20–30
9	3.76–3.83	10–20
10	3.32–3.76	0–10

Using the segmental linear function method, the OCV can be approximated to segmental linear functions. Taking the OCV at 10 equal SOC discontinuity points of the single battery as an example, the voltage interval and corresponding value of the SOC are shown in Table 3.2 The fitting equation for this method is shown in Equation 3.11.

$$\mathrm{SOC}=\mathrm{SOC}_l+\frac{\mathrm{OCV}-V_l}{V_h-V_l}(\mathrm{SOC}_h-\mathrm{SOC}_l)\times 100\% \tag{3.11}$$

V_l and V_h are the OCV of SOC_l and SOC_h, respectively. Taking Table 2.1 for example, $V_l = 4.10$ V and $V_h = 4.18$ V correspond to $\mathrm{SOC}_l = 90$ % and $\mathrm{SOC}_h = 100$ %, respectively.

In the mathematical expression method, the electromotive force of batteries is expressed by the determined function formula, the mathematical expression for the electromotive force about OCV tested in the standing process and the standing time:

$$V_t = V_\infty - \frac{\Gamma\gamma}{t^\alpha \log^\delta(t)} e^{\varepsilon_t/2} \tag{3.12}$$

where $\gamma > 0, \alpha > 0, \delta > 0$ are all constants related to the rate of battery charging or discharging. If the battery terminal voltage increases with time (standing after the battery discharging), $\Gamma = +1$; if the battery decreases with time (standing after the battery charging), $\Gamma = -1$. V_∞ is the OCV when the battery is stable, that is the electromotive force of the battery; V_t is the OCV tested at time t; $\log^\delta(t)$ is the δ power of the natural logarithm of time t; ε_t is the stochastic error term.

The method of electromotive force is better for estimating SOC at the beginning and end of charging and is often used with the Ah counting method (see Section 3.2.4).

3.2.3 Resistance Method

The internal resistance of the battery can be classified into internal impedance and resistance, both of which are closely related to the SOC. By using electrochemical impedance spectroscopy (EIS) and the relationship between each electrical parameter of the equivalent circuit model and the SOC, the SOC can be estimated according to the equivalent circuit model parameter of the battery EIS in the practical application. However, the resistance method is rarely used as AC impedance is greatly affected by the temperature and there is some dispute about whether the measurement of AC impedance should be made in the balanced state of the battery or in the charge–discharge process [1].

3.2.4 Ampere-hour Counting Method

The Ah counting method is simple, reliable and easy to realize, so it is a commonly-used approach to estimate SOC [2]. If the original state of charging or discharging is SOC_0, then the current state is:

$$SOC_{end} = \frac{Q_{rem}}{Q_{max}} = SOC_0 + \frac{\Delta Q}{Q_{max}} = SOC_0 + \frac{\int_{t_{SOC_0}}^{t_{SOC_{end}}} i\mathrm{d}t}{Q_{max}} \tag{3.13}$$

Where, SOC_0 refers to the initial state of charging and SOC_{end} refers to the current state of charging, ΔQ refers to the capacitance change from the initial state to the current state, so the estimation of SOC needs knowledge of certain parameters, including ΔQ, SOC_0, and Q_{max}.

According to the definition of SOC, the maximum available capacity of the battery Q_{max} can be found from:

$$Q_{max} = \frac{\Delta Q}{SOC_2 - SOC_1} = \frac{\int_{t_1}^{t_2} i\mathrm{d}t}{SOC_2 - SOC_1} \tag{3.14}$$

The Ah counting method can be applied to the batteries of all electric vehicles, if the current measured is accurate and the data of the initial state is sufficient, then it is definitely a simple and reliable way to estimate SOC. The Ah integral cannot effectively solve problems such as accumulative error, self-discharging capacity loss and the changes in the maximal available capacity. With the usage time increasing, the estimation error of SOC will gradually increase, so it is important to revise SOC_0 regularly. The premise is to obtain U_{OCV}, and then to take advantage of SOC-OCV to obtain the new SOC_0.

For electric vehicles, the following three revision modes are commonly used, so SOC_0 can be corrected regularly. This solves the accumulative error resulting from the Ah integral and ensures the precision of the SOC estimation.

1. Vehicles out of operation for a long time. When vehicles are stopped, out of operation for a long time (such as at night or in non-working hours), the polarization voltage U_P can regress quickly during the period and, when the vehicles are charging again, it can be decided whether or not to revise the SOC, based on the current voltage by calculating the standing time.
2. Charging completion. When the battery's charging is complete (the voltage reaches the cutoff voltage and the current is small enough), because the battery is fully charged, on the one hand, the rate of OCV with regard to SOC changes a lot; every 1% SOC corresponds to much voltage variation. On the other hand, the charging current is small enough, so the

ohmic potential drop and polarization voltage can be ignored, and then the revision of the SOC_0 according to the external voltage of the battery can be realized directly.
3. Before charging. The method recognizing the polarization voltage quickly at the beginning of battery charging, then achieving U_{OCV} can revise the SOC_0.

3.2.5 Kalman Filter Method

The Kalman filter method uses a state space method to describe the system, whose solution is calculated by a recursive method and has the feature of self-adaptability [3–8]. For a linear system the Kalman filter is gradually stable. For the nonlinear characteristic of a power battery, the Taylor formula is used for linearization followed by calculation by the Kalman recursive method. This is known as the extended Kalman Filter (EKF). For a stationary stochastic nonlinear discrete system, the general form of the state equation and observation equation are:

$$\begin{cases} x_{k+1}=f(x_k,u_k)+\Gamma_k w_k \\ y_k=g(x_k,u_k)+v_k \end{cases} \tag{3.15}$$

where, x_k is the state vector, y_k the observation vector, u_k the control vector, w_k the system noise vector, v_k the measurement noise vector, A_k the system matrix, B_k the control input matrix, Γ_k the interference matrix, C_k the measurement matrix, Q_k the system noise variance matrix, R_k the measurement noise variance matrix. $E[w_k]=0$, $E\left[w_k w_j^{\mathrm{T}}\right]=Q_k\delta_{k,j}$, $E[v_k]=0$, $E\left[v_k v_j^{\mathrm{T}}\right]=R_k\delta_{k,j}$. A nonlinear system can be solved as a linear system through the Taylor formula:

$$f(x_k,u_k)\approx f(\hat{x}_k,u_k)+\frac{\partial f(x_k,u_k)}{\partial x_k}\Big|_{x_k=\hat{x}_k}(x_k-\hat{x}_k)$$

$$g(x_k,u_k)\approx g(\hat{x}_k,u_k)+\frac{\partial g(x_k,u_k)}{\partial x_k}\Big|_{x_k=\hat{x}_k}(x_k-\hat{x}_k)$$

Definition:

$$\hat{A}_k=\frac{\partial f(x_k,u_k)}{\partial x_k}\Big|_{x_k=\hat{x}_k},\ C_k=\frac{\partial g(x_k,u_k)}{\partial x_k}\Big|_{x_k=\hat{x}_k},$$

then:

$$\begin{cases} x_{k+1}\approx \hat{A}_k x_k+\left[f(\hat{x}_k,u_k)-\hat{A}_k\hat{x}_k\right]+\Gamma_k w_k \\ y_k\approx \hat{C}_k x_k+\left[g(\hat{x}_k,u_k)-\hat{C}_k\hat{x}_k\right]+v_k \end{cases} \tag{3.16}$$

The process of EKF of a nonlinear discrete system is from Equations 3.17 to 3.22, where, $\hat{x}_{k/k-1}$ is the predictive value of the estimative state, $\hat{x}_{k/k}$ is the filter value of the estimative state, K_k is the gain matrix of the Kalman filter, $P_{k/k}$ is the covariance matrix of the filter error, $P_{k/k-1}$ is the covariance matrix of the predictive error, and I is the unit matrix

$$\hat{x}_{0/0}=E(x_0), P_{0/0}=\mathrm{var}(x_0) \tag{3.17}$$

$$\hat{x}_{k/k-1} = f\left(\hat{x}_{k-1/k-1}, u_{k-1}\right) \tag{3.18}$$

$$P_{k/k-1} = A_{k-1}P_{k-1/k-1}A_{k-1}^{\mathrm{T}} + \Gamma_{k-1}Q_{k-1}\Gamma_{k-1}^{\mathrm{T}} \tag{3.19}$$

$$K_k = P_{k/k-1}C_k^{\mathrm{T}}\left(C_kP_{k/k-1}C_k^{\mathrm{T}} + R_k\right)^{-1} \tag{3.20}$$

$$\hat{x}_{k/k} = \hat{x}_{k/k-1} + K_k\left[y_k - g\left(\hat{x}_{k/k-1}, u_k\right)\right] \tag{3.21}$$

$$P_{k/k} = (I - K_kC_k)P_{k/k-1} \tag{3.22}$$

$$k = 1, 2, \ldots$$

To estimate the SOC through calculation of the EKF: first, build an appropriate battery equivalent model, generally a circuit model. Then recognize the circuit model parameter by using the experimental data of complex-pulse and build the state equation and observation equation of the circuit model by selecting a suitable state variable. Finally, estimate and filter the SOC by the Kalman filter method. The schematic diagram is shown in Figure 3.8 and the flowchart in Figure 3.9.

The filter process calculates using a recursive way of predicting and revising; the predictive value can be calculated by the filter value and the new filter value can be revised according to the information and Kalman gain matrix of the observation.

Recently, the unscented Kalman filtermethod (UKF) has been used to estimate the SOC. Based on an unscented transform, using the UKF to state and update the estimation, this method solves nonlinear problems by means of a sampling method approximating a nonlinear distribution. The precision is higher than EKF but closely related to sampling strategy, and the calculation is complicated, so it is mainly applied in the study of simulation instead of electric vehicles.

3.2.6 Neural Network Method

A battery is a highly nonlinear system and it is hard to build an exact mathematical model of the process of charging and discharging. A neural network has basic nonlinear character, parallel

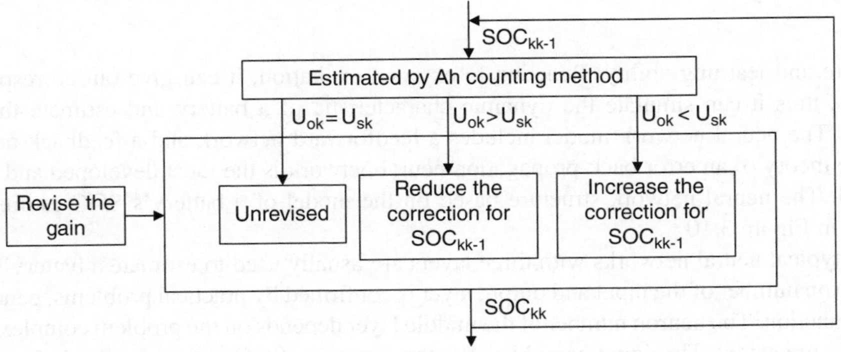

Figure 3.8 The theory of SOC estimation by the Kalman filter method.

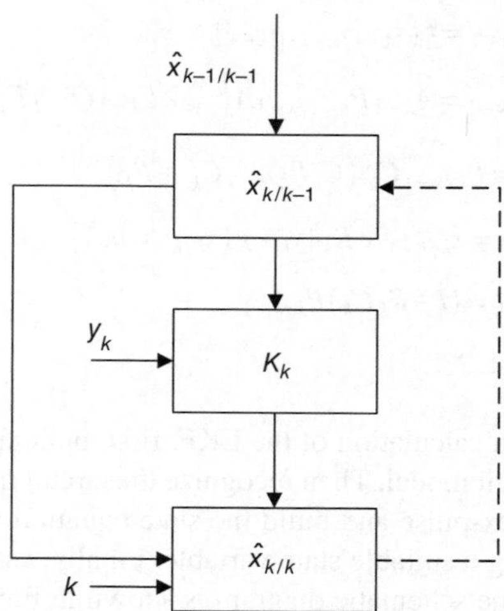

Figure 3.9 Flowchart for the Kalman filter method.

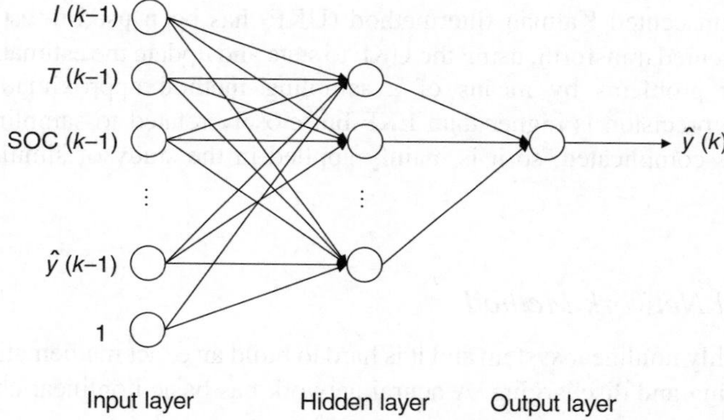

Figure 3.10 The structure of the neural network based on the cell model.

structure and learning ability. Reacting to external excitation, it can give out corresponding outputs, thus it can simulate the dynamic characteristic of a battery and estimate the SOC [9–14]. The neural network model includes a feedforward network and a feedback network, and the theory of an error back-propagation neural network is the most developed and widely applied. The neural network structure based on the model of a battery's SOC estimation is shown in Figure 3.10

The typical neural networks with three layers are usually used to estimate a battery's SOC; the neuron number of the input and output layer is confirmed by practical problems, generally a linear function. The neuron number of the middle layer depends on the problem complexity and analysis precision. The input variables for the battery's SOC estimation include voltage, current, accumulated releasing electricity, temperature, resistance and environment. Whether the choice or the number of input variables of the neural network is appropriate will directly influence the accuracy of the model and calculation. A neural network can be applied to all kinds of batteries but the disadvantage is that a large amount of reference data is needed for training and the estimation error is greatly affected by the training data and training method.

3.2.7 *Adaptive Neuro-Fuzzy Inference System*

Fuzzy inference does not rely on an exact mathematical model of the controlled object but on the knowledge of experts and the experience of operators, which involves a strong ability in knowledge representation and inference. This knowledge and experience can be described and refined by the form of rules; and through fuzzy logic inference a decision process similar to that of a human can be realized for the simulation of the battery output characteristic; but the disadvantage is the limited self-adaptability [15–20].

A neural network has the capability of approaching nonlinear mapping, learning ability and self-adaptability. It can find out the internal relations between input and output according to the past record in the system and obtain the estimated value of output variables. This process does not rely on any prior knowledge and rules of problems, so the neural network has a high adaptability and good application value in solving nonlinear problems of SOC estimation. Furthermore, a neural network has lots of parallel network structure of input and output and it can not only parallel process multivariable information effectively but also be easy for hardware realization. However, it has limited ability in achieving a qualitative expression of some uncertain knowledge.

The combination of a neural network and fuzzy inference can develop the superior properties of each and be mutually complementary. Using the technology of a neural network to solve fuzzy information can make the automatic extraction of fuzzy rule and automatic generation of a fuzzy membership function possible, thus making a fuzzy inference system adaptive. Meanwhile, the neural network is brought into the fuzzy inference, widening the scope and ability of the neural network to deal with both exact and fuzzy information. Thus, it can have a good effect on SOC estimation.

An adaptive neuro-fuzzy inference system (ANFIS) is an adaptive fuzzy inference system, with a five layers neural network structure; as shown in Figure 3.11. Taking a first-order

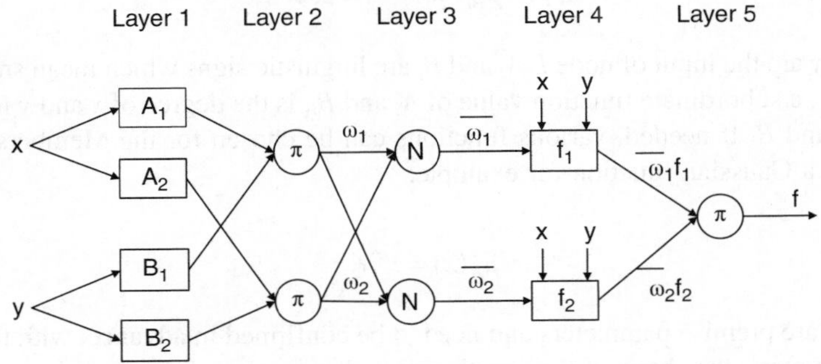

Figure 3.11 Structure diagram of the equivalent ANFIS system.

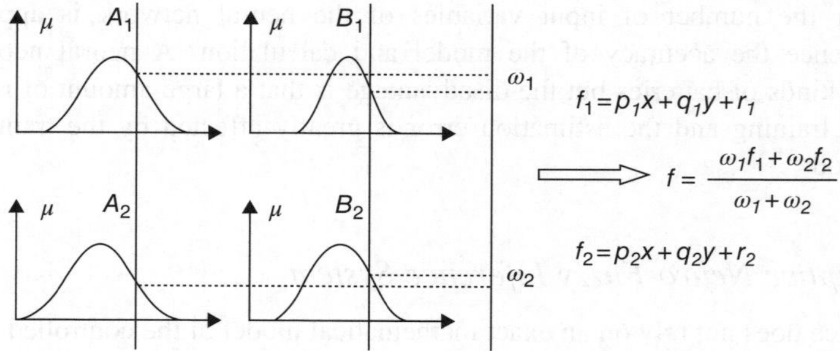

Figure 3.12 The setting method for ANFIS system with two inputs and one output.

Takagi–Sugeno fuzzy inference model of two inputs (x, y) and one output (f) as an example to illustrate the setting method of ANFIS fuzzy rules and the output expression of each layer, the setting process of fuzzy rule is shown in Figure 3.12.

Setting method of rules which have two "if-thens":

Rule 1: If x is A_1 and y is B_1, then

$$f_1 = p_1x + q_1y + r_1 \tag{3.23}$$

Rule 2: If x is A_2 and y is B_2, then

$$f_2 = p_2x + q_2y + r_2 \tag{3.24}$$

As is shown in Figure 3.11, the structure of ANFIS has five layers and the output expression of each layer is as follows:

The first layer: The nodes of this layer are adaptive, which is in charge of blurring of the input signal, node I has input function:

$$\begin{aligned} O_{1,i} &= \mu_{Ai}(x), \quad i=1,2 \text{ or} \\ O_{1,i} &= \mu_{Bi-2}(y), \quad i=3,4 \end{aligned} \tag{3.25}$$

Where x and y are the input of node I, A_i and B_i are linguistic signs which mean small, medium and large; $O_{1,i}$ a subordinate function value of A_i and B_i, is the degree of x and y indicating the degree of A and B. If needed, various functions can be chosen for the Membership function value, taking a Gaussian function for example:

$$\mu_A(x) = \mathrm{e}^{\frac{-(x-c)^2}{2\delta^2}} \tag{3.26}$$

Where $\{c, \delta\}$ are premise parameters and need to be confirmed in advance; with the change of premise parameters, the shape of the function will change; thus, it is necessary to set premise parameters according to different mappings.

The second layer: The nodes of this layer are fixed, marked with Π, which is in charge of multiplication of the membership function value input. The output value of every node is equal to the summation of all the nodes input.

$$O_{2,i} = \omega_i = \mu_{Ai}(x)\mu_{Bi}(y),\ i = 1,2 \tag{3.27}$$

The output value of every node means the credible strength of the rule, generally available for the operator T of the conducting AND algorithm.

$$O_{3,i} = \bar{\omega}_i = \frac{\omega_i}{\omega_1 + \omega_2},\ i = 1,2 \tag{3.28}$$

The third layer: the nodes of this layer are also fixed, marked with N; node *i* can calculate the ratio of the credible strength of this rule and all rules. The output of this layer is known as the normalization credible strength.

The fourth layer: the output of each node of this layer is an adaptive result, the output function of the nodes is:

$$O_{4,i} = \bar{\omega}_i f_i = \bar{\omega}_i(p_i x + q_i y + r_i),\ i = 1,2 \tag{3.29}$$

$\bar{\omega}_i$ is the normalization credible strength of the third layer output and $\{p_i, q_i, r_i\}$ is the setting parameter of the *i* node. The parameter of this layer is called the consequent parameter.

The fifth layer: there is only one fixed node in this layer marked with $\sum$; the overall output of all input signals is:

$$\text{overall output} = O_{5,1} = \sum_i \bar{\omega}_i f_i = \frac{\sum_i \omega_i f_i}{\sum_i \omega_i} \tag{3.30}$$

ANFIS uses the learning method of back propagation (BP). The disadvantages of standard BP are its slow convergence speed, it is prone to the local minimum, and so on. Thus, there are some improved methods to make up for the shortcomings, such as the additional momentum method, the auto-adaptive learning rate, the conjugate gradient algorithm, the quasi-Newton method, the Levenberg–Marquart (LM) method, the genetic algorithm, and so on. The additional momentum method and the auto-adaptive learning rate are simple and easy to add, but the other methods require a large amount of calculation which will halt the training system. In the 1990s, Jyh-Shing Roger Jang, the founder of ANFIS, came up with a comprehensive training method combining BP and least squares estimation (LSE), nonlinear parameters with BP and linear parameters with LSE. It not only achieves the expected training effect but also reduces the calculation amount. Both offline and online use is viable.

The specific process of an offline learning method inputs the discrete data into the model all at once for learning training. The following formula can be applied if there is an adaptive input variable learning system.

$$o = F(i, S) \tag{3.31}$$

where i is an input variable (vector), o an output variable, and S is the parameter set, which can be decomposed into the nonlinear parameters S_1 and linear parameters S_2. $d_\infty(f_1, f_2) = \sup_{x\in U}(|f_1(x) - f_2(x)|)$, and F is the function calculation of the adaptive network. Supposing that there exists the function computing H, the relationship between the synthesis function $H \circ F$ and the S_2 of S is linear. This part of the parameter can be identified by the LSE method. Given the parameters S_1, inputting n sets of training data, then we can get the matrix equation:

$$A\theta = y \tag{3.32}$$

This is a standard calculation of LSE, obtaining the LSE operator θ^* by minimizing $||A\theta - y||^2$,

$$\theta^* = \left(A^{\mathrm{T}}A\right)^{-1}A^{\mathrm{T}}y \tag{3.33}$$

Here, A^{T} is the transpose of A. If $A^{\mathrm{T}}A$ is singular, $(A^{\mathrm{T}}A)^{-1}A^{\mathrm{T}}$ is the inverse of A; if $A^{\mathrm{T}}A$ is non-singular, $(A^{\mathrm{T}}A)^{-1}A^{\mathrm{T}}$ is a pseudo-inverse of A. The i th row vector of a matrix A is defined as $A_i{}^{\mathrm{T}}$, the i th element of y is defined as $y_i{}^{\mathrm{T}}$, thus θ can be calculated in an iterative manner. P_i is proportional to the variance of the estimation values.

$$\begin{cases} \theta_{i+1} = \theta_i + P_{i+1}a_{i+1}\left(y_{i+1}^{\mathrm{T}} - a_{i+1}^{\mathrm{T}}\theta_i\right) \\ P_{i+1} = P_i - \dfrac{P_i a_{i+1} a_{i+1}^{\mathrm{T}} P_i}{1 + a_{i+1}^{\mathrm{T}} P_i a_{i+1}} \end{cases} \quad i = 0, 1, 2 \cdots N-1 \tag{3.34}$$

The initial condition is $\theta_0 = 0$, $P_0 = \beta I$. β is a relatively large positive number and I is a characteristic matrix. Finally, $\theta^* = \theta_N$.

Through online learning, we can update the training parameters. For a nonlinear parameter S_1, the decline in the strict sense is not a global gradient. For a linear parameter S_2, the LSE method is used for regression analysis; when new data are input, processing attenuation needs to be done with the old data. An effective way is to calculate the mean square error as a weight to the current data with the high weight on the introduction of a parameter, forgetting factor λ. Generally, λ takes a number from 0.9 to 1; the smaller λ, the faster the attenuation of the old data; this also produces a mathematical instability, and therefore λ should be adjusted according to the actual situation. At present, the study on SOC estimation using ANFIS is still in its infancy, and the estimation accuracy is greatly influenced by training methods and training data.

3.2.8 Support Vector Machines

The support vector machine (SVM) is a new kind of learning machine developed by statistical learning theory, which is based on the principle of structural risk minimization and has strong learning and generalization ability, especially for small sample data mode recognition and function estimation. It overcomes the shortcomings of the artificial neural network structure depending on the designers' experience, and solves, in a better way, a number of problems such as high-dimensional, local minima, the small sample machine learning, and so on [21, 22]. Battery SOC estimation based on SVM has been receiving more and more attention in recent years. It is focused on support vector regression and least squares SVM in the application for SOC estimation. The structure of the support vector regression method SOC estimation model is shown in Figure 3.13, which is a three-layer network with input, kernel function and output. The input is composed of battery capacity, temperature, current and voltage. The kernel function is a nonlinear transformation function, which makes the number of low-dimensional space in a high-dimensional space become linearly separable after the conversion. The output is usually the estimated value of the battery SOC.

The above SOC estimation methods can be summarized as direct measurement, time measurement methods and self-adaptive algorithms. Load voltage, electromotive force and internal resistance are direct measurement methods. Kalman filter estimates, the neural network method, neuro-fuzzy forecasts and SVM are self-adaptive algorithms. A comparison of various SOC estimation methods is shown in Table 3.3.

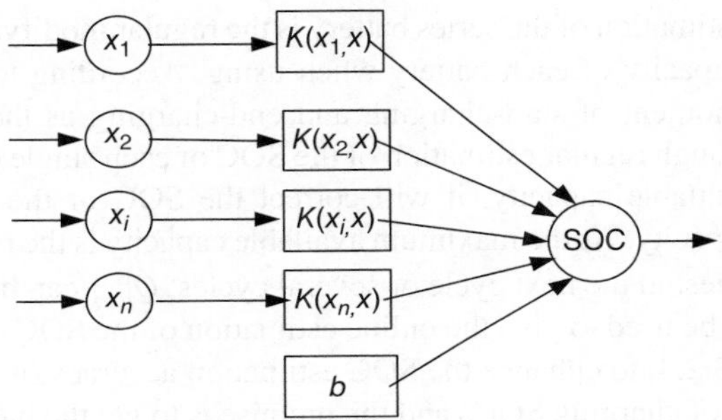

Figure 3.13 Model structure of support vector regression.

Table 3.3 Comparison of SOC estimation methods.

Estimation method	Application	Advantages	Disadvantages
Load voltage	All battery systems	Real time estimation, high estimation precision under constant current conditions	Large amounts of data; low estimation precision under variable current conditions
Electromotive force	Lithium-ion battery	Online battery electromotive force estimation	Long standing time
AC impedance	All batteries	SOH and battery quality information is given	More sensitive to temperature; expensive
Ah counting	All batteries	Preferably SOC estimation accuracy under conditions of high current detection precision	Greatly influenced by initial SOC
Neural networks	All batteries	Online estimation	Much training data arequired and influenced by the training method
Neuro-fuzzy forecast	All batteries	Online estimation	Needs complex computation and the adaptation is poor
Kalman filter	All batteries, dynamic system	Online dynamic estimation	High demand on the accuracy of the model

3.3 Battery SOC Estimation Algorithm Application

3.3.1 The SOC Estimation of a PEV Power Battery

In the process of single or successive charging and discharging, the maximum available capacity change of the battery can be ignored. In the process of battery pack charging and discharging, the remaining capacity $Q_{\text{rem}}^{\text{B}}$ will increase (or decrease) with the process of charging (discharging), but the battery pack maximum available capacity $Q_{\max}^{\text{B}}$ does not change. For the batteries in the initial charging state $\text{SOC}_0^{\text{B}} = Q_{\text{rem}}0^{\text{B}}/Q_{\max}^{\text{B}}$, after releasing the capacity of ΔQ^{B}, the charging state of the battery pack changes to SOC_1^{B} where:

$$\text{SOC}_1^{\text{B}} = \frac{Q_{\text{rem0}}^{\text{B}} - \Delta Q^{\text{B}}}{Q_{\max}^{\text{B}}} = \frac{Q_{\text{rem0}}^{\text{B}}}{Q_{\max}^{\text{B}}} - \frac{\Delta Q^{\text{B}}}{Q_{\max}^{\text{B}}} = \text{SOC}_0^{\text{B}} - \frac{\Delta Q^{\text{B}}}{Q_{\max}^{\text{B}}} \tag{3.35}$$

The core of the SOC estimation of the series battery is the regular modifying of the SOC and the maximum available capacity of each battery when using. According to the PEV operational features, it uses the moment of start-charging and end-charging as the point of penetration of SOC correction; through regular estimation of the SOC of each single cell in the battery pack and the maximum available capacity, it will correct the SOC of the series battery and its maximum available capacity. As the maximum available capacity is the result of the cumulative effects of multiple cycles, in the next cycle or several cycles, $Q_{\max}^{\text{B}}$ can be considered constant, and Equation 3.35 can be used to give the online estimation of the SOC of the battery pack. As mentioned in Section 3.2.4, to enhance the SOC estimation accuracy, it is necessary to correct the battery initial state of charging SOC_0 and the premise is to get the battery OCV. Herein the U_{OCV} method is used, and obtained by quickly identifying the polarizing voltage at the time of the initial battery charging. Therefore, we will first describe the online method for identification of the polarization voltage of the battery.

3.3.1.1 The Online Method for Identification of the Polarization Voltage

According to the operational features of the PEV, when the vehicle is to be charged after running, the polarization voltage of the battery renders the discharge polarization state. So in the early stage of charging, the discharge polarization fades, and the polarization depth gradually declines; then the polarization voltage gradually establishes and the polarization depth gradually increases. So there is a time of shallow polarization depth. If charging is stopped at this time and the polarization voltage of the battery is identified in the way described in the previous section it can greatly shorten the identification time to meet the needs of online identification.

Depolarization Capacity Calculation

In the initial stage of charging, the polarization depth changes first from shallow to deep, and then increases progressively, so knowing how much power can be charged into the battery in the initial stage to achieve the goal that the polarization depth of the battery can be completely dissipated or dropped to a lower level, is the key to achieving quick recognition of polarization voltage.

1. Time and capacity of depolarization
During charging, the expression of any first-order polarization voltage is:

$$u_P(t) = U_P(0)e^{-t/R_PC_P} + I_{ch}R_P\left(1 - e^{-t/R_PC_P}\right) \tag{3.36}$$

Then, in order to remove the polarization voltage $U_P(0)$ in the discharging process we charge the battery with a charging current I_{ch} until the polarization voltage is exactly removed:

$$u_P(t) = U_P(0)e^{-t/R_PC_P} + I_{ch}R_P\left(1 - e^{-t/R_PC_P}\right) = 0 \tag{3.37}$$

The time taken to depolarize is:

$$t = R_PC_P\ln\left(1 + \frac{U_P(0)}{I_{ch}R_P}\right) = R_PC_P\ln\left(1 + \frac{U_P(0)}{I_{ch}R_P}\right) \tag{3.38}$$

The corresponding charging capacity of the battery is:

$$Q_P = f(I_{ch}) = I_{ch}t = I_{ch}R_PC_P\ln\left(1 + \frac{U_P(0)}{I_{ch}R_P}\right) \tag{3.39}$$

Obviously the time and the capacity of depolarization is related to the charging current I_{ch}, the polarized parameters R_P, C_P and polarizing voltage $U_P(0)$ building up during discharging. Taking into account the complexity of obtaining effectively the parameters of the cell polarization in the operating vehicle, the above equation will be transformed and processed accordingly.

The discrete expression of the polarization voltage for the Thevenin model (shown in Figure 2.14) can be given by:

$$U_P(k) = \left(1 - \frac{T}{R_PC_P}\right)U_P(k-1) + \frac{I_{Dch}(k)T}{C_P} \tag{3.40}$$

where T is the sampling period, $I_{Dch}(k)$ the discharging current, $U_P(k-1)$ is the polarizing voltage of the last time, $U_P(k)$ is the polarizing voltage and R_PC_P is the polarizing parameter.

If

$$\alpha = 1 - \frac{T}{R_PC_P} < 1$$

then

$$\begin{aligned} U_P(k) &= \frac{I_{Dch}(k)T}{C_P} + \alpha U_P(k-1) \\ &= \frac{I_{Dch}(k)T}{C_P} + \alpha\left(\frac{I_{Dch}(k-1)T}{C_P} + \alpha U_P(k-2)\right) \end{aligned}$$

$$= \frac{I_{\mathrm{Dch}}(k)T}{C_{\mathrm{P}}} + \alpha\frac{I_{\mathrm{Dch}}(k-1)T}{C_{\mathrm{P}}} + \alpha^2 U_{\mathrm{P}}(k-2)$$

$$= \frac{I_{\mathrm{Dch}}(k)T}{C_{\mathrm{P}}} + \ldots + \alpha^{N-1}\frac{I_{\mathrm{Dch}}(k-N+1)T}{C_{\mathrm{P}}} + \alpha^N U_{\mathrm{P}}(k-N)$$

When N is large enough, then $\alpha^N \to 0$, so

$$U_{\mathrm{P}}(k) \approx \frac{I_{\mathrm{Dch}}(k)T}{C_{\mathrm{P}}} + \ldots + \alpha^{N-1}\frac{I_{\mathrm{Dch}}(k-N+1)T}{C_{\mathrm{P}}} \tag{3.41}$$

2. Analysis on a discharging model with constant current
   When the battery discharges with a constant current, then,

$$I_{\mathrm{Dch}}(k) = I_{\mathrm{Dch}}(k) = \ldots = I_{\mathrm{Dch}}(k-N+1) = I_{\mathrm{Dch}}$$

   Then,

$$U_{\mathrm{P}}(k) \approx \frac{I_{\mathrm{Dch}}T}{C_{\mathrm{P}}} + \ldots + \alpha^N\frac{I_{\mathrm{Dch}}T}{C_{\mathrm{P}}} = \frac{I_{\mathrm{Dch}}T}{C_{\mathrm{P}}}\left(1+\alpha+\ldots+\alpha^{N-1}\right)$$

   Meanwhile, $1+\alpha+\ldots+\alpha^N = \dfrac{1-\alpha^N}{1-\alpha} \approx \dfrac{1}{1-\alpha}$.
   Apply $\alpha = 1-\dfrac{T}{R_{\mathrm{P}}C_{\mathrm{P}}}$, then $U_{\mathrm{P}}(k) \approx \dfrac{I_{\mathrm{Dch}}T}{C_{\mathrm{P}}} \times \dfrac{R_{\mathrm{P}}C_{\mathrm{P}}}{T} = I_{\mathrm{Dch}} \times R_{\mathrm{P}}$, so $\dfrac{U_P(k)}{R_P} = I_{Dch}$.
   so the time and capacity of depolarizing in the stage separately are:

$$\begin{cases} t = R_{\mathrm{P}}C_{\mathrm{P}}\ln\left(1+\dfrac{I_{\mathrm{Dch}}}{I_{\mathrm{Ch}}}\right) \\ Q_{\mathrm{P}} = I_{\mathrm{Ch}}R_{\mathrm{P}}C_{\mathrm{P}}\ln\left(1+\dfrac{I_{\mathrm{Dch}}}{I_{\mathrm{Ch}}}\right) \end{cases} \tag{3.42}$$

3. Analysis of the changing-current mode
   When vehicles are in actual operation, the discharging current changes with the load changing, and their discharging modes with variable current are:

$$U_{\mathrm{P}}(k) \approx \frac{I_{\mathrm{Dch}}(k)T}{C_{\mathrm{P}}} + \ldots + \alpha^{N-1}\frac{I_{\mathrm{Dch}}(k-N+1)T}{C_{\mathrm{P}}}$$

$$= \frac{T}{C_{\mathrm{P}}}\left(I_{\mathrm{Dch}}(k) + \ldots + \alpha^{N-1}I_{\mathrm{Dch}}(k-N+1)\right)$$

$$= \frac{T}{C_{\mathrm{P}}}M_k(k) \tag{3.43}$$

   Meanwhile, $M_k(k) = I_{\mathrm{Dch}}(k) + \ldots + \alpha^{N-1}I_{\mathrm{Dch}}(k-N+1)$, this is called the cumulative current coefficient.

To simplify the calculation and reduce the consumption of internal memory, the formula can be obtained as:

$$\begin{aligned} M_k(k) &= I_{\mathrm{Dch}}(k) + \alpha(I_{\mathrm{Dch}}(k) + \ldots + \alpha^{N-1} I_{\mathrm{Dch}}(k-N+1)) \\ &= I_{\mathrm{Dch}}(k) + \alpha M_k(k-1) \end{aligned} \tag{3.44}$$

So the recurrence formula can be obtained.

It can be seen that the initial value of $M_k(0)$ will gradually decrease with time, so the selection scope is wide. Though a greater difference of the initial value exists, the difference will gradually decrease with time and thereby a better estimate of effect can be obtained. When $M_k(0)$ is confirmed, the present cumulative coefficient $M_k(k)$ can be obtained by resurrecting the real-time calculation in Equation 3.34.

Both sides of Equation 3.43 are divided by R_{P}, then:

$$\frac{U_{\mathrm{P}}(k)}{R_P} = \frac{T}{R_{\mathrm{P}} C_{\mathrm{P}}} \times M_k(k) \tag{3.45}$$

Applying Equations 3.38 and 3.39, then:

$$\begin{cases} t = R_{\mathrm{P}} C_{\mathrm{P}} \ln\left(1 + \dfrac{T M_k(k)}{I_{\mathrm{Ch}} R_{\mathrm{P}} C_{\mathrm{P}}}\right) \\ Q_{\mathrm{P}} = I_{\mathrm{Ch}} R_{\mathrm{P}} C_{\mathrm{P}} \ln\left(1 + \dfrac{T M_k(k)}{I_{\mathrm{Ch}} R_{\mathrm{P}} C_{\mathrm{P}}}\right) \end{cases} \tag{3.46}$$

Meanwhile, I_{Ch} and T are known data, $M_k(k)$ can be obtained by real-time calculation in Equation 3.44, then the time and capacity of depolarizing only relate to the time constant in the stage.

4. Influences of polarizing-time constant on the time and power

If

$$\frac{T M_k(k)}{I_{\mathrm{Ch}}} = k > 0,\ R_{\mathrm{P}} C_{\mathrm{P}} = x > 0, \text{ then:}$$

$$t = R_{\mathrm{P}} C_{\mathrm{P}} \ln\left(1 + \frac{T M_k(k)}{I_{\mathrm{Ch}} R_{\mathrm{P}} C_{\mathrm{P}}}\right) = x \ln\left(1 + \frac{k}{x}\right)$$

The first-order derivative between the depolarizing time and the polarizing time-constant:

$$\frac{\mathrm{d}t}{\mathrm{d}x} = \ln\left(1 + \frac{k}{x}\right) - \frac{k}{x+k}$$

The second-order derivative between the depolarizing time and the polarizing time-constant:

$$\frac{\mathrm{d}^2 t}{\mathrm{d}x^2} = \frac{k}{(x+k)^2} - \frac{k}{x^2 + kx} = k\left(\frac{1}{x^2 + 2kx + k^2} - \frac{1}{x^2 + kx}\right)$$

As $k>0$, $x>0$, then $x^2+2kx+k^2>x^2+kx$, so $\frac{\mathrm{d}^2t}{\mathrm{d}x^2}<0$

Hence , $\frac{\mathrm{d}t}{\mathrm{d}x}=\ln\left(1+\frac{k}{x}\right)-\frac{k}{x+k}$ is a monotonically decreasing function.

When $x\to+\infty$, $\frac{\mathrm{d}t}{\mathrm{d}x}\to 0$, for any x, $\frac{\mathrm{d}t}{\mathrm{d}x}>0$, so, $t=R_PC_P\ln\left(1+\frac{TM(k)}{I_{Ch}R_PC_P}\right)$ is a monotonically increasing function.

The depolarizing time increases with the polarization time-constant and the highest value of the polarizing time-constant decides the time length and capacity of depolarizing.

The battery achieves full settlement in 2 h, therefore for any value of the time constants, $R_PC_P\le\frac{2}{3}$ h.

Then,

$$\begin{cases} t\le\frac{2}{3}\ln\left(1+\frac{3TM_k(k)}{2I_{Ch}}\right)\ \mathrm{h} \\ Q_P\le\frac{2}{3}I_{Ch}\ln\left(1+\frac{3TM_k(k)}{2I_{Ch}}\right)\ \mathrm{Ah} \end{cases} \tag{3.47}$$

Effects Verification

In order to truly reflect the estimated effect of the battery of a PEV in actual operating conditions, to verify the feasibility and accuracy of the proposed methods, a battery module is taken. The battery module is composed of four lithium manganese oxide single cells of rated capacity 90 Ah. The test is conducted by using the Federal Urban Driving Schedule (FUDS) simulation operating mode; the current sampling period is 1 s, $T=1$ s. The test procedure is:

First, the battery is fully charged. Then 5% of the maximum available capacity is released (to prevent overvoltage happening during the testing process, which may lead to energy feedback). Using the continual N cyclical FUDS simulation operating mode, the test is continued until the electric discharge depth achieves the predetermined value. Settle for 10 min, simulating the charging-up time of the vehicles from start to stop. Charge a certain amount of power into the battery, using the calculation method as previously described.

Settle for 2 h; actually it does not need 2 h for the polarizing voltage fitting; the reason for doing this is to ensure the battery is fully settled. Compare the difference between the fitting polarizing voltage and the actual polarizing voltage, and then check the fitting effect. Fully discharge the battery and verify the effect of battery maximum available capacity estimation.

The plot of current and SOC of the battery versus time during the whole process is shown in Figure 3.14; an enlarged section of the plot is shown in Figure 3.15. Take the depth of discharging, DOD = 60% for battery 4 as an example, the plot of the current cumulative coefficient M_k versus time during discharging is shown in Figure 3.16. Select a different initial value $M_k(0)$ ($M_k(0)=0$ and $M_k(0)=105$), the plot of M_k is shown in Figure 3.17. Even though there is a large difference in the initial value of M_k, M_k will gradually converge after operating for a while.

According to the computed result in the figure, a value $M_k=144497$ is obtained at the time of the electric discharge, after settling for 10 min,

$$M_k=144497\left(1-\frac{T}{R_PC_P}\right)^{600}=112529$$

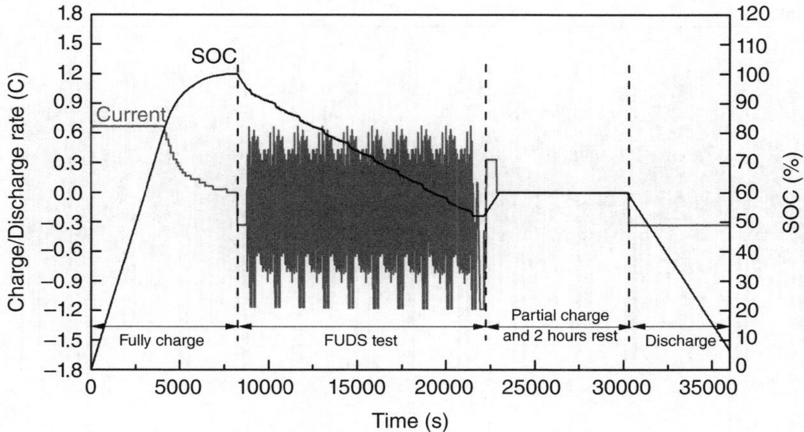

Figure 3.14 Testing curve of current and SOC when simulating the whole process.

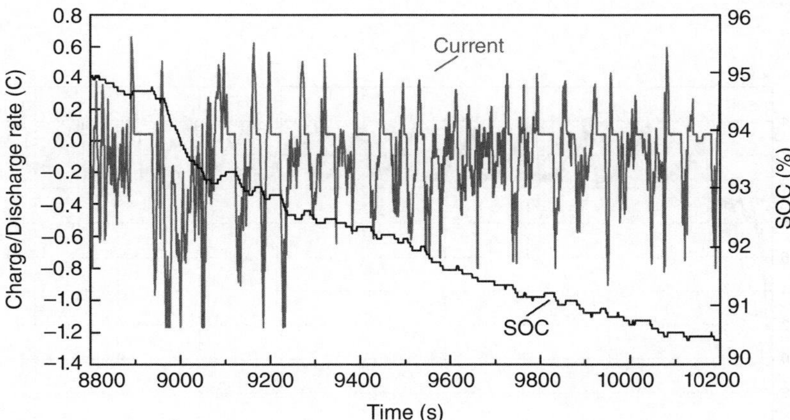

Figure 3.15 Expansion of Figure 3.14 for a period of the FUDS simulation process.

Therefore, when the charging current is 1/3 C (i.e. 120 A), the depolarizing time and capacity are:

$$\begin{cases} t \leq \dfrac{2}{3} \times \ln\left(1 + \dfrac{3 \times 112529}{2 \times 120 \times 3600}\right) \text{ h} = 0.22 \text{ h} = 13\text{min} \\ Q_{\text{P}} \leq \dfrac{2}{3} \times I_{\text{Ch}} \ln\left(1 + \dfrac{3 \times 112529}{2 \times 120 \times 3600}\right) \text{ Ah} = 26.4 \text{ Ah} \end{cases}$$

In accordance with the above-mentioned FUDS simulated test conditions, the preset depth of discharging is DOD = 60, 70, and 80%, respectively, after discharging in accordance

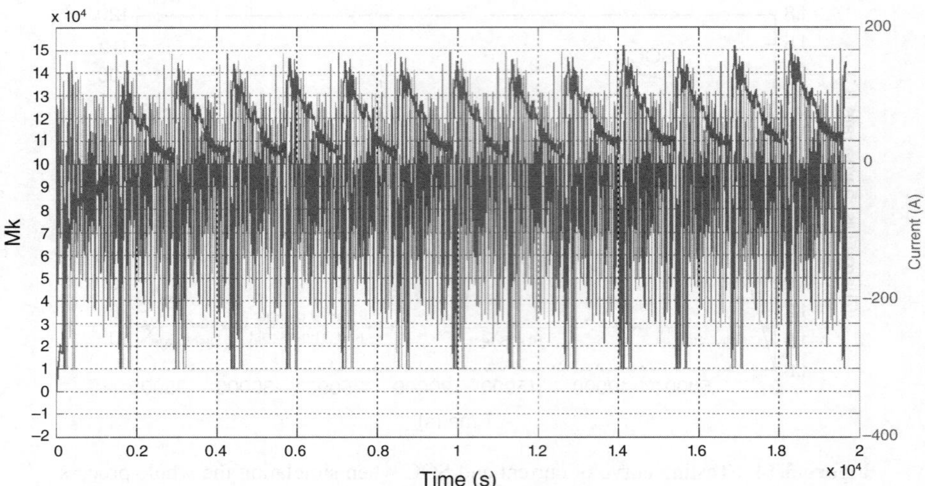

Figure 3.16 Curve of current and M_k during the FUDS simulation process.

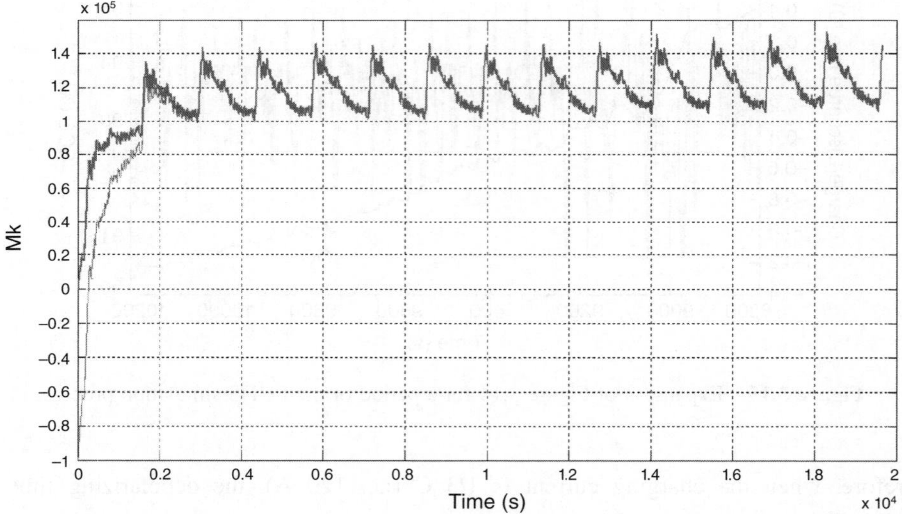

Figure 3.17 Curve of change in M_k with different initial values.

with 1/3 C (120 A) filled part of the capacity (see depolarization capacity calculation method), allowing to settle for 2 h, and making a least-squares fitting to the polarization voltage after settling, the effects can be seen in Figures 3.18–3.20, respectively. The main parameters of the testing process are shown in Table 3.4.

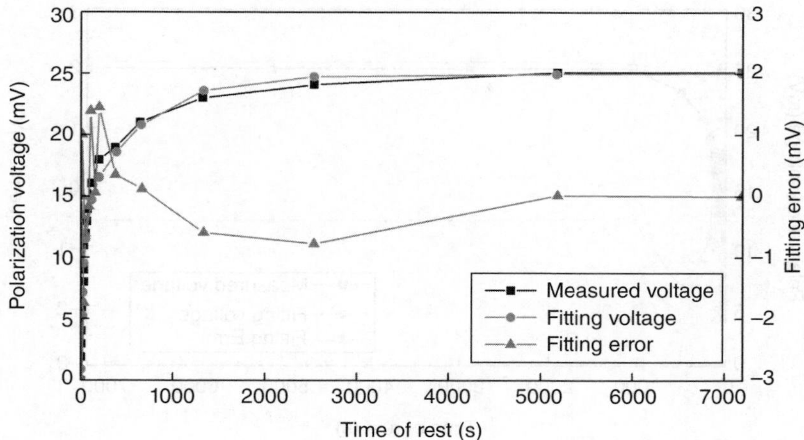

Figure 3.18 Fitting curve of polarizing voltage after discharging to DOD = 60% in the FUDS simulation conditions.

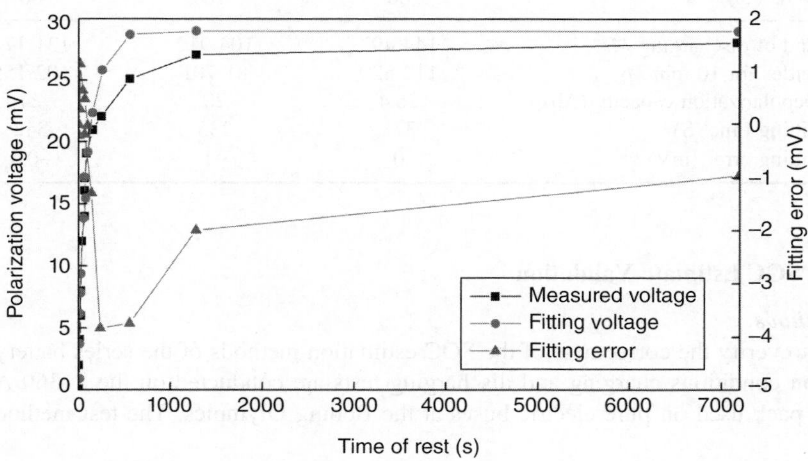

Figure 3.19 Fitting curve of polarizing voltage after discharging to DOD = 70% in the FUDS simulation conditions.

Obviously, using the least squares method to carry out the simulation to the battery polarizing voltage can obtain a good fit, the polarization voltage fitting time is significantly shortened, and the estimation errors are less than ±1 mV. That can meet the needs of the actual operation.

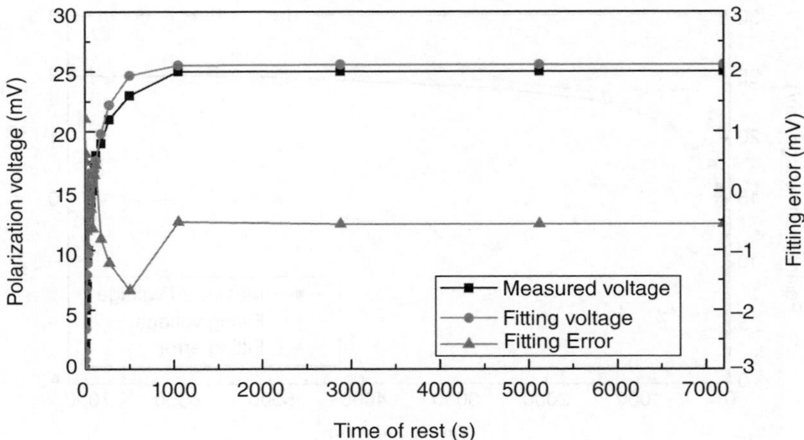

Figure 3.20 Fitting curve of polarizing voltage after discharging to DOD = 80% in the FUDS simulation conditions.

Table 3.4 Basic parameters of the FUDS test of the lithium manganese battery.

DOD (%)	60	70	80
End of discharging M_k	144 497	104 912	131 177
settles for 10 min M_k	112 529	81 701	102 155
Depolarization capacity (Ah)	26.4	20	24
Fitting time (S)	378	235	342
Fitting error (mV)	0	−1	−0.5

3.3.1.2 SOC Estimate Validation

Test Methods

In order to verify the correctness of the SOC estimation methods of the series battery, FUDS simulation conditions charging and discharging tests are conducted on the 8–360 Ah series batteries pack used on pure electric buses at the Beijing Olympics. The test methods are as follows:

1. Charge to the maximum battery voltage 4.23 V (end-of-charge voltage specified by the manufacturer) with constant current of 1/3 C (120 A), then keep the single battery maximumvoltage constant until the charging current reaches10 A (that the current is less than 1/30 C means that it is small enough and the battery is fully charged).
2. Discharge 5% of the maximum available capacity with constant current of 1/3 C (120 A) (to prevent overvoltage when energy feedback happens if the SOC is too high in the testing process).
3. Conduct tests with the FUDS simulated condition for continuous N cycles until the depth of discharging reaches a preset value of 302.4 Ah (this can also be any other value).
4. Rest for 10 min to simulate the process of vehicles from stopping running to charging-up.

the electrode is measured. The speed of the linear sweep has a large influence on the shape and value of the electrode polarization curves. When electrochemical reaction occurs in the process of charging and discharging, the higher the sweep speed, the higher the electrode polarization voltage. Only when the sweep speed is low enough can a stable volt–ampere characteristic curve be obtained. At this moment, the curve mainly reflects the relationship between the internal chemical reaction rate and the electrode potential of the battery. The volt–ampere curve reflects important characteristic information of the battery, but real-time measurement of the volt–ampere curve is not used in the actual engineering application. The main reason is that there is no condition of linear potential sweep in the process of battery charging and discharging, so it is not possible to obtain the volt–ampere curve of the battery directly.

Constant current-constant voltage (CC-CV) is a commonly used battery charging method. The potential in the potential sweep voltammetry always changes at a constant speed. The electrochemical reaction speed varies with the variation in the potential. The relation between the charging current i and the discharged electric quantity Q of the battery in a period $\{t_1\text{–}t_2\}$ is:

$$\Delta Q = \int_{t_1}^{t_2} i\mathrm{d}t \tag{3.48}$$

By online measurement of the voltage and current of the battery, constantly changing the voltage to the charging and discharging direction, a set of voltages ΔV can be obtained at equal intervals and thus a set of ΔQ can be obtained by integrating the currents at each interval of ΔV. The $\Delta Q/\Delta V$ curve based on online measurement can reflect the charge–discharge capacity of the battery at different electrode potential points. Figure 3.23 shows a $\Delta Q/\Delta V$ curve of the LFP battery with 20 Ah charging at a constant current 1/20 C.

At the charging current 1/20 C, it is usually considered that the polarization voltage of the battery is very small, and it is also suggested that the charging curve under this current stress is similar to the battery OCV curve. When the battery voltage increases in the charging process,

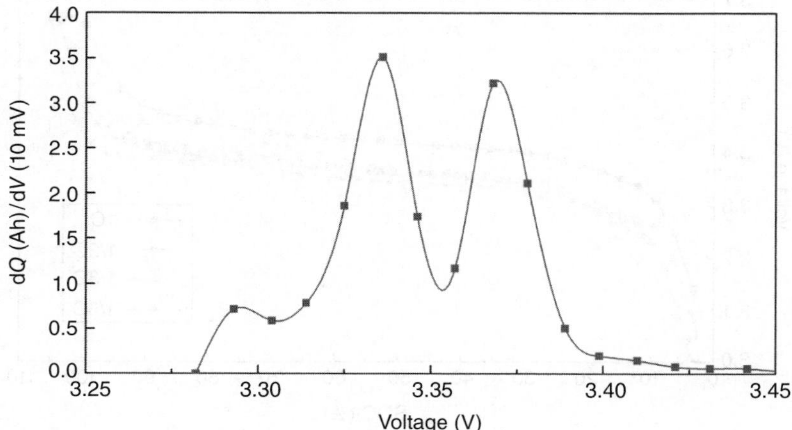

Figure 3.23 The $\Delta Q/\Delta V$ curve of the $LiFePO_4$ battery when charging at a constant current 1/20 C.

in two 10 mV periods corresponding to 3.34 and 3.37 V, the cumulative charged capacity is 3.5 and 3.2 Ah, respectively. The corresponding charged capacity begins to decline after passing the two maxima. The peak value corresponds to a high electrochemical reaction speed and, after the peak value, the concentration and flow of the reactants play a leading role. The reduction of the reactants involved in the chemical reaction causes the decrease in the charged capacity at the corresponding voltage interval.

Revising SOC by Peak Value Δ Q

The lithium-ion battery is a complex system. Considering the outer characteristics, the charging and discharging maximum allowable current (I) has an important relation with the battery capacity (Q), the temperature (T), the SOC, the state of health (SOH), and the consistency (EQ) [27], and shows a strong nonlinearity, expressed as:

$$I = f(Q, T, \mathrm{SOC}, \mathrm{SOH}, \mathrm{EQ}) \tag{3.49}$$

From the internal electrochemical analysis, the charged and discharged capacity corresponds to the intercalation/deintercalation amount of the lithium ions at the negative electrode. The speed of change of the charged capacity corresponding to the increment of voltage reflects the speed of the redox reaction of the battery system itself. The voltage platform of the LFP battery is formed within the $FePO_4$-$LiFePO_4$ phase change at the positive electrode and the intercalation/deintercalation of lithium ions at the negative electrode [28]. For the two redox reaction peaks of the LFP cell, the following analyzes the impact of aging of the charging and discharging current ratio and of battery aging on the revision of the battery SOC.

The Charging and Discharging Current Ratio

It is inappropriate to measure the battery performance in terms of the charging current because it will increase at large capacity. Figure 3.24 shows the charging curves of a single battery with 20 Ah at the charging–discharging current ratios of 1, 1/2, 1/3, and 1/5 C.

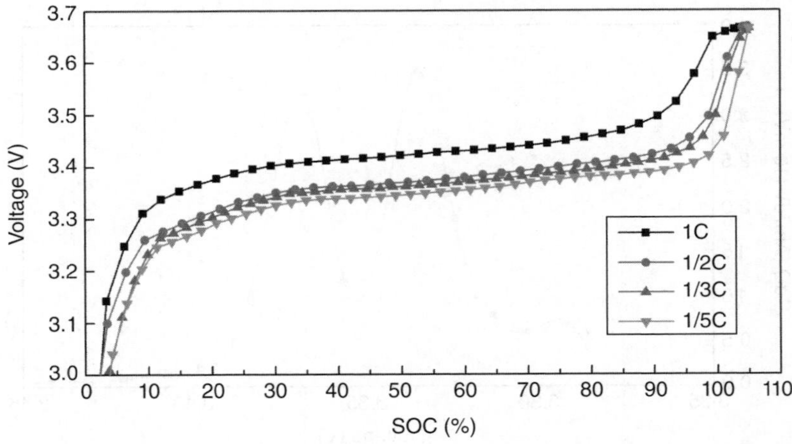

Figure 3.24 The battery voltage curves at different charging ratios.

The actual battery voltage measured on line is the outer voltage (U_O) across the two electrode pillars. The external voltage of the battery is equal to the OCV and the ohm drop (U_R) as well as the polarization voltage (U_P) of the battery. Different charging ratios will cause different U_R of the battery, and the different receiving capacities of the battery to current stress also lead to different U_P. If SOC revision is required, it is not practical to rely on a battery voltage curve.

When the charging and discharging current of the battery is 0, and after sufficient standing time, the U_R and U_P of the battery are both 0, then the OCV of the battery is equal to the terminal voltage U_O. But according to the OCV–SOC curve, the SOC of the LFP cell cannot be revised accurately.

Figure 3.25 depicts the ΔSOC/ΔV curves at different charging ratios. For a more intuitive understanding of the rate of change of the charged capacity, the vertical axis is expressed as the change in the value of the SOC. The four ratios corresponding to the peak curves of the SOC with the change in voltage have their own densities and peak positions, reflecting the battery's internal chemical reaction process at different charging ratios, and describing the current acceptance of the battery at different voltage points and different charging ratios. From Figure 3.25 it can be observed that: (i) at the ratios of 1/2, 1/3, and 1/5 C there appear two obvious peak positions, similar to the characteristic curve shown in Figure 3.2; (ii) the voltage values corresponding to the peak positions at the ratios of 1, 1/2, 1/3, and 1/5 C successively become larger; (iii) the battery capacity is charged when concentrated in the vicinity of the two peak values, and the voltage corresponding to the peak lies on the voltage platform of the battery.

The ohm drop and the polarization voltage of the batteries are mainly influenced by the current ratios. Without considering the accumulation of the polarization voltage, the greater the current ratio at the same SOC, the larger U_R and U_P. Figure 3.26 is obtained by changing the abscissa of Figure 3.25 into the SOC value.

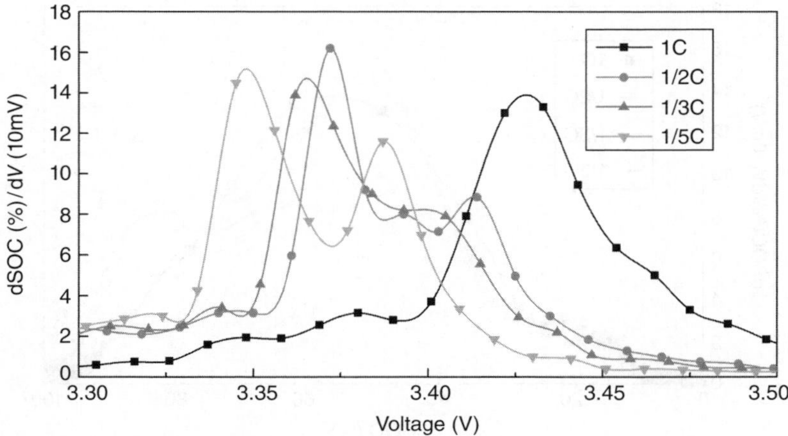

Figure 3.25 The ΔSOC/ΔV curves at different charging ratios. (Reproduced with permission from Wei Shi, Jiuchun Jiang, Suoyu Li, Rongda Jia "Research on SOC estimation for $LiFePO_4$ Li-ion batteries", Journal of Electronic Measurement and Instrument (in Chinese)., vol 24, no.8, 769–774, 2010.)

The data points shown in Figure 3.26 are still selected at every 10 mV of voltage. The SOC is derived from a precisely calibrated ampere-hour integral. It can be seen that the peak values at the charging ratios of 1/2, 1/3, and 1/5 C corresponding to the SOC are 50 and 85%. As can be seen with reference to Figure 3.3, at 1 C, the ohm drop and polarization voltage of the battery are larger. At the same time, with constant current charging, the internal resistance of the battery changes a little with the variation of SOC, that is, U_R changes a little. Therefore the disappearance of the second peak value at 1 C in Figures 3.25 and 3.26 is mainly due to the change in the polarization voltage, leading to difficulty in observing the higher charged capacity value at the same rate of change ofthe voltage. In addition, the charging rate of the usual energy-type battery is under 1 C, and therefore the characteristics of the battery under the conditions of normal charging ratio are mainly analyzed.

The ΔSOC/SOC curve at different discharging ratios is shown in Figure 3.27. It can be observed that the peak values at the discharging ratios of the 1/2, 1/3, and 1/5 C corresponding to the SOC are 80 and 55%. However, because the discharging current is not easy to keep stable in the actual application, the working condition is a little complex, and the produced changes in U_R and U_P are difficult to eliminate. The obtained ΔV value contains a larger error, which has an impact on the accuracy of the revised SOC from the $\Delta Q/\Delta V$ curve peak value.

If the influence of the internal resistance and the polarization is removed when obtaining the online-measuring battery voltage in the process of charging in the battery management system (BMS), a$\Delta Q/\Delta V$ curve like that in Figure 3.27 should be obtained. This also shows that the SOC value corresponding to the peak value of the $\Delta Q/\Delta V$ curve obtained at different ratios can be the condition of the accurate revision of the SOC. Especially under the condition of a very flat voltage platform of the $LiFePO_4$ cell, the peak amplitude is more apparent.

Aging of Battery

The aging of the battery mainly involves loss of capacity and increase in the internal resistance. Research has been carried out around the world into the reasons for this. It is generally believed

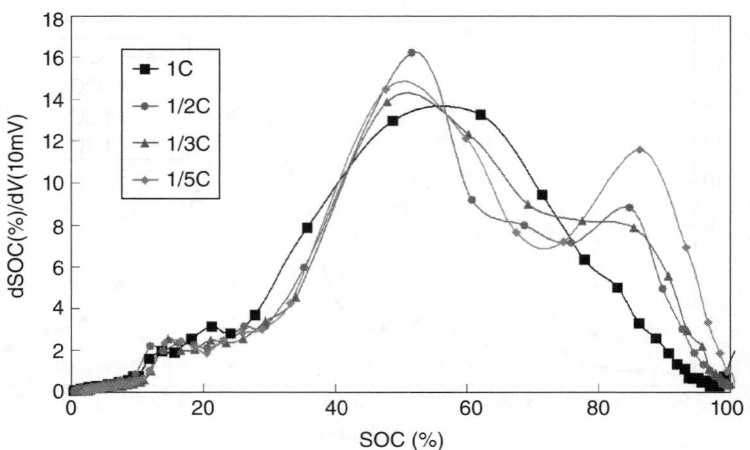

Figure 3.26 The ΔSOC/SOC curves at different charging ratios. (Reproduced with permission from Wei Shi, Jiuchun Jiang, Suoyu Li, Rongda Jia "Research on SOC estimation for $LiFePO_4$ Li-ion batteries", Journal of Electronic Measurement and Instrument (in Chinese)., vol 24, no.8, 769–774, 2010.)

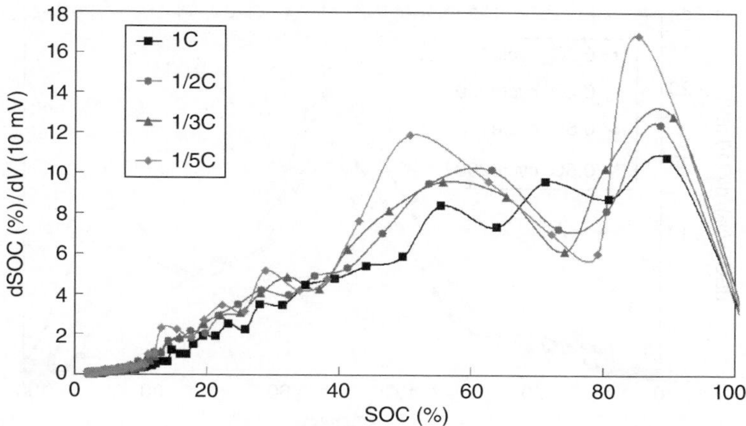

Figure 3.27 The ΔSOC/SOC curves at different discharging ratios. (Reproduced with permission from Wei Shi, Jiuchun Jiang, Suoyu Li, Rongda Jia "Research on SOC estimation for $LiFePO_4$ Li-ion batteries", Journal of Electronic Measurement and Instrument (in Chinese)., vol 24, no.8, 769–774, 2010.)

that the decrease in capacity is due to the loss of lithium ions involved in the reaction, caused by the irreversible chemical reaction that occurs in the charge–discharge process. The increase in the internal resistance of the battery is believed to be due to passivation of the internal structure of the battery, such as thickening of the SEI membrane and change in the structure at the positive and negative electrodes.

When battery aging starts, the application range of the OCV and ampere-hour integration method is not changed, but it has a great impact on the artificial neural network and Kalman filtering methods, because the parameters of the established battery model change with aging. Especially, the different aging paths caused by inconsistencies in the applied battery pack result in reduction of the applicability of the models, for example, the neural network needs to be re-trained, and the parameters of the model based on Kalman algorithms need to be changed. The revision of the SOC after battery aging is of great significance for improving the performance of the BMS and prolonging the life of the battery pack.

Because the $\Delta Q/\Delta V$ curve reflects the electrochemical characteristics inside the battery, a battery capacity below 80% of the rated capacity means, in general, the end of battery life in the application of electric vehicles. At this time, the main chemical reaction inside the battery depends on the concentration of the reactants and the internal structure of the battery system. Figure 3.28 depicts the ΔSOC/SOC characteristics of a $LiFePO_4$ cell after 200 cycles with the DOD of 100% and shows that its capacity declines to 95% of the rated capacity.

After 200 cycles, the capacity retention of the tested battery decreases, its internal structure also changes, and the increase in capacity is concentrated in a first peak value corresponding to the SOC value. It is found that the charged capacity corresponding to the second peak significantly reduces compared with a new battery, which indicates the decreased inserting ability of the lithium ions at the negative electrode, the reduced current acceptance, and the increased polarization voltage as well as the shortened life span [24].

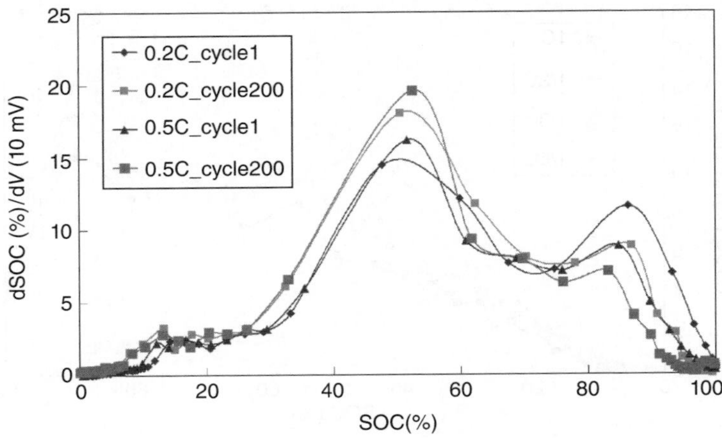

Figure 3.28 The comparison of ΔSOC/SOC curve before and after aging.

Revising SOC

The BMS system gathers the voltage and current of the single battery in real time, and calculates the internal resistance of the battery by analyzing the voltage change of the signal of the step-leap current. Eliminating the impact of the ohm drop U_R helps to obtain the voltage change ΔV with the changed current and other optimized charging methods (constant current charging has no effect), and then obtains the ampere-hour integral value ΔQ corresponding to the section of equal interval (for instance 10 mV). Mathematically, judging the extreme value of the $\Delta Q/\Delta V$ curve requires the first derivative of the function of the curve. In the actual application, the voltages at the two maximum values have a certain range. The method is: start charging the battery from the low SOC point and record a set of ΔQ values during charging, then obtain the two maxima to meet the requirements by a simple data processing (especially, at the charging ratio of 1 C with high charge rate there is only one maximum value). Compared with the voltage value appearing at the peak point, judge the position of the first peak point and record it. When the recorded peak points are the same for two or more processes of charging, and have a difference of 8% or more from the SOC value recorded with the BMS (usually electric vehicles require SOC accuracy of about 8%), then execute the revision of the SOC, and record the corrected events for debug analysis.

3.3.2 Power Battery SOC Estimation for Hybrid Vehicles

3.3.2.1 SOC Estimation Algorithm based on the EKF

Taking the power-type $LiMn_2O_4$ cell with rated capacity 8 Ah, with the Thevenin equivalent circuit model established in Chapter 2, based on the estimated SOC principle by EKF in Section 3.2, taking the polarization voltage U_p across both ends of the polarization capacitor C_p of the SOC and the model as the state variables, and by using the battery load voltage as the observable, then the state Equation 3.48 and observation Equation 3.49 can be obtained.

State equation:

$$\begin{bmatrix} \mathrm{SOC}_{k+1} \\ U_{\mathrm{p}}(k+1) \end{bmatrix} = \begin{bmatrix} 1 & 0 \\ 0 & \exp(-\Delta t/\tau) \end{bmatrix} \times \begin{bmatrix} \mathrm{SOC}_k \\ U_{\mathrm{p}}(k) \end{bmatrix} + \begin{bmatrix} \dfrac{k_{\mathrm{T}} k_{\mathrm{R}} \Delta t}{C_{\mathrm{N}}} \\ R_{\mathrm{P}}(1-\exp(-\Delta t/\tau)) \end{bmatrix} \times i_k + w_k \quad (3.48)$$

Observation equation:

$$U_{\mathrm{L}}(k) = U_{\mathrm{OC}} + i_k R_0 + U_{\mathrm{P}}(k) + v_k \quad (3.49)$$

In Equations 3.48 and 3.49, k_{T} is the correction coefficient of the temperature to SOC, k_{R} is the correction coefficient of the charge–discharge ratio to SOC. i_k is the battery load current at the sampling point k, making the charging direction as the positive. C_{N} is the rated capacity of the $LiMn_2O_4$ cell, SOC_k is the battery state of charging at the sampling point k, Δt is the experimental sampling period, $U_{\mathrm{p}}(k)$ is the estimated polarization voltage at the sample point k. τ is the time constant of the RC link, and is defined by $\tau = R_{\mathrm{P}} C_{\mathrm{P}}$. w_k is the process noise at the sampling point k, in the actual calculation process, generally set as constant. $U_{\mathrm{L}}(k)$ is the battery load voltage estimated value at the sample point k, U_{oc} is the OCV of the battery, R_0 is the ohm resistance of the battery, v_k is the observation noise at the sampling point k.

Suppose $x_k = [\mathrm{SOC}_k, U_{\mathrm{P}}(k)]^{\mathrm{T}}$, comparing with Equation 3.48, the matrices $\hat{A}_k$ and $\hat{C}_k$ can be obtained:

$$\hat{A}_k = \frac{\partial f(x_k, u_k)}{\partial x_k}\bigg|_{x_k = \hat{x}_k} = \begin{bmatrix} 1 & 0 \\ 0 & \exp(-\Delta t/\tau) \end{bmatrix} \quad (3.50)$$

$$\hat{C}_k = \frac{\partial g(x_k, u_k)}{\partial x_k}\bigg|_{x_k = \hat{x}_k} = \left[\left(\frac{\mathrm{d}U_{\mathrm{OC}}}{\mathrm{dSOC}} + \frac{\mathrm{d}R_0}{\mathrm{dSOC}} \times i_k \right)\bigg|_{\mathrm{SOC}_k = \hat{\mathrm{SOC}}_k}, 1 \right] \quad (3.51)$$

From Equation 3.51 the function relation between U_{OC}, R_0, and SOC can be calculated, and the differential relation between U_{OC}, R_0, and SOC can be obtained from the function relation. From Tables 3.6–3.8, with the least squares method, the polynomial function relation between U_{OC}, R_0, and SOC is obtained:

Table 3.6 Parameters of charging and discharging ohm resistance.

SOC	0.95	0.90	0.85	0.80	0.75	0.70	0.65	0.60	0.55
R_{d}/Ω	0.016 57	0.015 06	0.013 96	0.013 18	0.012 65	0.012 30	0.012 07	0.011 94	0.011 85
SOC	0.50	0.45	0.40	0.35	0.30	0.25	0.20	0.15	0.10
R_{d}/Ω	0.011 81	0.011 80	0.011 84	0.011 93	0.012 12	0.012 45	0.012 96	0.013 74	0.014 86
SOC	0.95	0.90	0.85	0.80	0.75	0.70	0.65	0.60	0.55
R_{c}/Ω	0.013 57	0.012 80	0.012 18	0.011 69	0.011 32	0.011 04	0.010 85	0.010 73	0.010 66
SOC	0.50	0.45	0.40	0.35	0.30	0.25	0.20	0.15	0.10
R_{c}/Ω	0.010 64	0.010 65	0.010 68	0.010 73	0.010 78	0.010 83	0.010 88	0.010 91	0.010 93

Table 3.7 Model parameters in the discharging direction.

SOC	U_{oc}/V	R_o/Ω	τ/s	R_p/Ω	C_p/kF	U_p/V
0.95	16.238	0.016 57	10.40	0.012 05	3.45	−0.5000
0.90	16.040	0.015 06	10.90	0.009 05	3.45	−0.5083
0.85	15.841	0.013 96	11.37	0.010 27	1.93	−0.4847
0.80	15.632	0.013 18	9.95	0.009 74	1.34	−0.4661
0.75	15.450	0.012 65	9.12	0.006 19	2.02	−0.4574
0.70	15.295	0.012 30	9.89	0.006 08	2.12	−0.3409
0.65	15.157	0.012 07	10.72	0.005 95	1.85	−0.3413
0.60	15.037	0.011 94	11.05	0.005 89	1.69	−0.3478
0.55	14.932	0.011 85	10.91	0.005 88	1.70	−0.3550
0.50	14.843	0.011 81	10.49	0.005 91	1.77	−0.3602
0.45	14.760	0.011 80	10.03	0.005 92	1.84	−0.3584
0.40	14.648	0.011 84	9.96	0.005 67	1.95	−0.3497
0.35	14.535	0.011 93	11.04	0.004 92	2.18	−0.3389
0.30	14.450	0.012 12	12.90	0.004 71	2.10	−0.1456
0.25	14.369	0.012 45	12.48	0.004 82	1.89	−0.1181
0.20	14.279	0.012 96	13.60	0.005 03	1.98	−0.1205
0.15	14.166	0.013 74	19.87	0.004 82	2.36	−0.1253
0.10	14.040	0.014 86	31.23	0.006 42	1.70	−0.1497

Table 3.8 Model parameters in the charging direction.

SOC	U_{oc}/V	R_o/Ω	τ/s	R_p/Ω	C_p/kF	U_p/V
0.95	16.184	0.013 57	38.25	0.004 95	5.05	0.1500
0.90	16.002	0.012 80	28.55	0.004 30	4.93	0.1573
0.85	15.818	0.012 18	24.58	0.004 16	3.65	0.1646
0.80	15.632	0.011 69	22.67	0.003 40	4.30	0.1308
0.75	15.463	0.011 32	19.16	0.002 72	2.85	0.1023
0.70	15.312	0.011 04	14.18	0.002 17	3.99	0.1695
0.65	15.178	0.010 85	11.00	0.002 08	4.33	0.2316
0.60	15.061	0.010 73	9.44	0.002 02	4.51	0.2320
0.55	14.962	0.010 66	9.16	0.001 98	4.96	0.2312
0.50	14.876	0.010 64	9.84	0.001 96	5.02	0.2283
0.45	14.803	0.010 65	9.80	0.001 98	4.64	0.2282
0.40	14.734	0.010 68	9.12	0.001 99	4.74	0.2362
0.35	14.654	0.010 73	9.02	0.002 08	5.29	0.4256
0.30	14.562	0.010 78	8.64	0.002 59	5.48	0.4509
0.25	14.457	0.010 83	7.75	0.002 52	7.95	0.4510
0.20	14.330	0.010 88	14.59	0.002 75	8.25	0.4452
0.15	14.200	0.010 91	15.19	0.003 07	8.00	0.4503
0.10	14.020	0.010 93	21.18	0.003 30	8.65	0.4504

Discharging direction:

$$U_{\mathrm{OC}} = 12.961 + 11.936\mathrm{SOC} - 33.191\mathrm{SOC}^2 + 41.842\mathrm{SOC}^3 - 17.149\mathrm{SOC}^4$$

$$R_{\mathrm{d}} = 0.0185 - 0.0479\mathrm{SOC} + 0.1306\mathrm{SOC}^2 - 0.1636\mathrm{SOC}^3 + 0.081\mathrm{SOC}^4$$

Thus the differential form of U_{OC} and R_0 with respect to SOC is obtained:

$$\left.\frac{\mathrm{d}U_{\mathrm{OC}}}{\mathrm{dSOC}}\right|_{\mathrm{SOC}_k=\hat{\mathrm{SOC}}_{k-1}} = 11.936 - 66.382\mathrm{SOC} + 125.526\mathrm{SOC}^2 - 68.596\mathrm{SOC}^3$$

$$\left.\frac{\mathrm{d}R_0}{\mathrm{dSOC}}\right|_{\mathrm{SOC}_k=\hat{\mathrm{SOC}}_{k-1}} = -0.0479 + 0.2612\mathrm{SOC} - 0.4908\mathrm{SOC}^2 + 0.324\mathrm{SOC}^3$$

From the calculated result we obtain:

Charging direction:

$$U_{\mathrm{OC}} = 12.468 + 17.767\mathrm{SOC} - 51.674\mathrm{SOC}^2 + 64.564\mathrm{SOC}^3 - 26.834\mathrm{SOC}^4$$

$$R_{\mathrm{C}} = 0.0109 + 0.0007\mathrm{SOC} - 0.0045\mathrm{SOC}^2 + 0.0008\mathrm{SOC}^3 + 0.0066\mathrm{SOC}^4$$

Thus the differential form of U_{OC} and R_0 with respect to SOC is obtained:

$$\left.\frac{\mathrm{d}U_{\mathrm{OC}}}{\mathrm{dSOC}}\right|_{\mathrm{SOC}_k=\hat{\mathrm{SOC}}_{k-1}} = 17.767 - 103.348\mathrm{SOC} + 193.692\mathrm{SOC}^2 - 107.336\mathrm{SOC}^3$$

$$\left.\frac{\mathrm{d}R_0}{\mathrm{dSOC}}\right|_{\mathrm{SOC}_k=\hat{\mathrm{SOC}}_{k-1}} = 0.0007 - 0.009\mathrm{SOC} + 0.0024\mathrm{SOC}^2 + 0.0264\mathrm{SOC}^3$$

Thus obtain:

$$\hat{C}_k = \left.\frac{\partial g(x_k,u_k)}{\partial x_k}\right|_{x_k=\hat{x}_k} = \left[\left.\left(\frac{\mathrm{d}U_{\mathrm{OC}}}{\mathrm{dSOC}} + \frac{\mathrm{d}R_0}{\mathrm{dSOC}}i_k\right)\right|_{\mathrm{SOC}_k=\hat{\mathrm{SOC}}_{k-1}},1\right]$$

$$= [(17.767 - 0.0007i_k) - (103.348 - 0.009i_k)\mathrm{SOC} + (193.692 - 0.0024i_k)\mathrm{SOC}^2 - (107.336 - 0.0264i_k)\mathrm{SOC}^3, 1]$$

The set of initial conditions is:

1. Initial SOC SOC_0. In practical applications, it is generally found through reading the last data results stored in the EEPROM of the BMS if, before this, the battery is allowed to rest for a sufficiently long time, the correction can be carried out by using the ($U_{\mathrm{OC}} \sim$ SOC) curve, the battery temperature and other conditions. In the simulation, generally any values between [0, 1] can be artificially set as the initial SOC value, for instance 0.5.

2. Polarization voltage U_P. When the filtering algorithm is initialized, the general polarization effect is not particularly obvious and, at this time, the circuit load voltage is zero, that is, $U_P(0) = 0$.
3. Error covariance. In order to set this parameter, it is necessary to select the appropriate value based on engineering practice. If selected properly, the convergence rate of the filtering algorithm can be speeded up.

3.3.2.2 Introduction of the Gain Factor

The biggest advantage of the EKF algorithm is to get the state value with an initial error quickly to the convergence of the exact value. In actual operation of the hybrid vehicle, due to its relatively complex operating conditions, there are a large number of repeated charge–discharge phenomena of the battery, and the pulse charge–discharge currents are relatively large with a shorter pulse time. In the conversion process of the charging and discharging, the fluctuation of the voltage is very large, which makes fluctuation of the output error of the model become large, resulting in oscillation of the SOC estimation value. For a power-type battery, its rated capacity is generally small, allowing a large charging and discharging current ratio. Taking the $LiMn_2O_4$ cell with 8 Ah used herein as an example, if the charge–discharge current is 20 C and the pulse time is10 s, the rate of change of SOC is 5.55%. When mutation arises in the system state, the true system state cannot be timely tracked, thus affecting the calculation accuracy of the algorithm; therefore, the filtering estimated actual track effect needs to be considered. In the iterative process of the algorithm, the single-step predictive value and the filter gain are the key factors in determining the convergence rate, of which, the SOC single-step predictive value formula is:

$$\mathrm{SOC}_k = \mathrm{SOC}_{k-1} + \frac{K_R \Delta t}{C} i_k \tag{3.52}$$

where SOC_{k-1} is the initial set value or the one read in the EEPROM of the management system, both of which are constant. K_R is the revised coefficient of the discharging ratio, Δt is the sampling time. C is the rated capacity of the battery, which is constant, therefore it is difficult to optimize the estimate of the single-step predictive value. Here, the optimization of EKF is done by setting the gain.

Kalman gain matrix formula:

$$K_k = P_{k/k-1} C_k^{\mathrm{T}} \left(C_k P_{k/k-1} C_k^{\mathrm{T}} + R_k \right)^{-1} \tag{3.53}$$

The updated formula of the state estimated measurement [10]:

$$\hat{x}_{k/k} = \hat{x}_{k/k-1} + K_k \left(y_k - g\left(\hat{x}_{k/k-1}, u_k \right) \right) \tag{3.54}$$

From Equation 3.54, it can be seen that the default coefficient of K_k is 1. After the experiment, it can be found that a good filtering adjusted effect cannot be achieved without the proper adjustments. For this reason, on the basis of the Kalman gain, on introducing the gain factor λ the Kalman gain matrix K_k' after adjustment is:

$$K_k' = \lambda K_k = \lambda P_{k/k-1} C_k^{\mathrm{T}} \left(C_k P_{k/k-1} C_k^{\mathrm{T}} + R_k \right)^{-1} \tag{3.55}$$

The formula of state estimated measurement after adjustment is:

$$\hat{x}_{k/k} = \hat{x}_{k/k-1} + K_k' \left(y_k - g\left(\hat{x}_{k/k-1}, u_k \right) \right) \tag{3.56}$$

After the experiment, during the charging and discharging period, the gain factor λ is larger, and its value is related to the charging and discharging currents. According to the different charging and discharging currents, when the value is generally taken from 40 to 60, the convergence rate is ideal. When the battery is at rest, the SOC value is unchanged, and at this time the polarization phenomenon of the battery voltage plays a main role. Because the polarization effect of the battery is relatively complex, using a larger gain factor will cause increase in the estimation error, and thus, while at rest, set the gain factor λ to the default value of 1. It is found through experiment that the gain factor λ during the charging and discharging period is mainly related to the value of the charging and discharging currents. In the filtering process, on the basis of the different charging and discharging currents, when the different values have been set, the filtering effect is relatively ideal.

3.3.2.3 Verification of the Estimated Effect

In order to simulate the working situation in the actual operation of hybrid vehicles, the self-defined compound pulse authentication is adopted in this book, with an EKF algorithm (improved EKF algorithm) estimated effect of the gain factor, and the experimental procedure is as follows:

1. Charge with 8 A current until any one of the single battery voltages reaches 4.2 V (SOC = 100%) and rest for 18 min;
2. Discharge with 8 A current for 18 min (SOC = 70%) and rest for 18 min;
3. Discharge with 840 W constant power for 30 s;
4. Charge with 140 W constant power for 2 min;
5. Discharge with 420 W constant power for 1 min;
6. Charge with 700 W constant power for 30 s and rest for 10 min;
7. Repeat procedure 3–6 9 times and the experiment is over.

Adjusting the battery SOC to 70% to simulate the normal working situation of hybrid vehicles (the normal working range of a hybrid vehicle with a lithium battery is between 30 and 70%), discharging with 840 W constant power for 30 s to simulate the initial stage of the vehicles, charging with 140 W constant power for 2 min to simulate the idling stage, discharging with 420 W constant power for 1 min to simulate the accelerating stage, charging with 700 W constant power for 30 s to simulate the braking stage. After one cycle, the drop in SOC is about 4.5%; after nine cycles, the drop in SOC is about 30%, consistent with the working range of the battery in a hybrid vehicle.

Figure 3.29 shows the changed waveforms of the voltage and current in the second four-working-condition experiment. Figure 3.30 shows the comparison before and after the

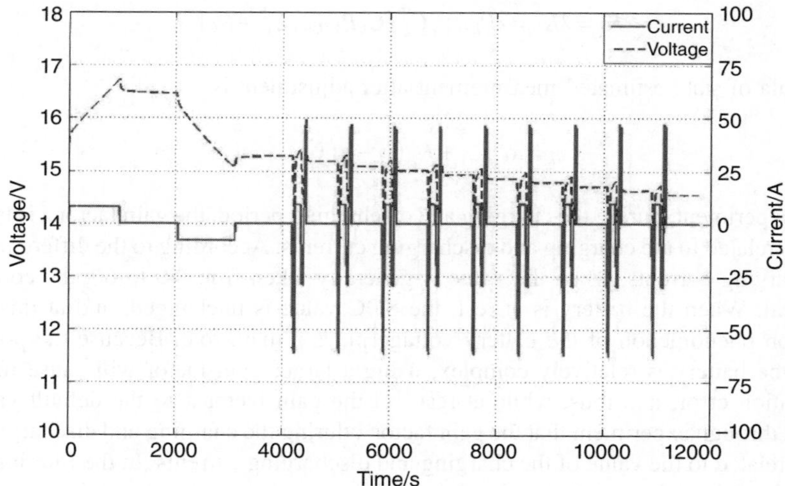

Figure 3.29 Self-defined compound pulse current and voltage wave forms.

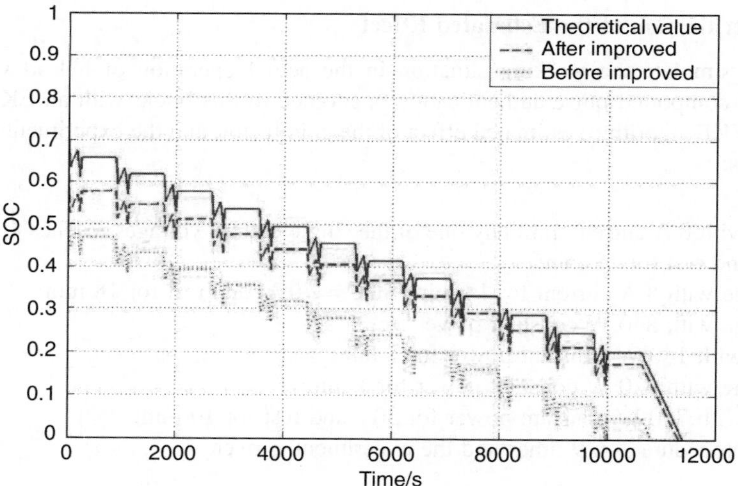

Figure 3.30 Simulation comparison before and after the improvement of the EKF algorithm under the working condition of the self-defined compound pulse.

improvement of the EKF algorithm in the second four-working-condition experiment. Figure 3.31 shows the estimated error curve before and after improvements of the EKF algorithm. From the error curve in Figure 3.31 it can be seen that, before improvement of the algorithm, the changing of the error curve is slow, and the error is maintained at above 15%, and both the convergence speed and accuracy cannot meet the actual requirements.

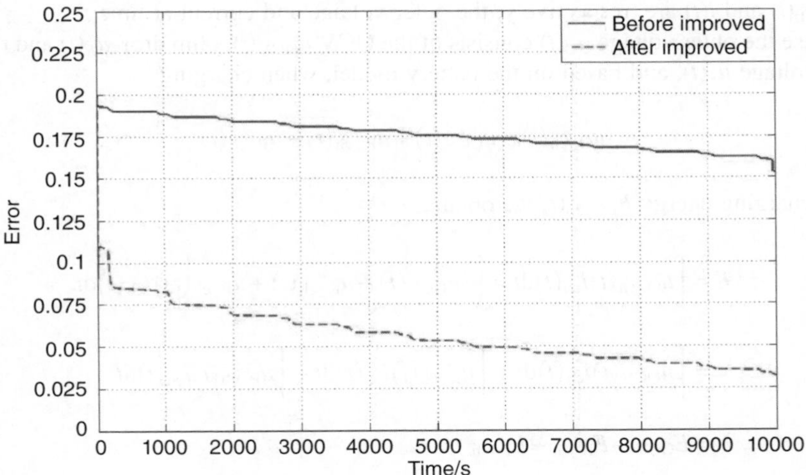

Figure 3.31 SOC estimated error curve.

After the improvement, the error quickly converges to 10% or less, the final error converges to about 3.5%, with better convergence speed and estimation accuracy to meet the actual requirements.

3.4 Definition and Estimation of the Battery SOE

3.4.1 Definition of the Single Battery SOE

The battery SOE is the remaining energy (E_{rem}) as a percentage of the maximum available energy (E_{max}), that is:

$$\mathrm{SOE} = \frac{E_{rem}}{E_{max}} \times 100\% \tag{3.57}$$

Remaining energy: the released energy of the battery to discharge with a small enough current from the present state to the power-up state.

Maximum available energy: the released energy of the battery to discharge with small enough current from the fully charged state to the power-up state.

The battery SOE is closer to the essential function of the battery to store and release energy than the SOC. In the actual application of electric vehicles, the energy directly affects the mileage of electric vehicles and the life of electrical equipment, and so on. Therefore analysis of the energy state of the battery has a practical significance in the charging and discharging process of the battery.

The formula for calculation of the battery charging and discharging energy is:

$$E = \int u_0(t)i(t)\mathrm{d}t \tag{3.58}$$

Where $u_O(t)$ and $i(t)$ are, respectively, the outer voltage and current at time t.

Because the outer voltage $u_O(t)$ consists of the OCV $u_{OCV}(t)$, ohm drop $u_R(t)$ and the polarization voltage $u_P(t)$, and based on the battery model, when charging:

$$u_{O\_ch}(t) = u_{OCV}(t) + u_{R\_ch}(t) + u_{P\_ch}(t) \tag{3.59}$$

For the charging energy $E_{O\_ch}(t)$ we obtain:

$$\begin{aligned} E &= \int u_{O\_ch}(t) i_{ch}(t) \mathrm{d}t = \int \left(u_{OCV}(t) + u_{R\_ch}(t) + u_{P\_ch}(t)\right) i_{ch}(t) \mathrm{d}t \\ &= \int u_{OCV}(t) i_{ch}(t) \mathrm{d}t + \int u_{R\_ch}(t) i_{ch}(t) \mathrm{d}t + \int u_{P\_ch}(t) i_{ch}(t) \mathrm{d}t \\ &= E_{OCV} + E_{R\_ch} + E_{P\_ch} \end{aligned} \tag{3.60}$$

where the energy $E_{OCV} = \int u_{OCV}(t) i_{ch}(t) \mathrm{d}t$ corresponding to $u_{OCV}(t)$ is absorbed by the battery and stored, and is called the stored energy. The energy $E_{R\_ch} = \int u_{R\_ch}(t) i_{ch}(t) \mathrm{d}t$ corresponding to $u_R(t)$ is consumed in the charging process and is called the charging ohm loss. The energy $E_{P\_ch} = \int U_{P\_ch}(t) i_{Ch}(t) \mathrm{d}t$ corresponding to $u_P(t)$ is consumed in the charging process, and is called the charging polarization loss.

Similarly, when discharging:

$$u_{O\_dch}(t) = u_{OCV}(t) - u_{R\_dch}(t) - u_{P\_dch}(t) \tag{3.61}$$

So the discharging energy $E_{O\_dch}(t)$ can be written as:

$$\begin{aligned} E_{O\_dch} &= \int u_{O\_dch}(t) i_{dch}(t) \mathrm{d}t = \int \left(u_{OCV}(t) - u_{R\_dch}(t) - \mathrm{u}_{P\_dch}(t)\right) i_{dch}(t) \mathrm{d}t \\ &= \int u_{OCV}(t) i_{dch}(t) \mathrm{d}t - \int u_{R\_dch}(t) i_{dch}(t) \mathrm{d}t - \int u_{P\_dch}(t) i_{dch}(t) \mathrm{d}t \\ &= E_{OCV} - E_{R\_dch} + E_{P\_dch} \end{aligned} \tag{3.62}$$

Where the energy $E_{OCV} = \int u_{OCV}(t) i_{dch}(t) \mathrm{d}t$ corresponding to $u_{OCV}(t)$ is the actual released energy inside the battery, called the released energy. The energy $E_{R\_dch} = \int u_{R\_dch}(t) i_{dch}(t) \mathrm{d}t$ corresponding to $u_R(t)$ is consumed in the discharging process, and is called the discharging ohm loss. The energy $E_{P\_dch} = \int u_{P\_dch}(t) i_{dch}(t) \mathrm{d}t$ corresponding to $u_P(t)$ is called the discharging polarization loss.

When the charge–discharge current is not small enough, the $u_R(t)$ and $u_P(t)$ of the battery cannot be ignored; therefore, during the charging process, $u_O(t) > u_{OCV}(t)$, $E_{O\_ch} > E_{OCV}$, and the extra parts are E_{R_ch} and E_{P_ch}; or vice versa in the discharging process. Therefore, in the charge–discharge process, $E_{O\_ch} > E_{OCV} > E_{dch}$, and is described by using the efficiency η_E of the battery charging and discharging energy:

$$\eta_{\mathrm{E}} = \frac{E_{\mathrm{O\_dch}}}{E_{\mathrm{O\_ch}}} = \frac{\int u_{\mathrm{o\_dch}}(t) i_{\mathrm{dch}}(t) \mathrm{d}t}{\int u_{\mathrm{o\_ch}}(t) i_{\mathrm{ch}}(t) \mathrm{d}t} = \frac{E_{\mathrm{OCV}} - E_{\mathrm{R\_dch}} - E_{\mathrm{P\_dch}}}{E_{\mathrm{OCV}} + E_{\mathrm{R\_ch}} + E_{\mathrm{P\_ch}}}$$

$$= \frac{\int u_{\mathrm{OCV}}(t) i_{\mathrm{dch}}(t) \mathrm{d}t - \int u_{\mathrm{R\_ch}}(t) i_{\mathrm{dch}}(t) \mathrm{d}t - \int u_{\mathrm{P\_dch}}(t) i_{\mathrm{dch}}(t) \mathrm{d}t}{\int u_{\mathrm{OCV}}(t) i_{\mathrm{ch}}(t) \mathrm{d}t + \int u_{\mathrm{R\_ch}}(t) i_{\mathrm{ch}}(t) \mathrm{d}t + \int u_{\mathrm{P\_ch}}(t) i_{\mathrm{ch}}(t) \mathrm{d}t} \quad (3.63)$$

As seen, during the charging and discharging process, $u_{\mathrm{R}}(t)$ and $u_{\mathrm{P}}(t)$ will cause a certain energy loss, so the efficiency of the battery charging and discharging energy must be less than 1.

In the charging and discharging process, the energy E_{OCV} corresponding to $u_{\mathrm{OCV}}(t)$ is the stored or released energy of the battery. When the charge–discharge capacity changes are small enough, $u_{\mathrm{OCV}}(t)$ can be considered almost constant. So the charge–discharge capacity Q of the period $[t_1, t_2]$ is divided into n equal parts, and it is assumed that the corresponding time slice is, respectively, $t_1 \sim t_{a1}, t_{a1} \sim t_{a2}, \ldots, t_{a(n-2)} \sim t_{a(n-1)}, t_{a(n-1)} \sim t_2$ and the $u_{\mathrm{OCV}}(t)$ is, respectively, $u_{\mathrm{OCV}}(1) \ldots, u_{\mathrm{OCV}}(n)$. When n is large enough:

$$\int_{t_1}^{t_{a1}} i(t) dt = \int_{t_{a1}}^{t_{a2}} i(t) \mathrm{d}t = \ldots = \int_{t_{a(n-1)}}^{t_2} i(t) \mathrm{d}t = \frac{Q}{n} \quad (3.64)$$

Meanwhile:

$$\begin{aligned}
E_{\mathrm{OCV}} &= \int_{t1}^{t2} u_{\mathrm{OCV}}(t) i(t) \mathrm{d}t \\
&= \int_{t_1}^{t_{a1}} U_{\mathrm{OCV}}(1) i(t) \mathrm{d}t + \int_{t_{a1}}^{t_{a2}} U_{\mathrm{OCV}}(2) i(t) \mathrm{d}t + \ldots + \int_{t_{a(n-1)}}^{t_2} U_{\mathrm{OCV}}(n) i(t) \mathrm{d}t \\
&= U_{\mathrm{OCV}}(1) \times \int_{t_1}^{t_{a1}} i(t) \mathrm{d}t + U_{\mathrm{OCV}}(2) \times \int_{t_{a1}}^{t_{a2}} i(t) \mathrm{d}t + \ldots + U_{\mathrm{OCV}}(n) \int_{t_{a(n-1)}}^{t_2} U_{\mathrm{OCV}}(n) i(t) \mathrm{d}t \\
&= U_{\mathrm{OCV}}(1) \frac{Q}{n} + U_{\mathrm{OCV}}(2) \frac{Q}{n} + \ldots + U_{\mathrm{OCV}}(n) \frac{Q}{n} \\
&= \frac{Q}{n} \sum_{i=1}^{n} U_{\mathrm{OCV}}(i) = Q U_{\mathrm{OCV}}\big|_{t=[t_1 \sim t_2]}
\end{aligned}$$

Where, $U_{\mathrm{OCV}}\big|_{t=[t_1 \sim t_2]}$ is the average of the OCV after the equal-capacity section in the period $[t_1 \sim t_2]$.

As is seen, in the charging and discharging process, the stored or released energy of the battery is equal to the product of the charging and discharging capacity and the average of the OCV corresponding to this capacity. Also, the OCV is closely related to *SOC*, battery type and battery formula, so:

1. The stored or released energy of the battery is related to not only the charging and discharging capacity, but also the current state of charging of the battery. As shown in Figure 3.32, at the low-end of the SOC, the energy corresponding to the same capacity change is less (EL < EH); on the contrary, at the high-end of the SOC, the energy corresponding to the same capacity change is more (EL > EH). Hence, in a different state of charging, when the battery is charged or releases the same capacity, there are differences between the charging and discharging energy (corresponding to the life of the equipment or the mileage of the vehicles).
2. The voltage levels of different types of batteries (such as lead-acid, nickel hydrogen and lithium-ion batteries) have large differences, so even with the same capacity the stored energy of the battery is bound to be very different, which causes a corresponding difference in the useful life or mileage of vehicles.
3. Even for the same lithium-ion battery, because the formula is not the same, the OCV–SOC curve of the battery has differences, and the corresponding battery energy is bound to produce a difference. But from the aspect of the capacity, it cannot effectively reflect the problem. Therefore, the measuring of energy use has a practical significance.

When the charge–discharge current is small enough, the ohm voltage drop u_R and the polarization voltage drop u_P can be ignored, then the terminal voltage u_0 is equal to the OCV u_{OCV}, and the charging and discharging energy efficiency is equal to 1. The combination of the above analysis shows that:

$$\begin{cases} E_{OCV} = Q_{max} U_{OCV|SOC=[0\%,100\%]} \\ E_{rem} = Q_{rem} U_{OCV|SOC=[0\%,SOC_0]} = Q_{max} SOC_0 U_{OCV|SOC=[0\%,100\%]} \end{cases}$$

Q_{max} is the maximum available capacity of the battery, $U_{OCV|SOC=[0\%,100\%]}$ is the average of the OCV corresponding to the whole capacity section, Q_{rem} is the remaining capacity of the

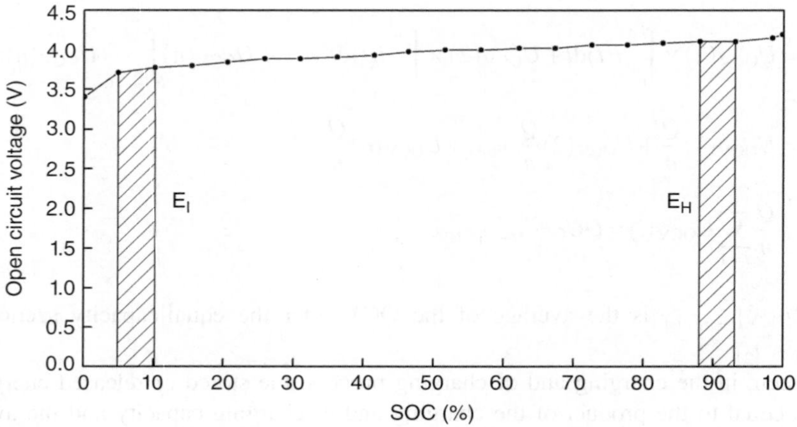

Figure 3.32 Discharging energy difference at different SOCs.

battery, SOC_0 is the current state of charging, $U_{OCV|SOC=[0\%,SOC_0]}$ is the average of the OCV of the state of charging from SOC_0 to the power-up process.

So :

$$SOE = \frac{E_{rem}}{E_{max}} \times 100\% = SOC_0 \frac{U_{OCV|SOC=[0\%,SOC_0]}}{U_{OCV|SOC=[0\%,100\%]}} \tag{3.65}$$

As is seen, for the same battery, OCV – SOC is constant; therefore E_{max} and E_{rem} are only related to Q_{max} and SOC. Because both of them are decoupling, the SOE of the battery also achieves decoupling with the operation conditions.

When the battery with initial SOE SOE_0, state of charging SOC_0, releases a certain energy ΔE, the SOE is changed to SOE_1 and the SOC to SOC_1, and then:

$$\begin{aligned} SOE_1 &= \frac{E_{rem0} - \Delta E}{E_{max}} \times 100\% = \left(\frac{E_{rem0}}{E_{max}} - \frac{\Delta E}{E_{max}}\right) \times 100\% \\ &= SOE_0 - \frac{Q_{max}|SOC_0 - SOC_1| U_{OCV|SOC=[SOC_0,SOC_1]}}{Q_{max} U_{OCV|SOC=[0\%,SOC_0]}} \\ &= SOE_0 - \frac{\Delta SOC\, U_{OCV|SOC=[SOC_0,SOC_1]}}{Q_{max} U_{OCV|SOC=[0\%,SOC_0]}} \end{aligned} \tag{3.66}$$

Similar to the estimation method of SOC, this makes the SOE of the battery recursive.

3.4.2 SOE Definition of the Battery Groups

For a single battery, when the SOC changes from SOC_0 to SOC_1, the change in the capacity Q and energy E of the battery have the following relation:

$$E = QU_{OCV}\Big|_{SOC=[SOC_0,SOC_1]} = Q_{max}|SOC_1 - SOC_0| U_{OCV}\Big|_{SOC=[SOC_0,SOC_1]} \tag{3.67}$$

When charging, the maximum charging energy of the battery is:

$$\begin{aligned} E_{ch\_max} &= Q_{max}(100\% - SOC_0) U_{OCV}\Big|_{SOC=[SOC_0,100\%]} \\ &= Q_{ch\_max} U_{OCV}\Big|_{SOC=[SOC_0,100\%]} \end{aligned}$$

When discharging, the maximum discharging energy of the battery, also called the remaining energy is:

$$\begin{aligned} E_{rem} &= E_{dis\_max} = Q_{max} SOC_0 U_{OCV}\Big|_{SOC=[SOC_0,0\%]} \\ &= Q_{dis\_max} U_{OCV}\Big|_{SOC=[SOC_0,0\%]} = Q_{rem} SOC_0 U_{OCV}\Big|_{SOC=[SOC_0,0\%]} \end{aligned}$$

If n batteries are collected in series to form a group, its maximum available capacity is, respectively, $Q_{max}[1], \ldots Q_{max}[n]$, the corresponding current SOC is: $SOC_0[1], \ldots SOC_0[n]$. If we start changing this battery group, when it has been fully charged, the SOC of each battery is, respectively:

$$SOC_0[1] + \frac{Q^B_{ch\_max}}{Q_{max}}[1], \ldots SOC_0[n] + \frac{Q^B_{ch\_max}}{Q_{max}[n]}$$

The stored energy of any battery m in this battery group is:

$$E_{ch}[m] = Q^B_{ch\_max} U_{OCV}\Big|_{SOC=\left[SOC_0[m], SOC_0[m] + \frac{Q^B_{ch\_max}}{Q_{max}[m]}\right]} \tag{3.68}$$

The total stored energy E^B_{ch} of the battery group is the sum of the stored energy of each battery.

If this battery pack is discharging, at the end of discharging, the SOC of each battery is, respectively:

$$SOC_0[1] - \frac{Q^B_{rem}}{Q_{max}[1]}, \ldots SOC_0[n] - \frac{Q_{rem}}{Q_{max}[n]}$$

The stored energy of any battery m in this battery pack is:

$$E_{dch}[m] = Q^B_{rem} U_{OCV}\Big|_{SOC=\left[SOC_0[m] - \frac{Q^B_{rem}}{Q_{max}[m]}, SOC_0[m]\right]} \tag{3.69}$$

The releasing energy of the battery E^B_{dch}, also called the remaining energy of the battery E^B_{rem} is the sum of the released energy of each battery.

The above analysis shows that the SOC available section of any battery m in this battery pack is:

$$\left[SOC_0[m] - \frac{Q^B_{rem}}{Q_{max}[m]}, SOC_0[m] + \frac{Q^B_{ch\_max}}{Q_{max}[m]}\right]$$

The corresponding available energy $E[m]$ is:

$$\begin{aligned} E[m] &= Q_{max}[m]\left(SOC_0[m] + \frac{Q^B_{ch\_max}}{Q_{max}[m]} - SOC_0[m] + \frac{Q^B_{rem}}{Q_{max}[m]}\right) U_{OCV}\Big|_{SOC=\left[SOC_0[m] - \frac{Q^B_{rem}}{Q_{max}[m]}, SOC_0[m] + \frac{Q^B_{ch\_max}}{Q_{max}[m]}\right]} \\ &= \left(Q^B_{ch\_max} + Q^B_{rem}\right) U_{OCV}\Big|_{SOC=\left[SOC_0[m] - \frac{Q^B_{rem}}{Q_{max}[m]}, SOC_0[m] + \frac{Q^B_{ch\_max}}{Q_{max}[m]}\right]} \\ &= Q^B_{max} U_{OCV}\Big|_{SOC=\left[SOC_0[m] - \frac{Q^B_{rem}}{Q_{max}[m]}, SOC_0[m] + \frac{Q^B_{ch\_max}}{Q_{max}[m]}\right]} \end{aligned} \tag{3.70}$$

Where Q^B_{max} is the maximum available capacity of this battery pack.

The maximum available energy of this battery pack is the sum of the available energy of each battery:

$$E^{\mathrm{B}}_{\max} = \sum_{m=1}^{n} E[m] = Q^{\mathrm{B}}_{\max} \sum_{m=1}^{n} U_{\mathrm{OCV}} \Big|_{\mathrm{SOC}=\left[\mathrm{SOC}_0[m]-\frac{Q^{\mathrm{B}}_{\mathrm{rem}}}{Q_{\max}[m]},\ \mathrm{SOC}_0[m]+\frac{Q^{\mathrm{B}}_{\mathrm{ch\_max}}}{Q_{\max}[m]}\right]} \tag{3.71}$$

As is seen, the maximum available energy of the series battery pack is equal to the product of the maximum available capacity of the battery in the pack and the sum of the average OCV of all single batteries in the pack in the state of charging section.

Similar to the definition of the SOE of the single battery, the definition of $\mathrm{SOE}^{\mathrm{B}}$ of the series battery pack is the remaining energy of the battery pack, which is the ratio between $E^{\mathrm{B}}_{\mathrm{rem}}$ and the maximum available energy $E^{\mathrm{B}}_{\max}$.

$$\mathrm{SOE}^{\mathrm{B}} = \frac{E^{\mathrm{B}}_{\mathrm{rem}}}{E^{\mathrm{B}}_{\max}} \times 100\% = \frac{Q^{\mathrm{B}}_{\mathrm{rem}} \sum_{m=1}^{n} U_{\mathrm{OCV}} \Big|_{\mathrm{SOC}=\left[\mathrm{SOC}_0[m]-\frac{Q^{\mathrm{B}}_{\mathrm{rem}}}{Q_{\max}[m]},\ \mathrm{SOC}_0[m]\right]}}{Q^{\mathrm{B}}_{\max} \sum_{m=1}^{n} U_{\mathrm{OCV}} \Big|_{\mathrm{SOC}=\left[\mathrm{SOC}_0[m]-\frac{Q^{\mathrm{B}}_{\mathrm{rem}}}{Q^{\mathrm{B}}_{\max}[m]},\ \mathrm{SOC}_0[m]+\frac{Q^{\mathrm{B}}_{\mathrm{ch\_max}}}{Q^{\mathrm{B}}_{\max}[m]}\right]}} \times 100\%$$

$$= \mathrm{SOC}^{\mathrm{B}} \frac{U^{\mathrm{B}}_{\mathrm{OCV}} \Big|_{\mathrm{SOC}^{\mathrm{B}}=\left[\mathrm{SOC}^{\mathrm{B}}_0,0\%\right]}}{U^{\mathrm{B}}_{\mathrm{OCV}} \Big|_{\mathrm{SOC}^{\mathrm{B}}=[100\%,0\%]}}$$

Where $\mathrm{SOC}^{\mathrm{B}}$ is the state of charging of the battery pack, $U^{\mathrm{B}}_{\mathrm{OCV}} \Big|_{\mathrm{SOC}^{\mathrm{B}}=\left[\mathrm{SOC}^{\mathrm{B}}_0,0\%\right]}$ the average of the OCV of the battery pack from the state of charging $\mathrm{SOC}^{\mathrm{B}}_0$ to the power-up process, then:

$$U^{\mathrm{B}}_{\mathrm{OCV}} \Big|_{\mathrm{SOC}^{\mathrm{B}}=\left[\mathrm{SOC}^{\mathrm{B}}_0,0\%\right]} = \sum_{m=1}^{n} U_{\mathrm{OCV}} \Big|_{\mathrm{SOC}=\left[\mathrm{SOC}_0[m]-\frac{Q^{\mathrm{B}}_{\mathrm{rem}}}{Q_{\max}[m]},\ \mathrm{SOC}_0[m]\right]} \tag{3.72}$$

where $U^{\mathrm{B}}_{\mathrm{OCV}} \Big|_{\mathrm{SOC}^{\mathrm{B}}=[100\%,0\%]}$ is the average of the OCV of the battery pack from fully charged to the power-up process, and then:

$$U^{\mathrm{B}}_{\mathrm{OCV}} \Big|_{\mathrm{SOC}^{\mathrm{B}}=[100\%,0\%]} = \sum_{m=1}^{n} U_{\mathrm{OCV}} \Big|_{\mathrm{SOC}=\left[\mathrm{SOC}_0[m]-\frac{Q^{\mathrm{B}}_{\mathrm{rem}}}{Q_{\max}[m]},\ \mathrm{SOC}_0[m]+\frac{Q^{\mathrm{B}}_{\mathrm{ch\_max}}}{Q_{\max}[m]}\right]} \tag{3.73}$$

The following three points claim attention:

1. Because of the inconsistencies in the state of charging of the batteries in a battery pack, the OCV of the battery pack is inconsistent with the OCV of a single battery, which can effectively correspond to the SOC and SOE of the battery. This is the problem based on the estimation of the terminal voltage to the SOC and SOE of the battery pack.

2. Due to little changes in the state of charging of the battery once or over several consecutive charge–discharge cycle processes, the $U^{B}_{OCV}\big|_{SOC^{B}=[100\%,0\%]}$ is considered to be unchanged. But it changes with the variation in the SOC of the battery. Because it takes the average of the OCV at the lower end of the battery, $U^{B}_{OCV}\big|_{SOC^{B}=\left[SOC^{B}_{0},0\%\right]} \le U^{B}_{OCV}\big|_{SOC^{B}=[100\%,0\%]}$, that is $SOE^{B} \le SOC^{B}$, which also explains why the remaining energy percentage of the battery is less than its remaining capacity percentage. This is the reason why the energy is less when the same capacity is discharged at the lower end of the battery.
3. In the charge–discharge process, the energy variation of the battery:

$$\Delta E^{B} = E^{B}_{max}\Delta SOE^{B} = E^{B}_{max}\Delta SOC^{B}k$$

where

$$k = \frac{U^{B}_{OCV}\big|_{SOC^{B}=\left[SOC^{B}_{0},0\%\right]}}{U^{B}_{OCV}\big|_{SOC^{B}=[100\%,0\%]}}.$$

Because k is reduced with decreasing SOC^{B}, when the battery is charged with the same energy, the SOC^{B} of the battery is larger as well and the capacity variation of the battery is greater at the lower end of the state of charging. In other words, the released energy of the battery is less when charging the same capacity.

The battery electric quantity description method based on the capacity comes from the electrochemical mechanism, that is, the process of gaining and losing electrons in the battery's internal redox reaction. In the charge–discharge process, the electron flow is formed in the external circuit, and the ion flow is formed inside the battery, expressed as the current with certain amplitude, whose product with time is the coulomb quantity, namely the capacity of the battery. Due to the fixed number of substances involved in the reaction inside the battery, when the battery's internal substances are sufficiently reacted within the allowable range, the number of electrons emitted is the maximum available capacity of the battery. At a certain moment, inside the battery the remaining amount of substances is able to participate in the reaction determining the number of electrons generated in the remaining battery reaction, that is, the remaining capacity of the battery. The ratio of the remaining capacity and the maximum available capacity is the SOC of the battery. Therefore capacity is the macro-performance of the micro-chemical properties of the internal battery, and the basis of using the battery, reflecting the essential features of the internal battery. However, when in use, the role of the battery is storing energy and releasing energy, and the amount of external work is directly related to the output energy. Therefore, in terms of use, it is better to describe the battery state from the point of view of energy to establish a simpler (linear) relation with the time of use and the mileage of electric vehicles, and thus to make it easy to explain the related phenomena. From the energy of the battery and the definition of SOE and the estimation formula it can be seen that the capacity of the battery and SOC are the basis of the energy and the SOE definition and estimation.

3.5 Method for Estimation of the Battery Group SOE and the Remaining Energy

Estimation of the SOE of the battery requires estimation of the SOC and the average OCV from the current state and fully charged state to the power-up state. This can be done by calculating $U^{B}_{OCV}\big|_{SOC^{B}=[SOC^{B}_{0},0\%]}$ and $U^{B}_{OCV}\big|_{SOC^{B}=[100\%,0\%]}$ and then estimating the SOE.

For the battery under consideration a table of corresponding values of the OVC and SOC can be obtained by testing. From the initial SOC of the battery, SOC_0, the OCV is measured at intervals of equal capacity change (the more measurements, the better the accuracy) until SOC = 0%. Taking the average of the above data, then $U_{OCV}\big|_{SOC=[0\%,SOC_0]}$ can be obtained. When $SOC_0 = 100\ \%$, $U_{OCV}\big|_{SOC=[0\%,SOC_0]}$ can be obtained.

The OCV–SOC relation can be determined for a particular battery, so $U_{OCV}\big|_{SOC=[0\%,100\%_0]}$ is a constant and only associated with the SOC of the battery. Thus, there is a one-to-one relationship between the SOC and SOE of the battery. Then a table of corresponding SOC and SOE values can be established. Thus by estimation of the SOC, the battery SOE can be obtained by looking up the table, simplifying online processing.

Taking the OCV–SOC correspondence table of the I-type battery as an example, the OCV–SOC–SOE correspondence table (see Table 3.9) can be obtained.

Table 3.9 OCV–SOC–SOE correspondence table of LMO cell.

U_{OCV}(V)	SOC(%)	$U_{OCV}\big\|_{SOC=[0\%,SOC_0]}$ (V)	SOE(%)
3.300	0.0	3.300	0.0
3.695	5.2	3.498	4.6
3.775	10.4	3.616	9.5
3.812	15.6	3.675	14.5
3.849	20.7	3.714	19.5
3.877	25.9	3.744	24.5
3.904	31.1	3.768	29.6
3.925	36.3	3.789	34.8
3.946	41.5	3.808	39.9
3.967	46.7	3.824	45.1
3.988	51.9	3.839	50.3
4.003	57.1	3.854	55.6
4.018	62.2	3.867	60.8
4.036	67.4	3.879	66.1
4.057	72.6	3.891	71.4
4.070	77.8	3.903	76.7
4.083	83.0	3.913	82.1
4.096	88.2	3.924	87.4
4.121	93.4	3.934	92.8
4.167	98.6	3.945	98.3
4.210	100.0	3.957	100.0

The maximum available energy of the battery $E_{max} = Q_{max}U_{OCV}|_{SOC=[0\%,100\%]}$, so the E_{max} of the battery can be converted to the estimation of Q_{max} and $U_{OCV}|_{SOC=[0\%,100\%]}$, and these two values can be obtained by the preceding analysis. Since $U_{OCV}|_{SOC=[0\%,100\%]}$ is a constant, the maximum available energy is proportional to the maximum available capacity, and the coefficient is $U_{OCV}|_{SOC=[0\%,100\%]}$. Then the remaining energy of the battery $E_{rem} = E_{max}SOE$ can be obtained.

3.6 Method of Estimation of the Actual Available Energy of the Battery

Comparing the OCV with the EOC or EOD voltage to determine whether the battery is fully charged or discharged to achieve the decoupling between the working condition and the remaining energy or the SOE is only possible when the charging and discharging current is small enough. In actual use, the working current of the battery cannot continuously be maintained sufficiently small, so there is bound to be a certain ohm voltage drop U_R and polarization voltage U_P, to make the external voltage larger than the OCV when charging, and smaller when discharging. Meanwhile, in order to ensure the safety and longevity of the battery in use, when reaching the EOD voltage, the battery cannot continue to work with this current, and needs to work by dropping the power or terminating the discharging, which eventually leads to the actual discharging energy becoming less than the maximum available energy. So estimation of the actual releasing energy of the battery in the current working conditions, the actual available energy, has a practical significance for users.

The actual discharging energy E_{dch} of the battery is equal to the product of the discharging capacity Q_{dch} and the corresponding average discharging voltage U_{avg}, and U_{avg} is related to the corresponding OCV, the ohm voltage drop and the polarization voltage, that is, $E_{dch} = Q \times U_{avg} = Q(U_{OCV\_avg} - U_{R\_avg} - U_{p\_avg})$. Similar to the fact that the actual capacity is less than the maximum available capacity, the presence of U_R and U_P again explain why the actual available energy is less than the maximum available energy, and the discharging energy is less than the remaining energy. Running in the working condition, the over-voltage from the two aspects affects the actual discharging energy of the battery:

1. The actual discharge capacity decreases.
2. The utilization rate of the voltage decreases, and the corresponding charging and discharging energy efficiency of the battery decreases.

For capacity loss, assume that the SOC of the battery is SOC_H when the battery charge is completed and SOC_L, when the discharge is ended, the actual available capacity of the battery Q is $Q = Q_{max}(SOC_H - SOC_L)$, and the maximum available energy E of the battery is:

$$\begin{aligned} E &= Q_{max}(SOC_H - SOC_L)U_{OCV}\Big|_{SOC=[SOC_H, SOC_L]} \\ &= Q_{max}SOC_H U_{OCV}\Big|_{SOC=[SOC_H, 0\%]} - Q_{max}SOC_L U_{OCV}\Big|_{SOC=[SOC_L, 0\%]} \end{aligned} \tag{3.74}$$

As is seen, in the working condition, the capacity loss makes the available energy of the battery limited to the middle section of the SOC, the loss is from the fully charged state of the battery to SOC_H and SOC_L to power-up, and the loss of available energy is:

$$\begin{aligned} E_{loss} &= Q_{max}(100\% - SOC_H)U_{OCV}\big|_{SOC\in[100\%, SOC_H]} + Q_{max}SOC_L U_{OCV}\big|_{SOC\in[SOC_L, 0\%]} \\ &= E_{max} + Q_{max}SOC_L U_{OCV}\big|_{SOC\in[SOC_L, 0\%]} - Q_{max}SOC_H U_{OCV}\big|_{SOC\in[SOC_H, 0\%]} \end{aligned} \tag{3.75}$$

For the utilization rate of the voltage, U_R and U_P result in $U_0 \geq U_{OCV}$ when charging, and then $E_{0\_ch} \geq E_{OCV}$; $U_0 \leq U_{OCV}$ when discharging, and then $E_{0\_dch} \leq E_{OCV}$, which causes a problem with energy utilization efficiency in the charge–discharge process of the battery. When the battery discharges from SOC_0 to SOC_1, the energy ΔE_{OCV} corresponding to U_{OCV} is changed to:

$$\Delta E_{OCV} = Q_{max}(SOC_0 - SOC_1)U_{OCV}\big|_{SOC=[SOC_0, SOC_1]} \tag{3.76}$$

The discharging energy corresponding to:

$$\Delta E_0 = \int_{SOC_1}^{SOC_0} u_0 i \mathrm{d}t$$

Because u_0 and the current are measurable, and there is a dedicated integrated circuit integrating their product to obtain ΔE_0, which is able to guarantee the real time and accuracy of the calculation, the corresponding energy efficiency when the battery is discharging from SOC_0 to SOC_1 is:

$$\eta_{E\_dch} = \frac{\Delta E_0}{\Delta E_{OCV}} = \frac{\int_{SOC_1}^{SOC_0} u_0 i \mathrm{d}t}{Q_{max}(SOC_0 - SOC_1)U_{OCV}\big|_{SOC=[SOC_0, SOC_1]}} \tag{3.77}$$

Similarly, the corresponding charging energy efficiency when the battery SOC is charged from SOC_0 to SOC_1 is:

$$\eta_{E\_ch} = \frac{E_{OCV}E_0}{E_0} = \frac{Q_{max}(SOC_1 - SOC_0)U_{OCV}\big|_{SOC=[SOC_0, SOC_1]}}{\int_{SOC=SOC_0}^{SOC=SOC_1} u_0 i \mathrm{d}t} \tag{3.78}$$

So the relation between the maximum available and actual available energy of the single battery is shown in Figure 3.33.

The above discussions relate to a single battery. The analysis and estimation method from the actual and maximum available energy of a single battery to those of the battery pack will not be discussed here.

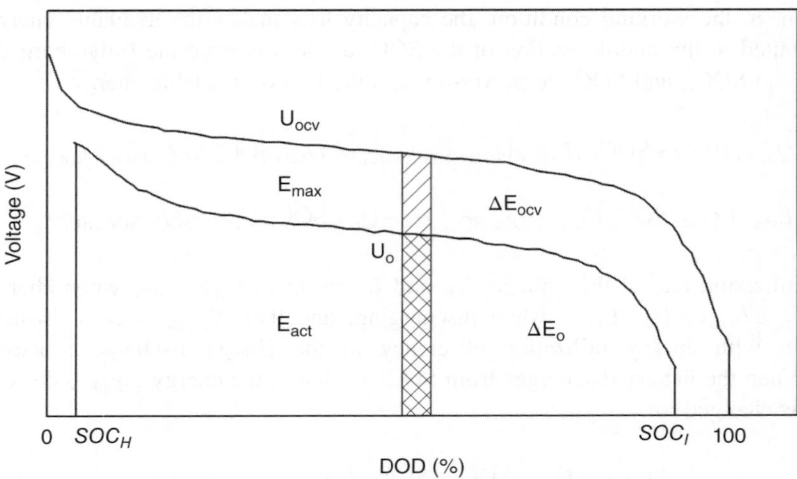

Figure 3.33 Maximum available and actual available energy of the battery. (Reproduced with permission from Feng Wen, "Study on basic issues of the Li-ion battery pack management technology for pure electric vehicles.", ©2009.)

References

[1] Gu, W.B. and Wang, C.Y. (2000) Thermal–electrochemical modeling of battery systems. *Journal of the Electrochemical Society*, **147**(8), 2910–2922.

[2] Piller, S., Perrin, M., and Jossen, A. (2001) Methods for state-of-charge determination and their applications. *Journal of Power Sources*, **96**, 113–120.

[3] Plett, G.L. (2004) Extended Kalman filtering for battery management systems of LiPB-based HEV battery packs Part 1. Background. *Journal of Power Sources*, **134**, 252–261.

[4] Plett, G.L. (2004) Extended Kalman filtering for battery management systems of LiPB-based HEV battery packs Part 2. Modeling and identification. *Journal of Power Sources*, **134**, 262–276.

[5] Plett, G.L. (2004) Extended Kalman filtering for battery management systems of LiPB-based HEV battery packs Part 3. State and parameter estimation. *Journal of Power Sources*, **134**, 277–292.

[6] Plett, G.L. (2006) Sigma-point Kalman filtering for battery management systems of LiPB-based HEV battery packs Part 1: Introduction and state estimation. *Journal of Power Sources*, **161**, 1356–1368.

[7] Plett, G.L. (2006) Sigma-point Kalman filtering for battery management systems of LiPB-based HEV battery packs Part 2: Simultaneous state and parameter estimation. *Journal of Power Sources*, **161**, 1369–1384.

[8] Vasebi, A., Bathaee, S.M.T., and Partovibakhsh, M. (2008) Predicting state of charge of lead-acid batteries for hybrid electric vehicles by extended Kalman filter. *Energy Conversion and Management*, **49**, 75–82.

[9] Jinchun, P., Yaobin, C., and Eberhart, R. (2000) Battery Pack State of Charge Estimator Design Using Computational Intelligence Approaches in Proceedings of The Fifteenth Annual Battery Conference on Applications and Advances. Long Beach, CA, 11–14 January 2000, 173–177.

[10] Gerard, O., Patillon, J.N., and D'Alche-Buc, F. (1999) Discharge prediction of rechargeable batteries with neural networks. *Integrated Computer Aided Engineering*, **6**(1), 41–52.

[11] Cheng-Hui, C., Dong, D., Zhi, Y., *et al.* (2002) Artificial Neural Network in Estimation of Battery State-of-Charge (SOC) with Non-Conventional Input Variables Selected by Correlation Analysis. Proceedings of International Conference on Machine Learning and Cybernetics. Beijing, China, 4–5 November 2002, 1619–1625.

[12] Shen, W.X. (2007) State of available capacity estimation for lead-acid batteries in electric vehicles using neural network. *Energy Conversion and Management*, **48**(2), 433–442.

[13] Lee, D.T., Shiah, S.J., Lee, C.M. *et al.* (2007) State-of-charge estimation for electric scooters by using learning mechanisms. *IEEE Transactions on Vehicular Technology*, **56**(2), 544–556.

[14] Cho, I., Kim, D., Jung, D., *et al.* (2007) Apparatus and method for testing state of charge in battery US Patent 2,007,005,276-A1, January **4**, 2007.

[15] Singh, P. and Fennie, C.J. (2000) A method for determining battery SOC using an intelligent system. US Patent 6,011,379, January 4, 2000.

[16] Reddy, V., Arey, S., Singh, P., *et al.* (1999) Preliminary Design SOC Meter for $LiIS0_2$ Cells based on Fuzzy Logic Methodology. Proceedings of the 14th Annual Battery Conference on Applications and Advances. Long Beach, CA, 12–15 January 1999, 237–239.

[17] Lee Y.S. and Jao, C.W. (2003) Fuzzy Controlled Lithium-Ion Battery Equalization with State-of-Charge Estimator. IEEE International Conference on System, Man & Cybernetics, 5–8 October 2003, 4431–4438.

[18] Yuang-Shung, L., Jao, J., and Tsung-Yuang, L. (2002) Lithium-Ion Battery Model and Fuzzy Neural Approach for Estimating Battery State-of-Charge. The 19th International Battery, Hybrid and Fuel Cell Electric Vehicle Symposium & Exhibition. EVS 19, Korea, 19–23 October, 1879–1890.

[19] Cai, C.H., Du, D., and Liu, Z.Y. (2003) Battery State-of-Charge (SOC) Estimation Using Adaptive Neuro-Fuzzy Inference System (ANFIS). *Proceedings of the 12th IEEE International Conference on Fuzzy Systems. St Louis, MO, 25-28 May 2003*, IEEE, pp 1068–1073.

[20] Yuang-Shung, L., Tsung-Yuan, K., and Wang, W.Y. (2004) Fuzzy Neural Network Genetic Approach to Design the SOC Estimator for Battery Powered Electric Scooter. *Proceedings of 2004 35th Annual IEEE Power Electronics Specialists Conference. Aachen, Germany*, IEEE, pp 2759–2765.

[21] Fu-Hua, S. (2008) A prediction model for remaining capacity in batteries based on least square support vector machine. *Chinese Journal of Power Sources, 2008*, **32**(7), 452–455.

[22] Sheng, P., Quan-Shi, C., Cheng-Tao, L. (2007) Study on estimating method for battery state of charge based on support vector regression *Chinese Journal of Power Sources*, 2007, **31**(3): 242–243, 252.

[23] Zhi-hua, W. and Cheng-liang, Y. (2008) Safety performance analysis of $LiFePO_4$ Li-ion batteries for electric vehicles. *Chinese Battery Industry*, **3**.

[24] Guo, H. and Yao, S. (2009) *The Basis of Electrochemical Measurement*, Chemical Industry Press, Beijing.

4

The Prediction of Battery Pack Peak Power

4.1 Definition of Peak Power

4.1.1 Peak Power Capability of Batteries

The battery instantaneous power refers to the product of the terminal voltage of the battery and the current flowing through the electrode in the specific state of the battery. Suppose that the open-circuit voltage (OCV) is U_{OCV}, the internal resistance R, the current controlled by the exterior power supply I_0 and the terminal voltage of the battery U_1. As Figure 2.13 shows:

$$P = U_1 I_0 = (U_{\mathrm{OCV}} - I_0 R) I_0 \tag{4.1}$$

Here, U_{OCV} can be obtained from the plot of the battery OCV versus its SOC (Figure 4.1).

If the battery is discharged with power P for t s, assuming that the SOC of the battery at time t is SOC′, the open-circuit voltage OCV and internal resistance of the battery are, respectively, U_{OCV}' and R, and the current I_t at time t then Equation 4.2 is satisfied:

$$P = U_1' I_t = (U_{\mathrm{OCV}}' - I_t R) I_t \tag{4.2}$$

The terminal voltage of the battery keeps changing because of its charging and discharging with constant power, so the current keeps changing inversely with the voltage. If the ampere hour efficiency of battery charging and discharging is 1, the battery can satisfy the following equation with the initial SOC and SOC at time t:

$$\mathrm{SOC}_0 - \mathrm{SOC} = \int_0^t I_t \mathrm{d}_t \tag{4.3}$$

Generally, manufacturers of batteries can provide the limits of voltage according to the type of battery, as constraints on battery use to prevent over charged or over discharged, and thus to

Fundamentals and Applications of Lithium-ion Batteries in Electric Drive Vehicles, First Edition.
Jiuchun Jiang and Caiping Zhang.

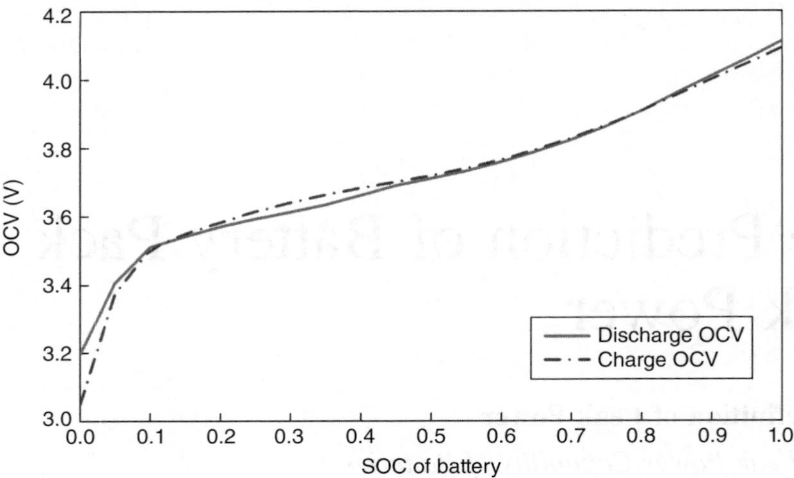

Figure 4.1 Plot of OCV versus SOC for lithium manganese oxide batteries.

avoid adversely influencing the service life of the battery. According to the constraints on the battery's voltage utilization, the definition of the battery charging peak power [1] is the value of the constant power when the terminal voltage reaches the maximum working voltage of the battery, while charging for t s at that constant power. Similarly, the discharging peak power of battery is the value of constant power when the terminal voltage declines to a minimum working voltage when discharged for t s at the constant power of the battery.

If we know the current SOC, the internal resistance R and the voltage working range of the battery $[U_{\min}, U_{\max}]$, the charging peak power $P_{\max}^{\mathrm{chr}}$ and the discharging peak power $P_{\max}^{\mathrm{dis}}$ cannot be obtained via Equations 4.1 and 4.2 because not only the battery SOC at a known time t and current I in the battery operation are coupling with each other, but also the current I and unsolved battery powers $P_{\max}^{\mathrm{chr}}$ or $P_{\max}^{\mathrm{dis}}$ are coupling and so each factor of the equation has a strong-coupling relation, this means that the charging peak power and discharging peak power at the present state of a battery cannot be obtained with a physical formula.

4.1.2 *Battery Power Density*

Battery power density is the ratio of battery output power to battery volume, namely specific power. There are two key factors which restrict the popularization of electric vehicles. One is that batteries can bear the lower charging rate but cannot achieve the minimum charging waiting time that oil-fueled automotives do. The other is that the battery has low power density and huge volume, and the ratio of gross mass of a pure electric vehicle to battery weight is 5:2, so some of the electric energy stored in batteries is consumed by the weight of the batteries, which impacts negatively on the performance of electric vehicles in braking, climbing and accelerating. In order to take advantage of electronic quality and ensure the service life of batteries, batteries cannot be over used as this will cause working fatigue. So the exact power density of batteries will need to be known in order to propose the recommendations for driving and maintenance of electric vehicles with regard to improving the maximum energy utilization ratio and ensuring the service life of the batteries. Thus, the driving conditions in electric vehicles can be improved and the cost of using batteries can be reduced.

4.1.3 *State of Function of Batteries*

The transient in a battery is defined as the state of function (SOF) of the battery [2]. The SOF can be classified into start-up SOF1 and charge acceptance SOF2. The major role of the SOF1 is to predict whether the batteries have the ability to drive motors with constant discharge and the major role of SOF2 is to predict whether the batteries have the ability to charge after constant charge for a while. Research is also being carried out to determine SOC and SOH (state of health) accurately online. However, it is obvious that the management system of the battery energy cannot reach the identification accuracy, and SOF, SOH, SOC and temperature T are highly correlated. If the transient of batteries can be predicted, and the expected parameters of the battery are input to the battery management system, then we can simplify greatly the internal mode and the algorithm of the battery management system and improve its precision management and reaction rate. So the SOF of batteries is one of the key technologies for studying batteries and their application.

4.2 Methods for Testing Peak Power

The power characteristic of a battery is measured by its power density and is meant to provide the maximum power corresponding to per unit battery quality in a short period of time. The power density of a battery is affected by internal resistance, SOC, temperature and pulse duration, which is one of key technical indexes for measuring the cell performance, especially under the condition of high power, such as in hybrid power electric vehicles and electric tools.

For hybrid power electric vehicles, in practice, in order to ensure an excess margin of battery power to power and satisfy the requirements for power in different situations, it is necessary to use more batteries, leading to high cost and battery weight, which influences the vehicle performance. If the power of the batteries is smaller, they may be over-charged and over-discharged in the process and it is necessary for the battery management system to provide instantaneous input and output power in order to protect the batteries and use them efficiently. Thus, it is necessary to test the peak power of batteries and obtain the maximal input and output power in different states.

4.2.1 *Test Methods Developed by Americans*

Some experts of the US Advanced Battery Consortium (USABC) and the United States Department of Energy compiled a test manual in 1996; furthermore, the United States Department of Energy organized some experts to compile the *Freedom CAR Battery Test Manual*, which was published in 2003. The test methods of peak power and hybrid pulse power characterization (HPPC) of the USABC focus on these two test manuals, the former focusing on pure electric vehicles and the latter on hybrid power electric vehicles. The two methods are described below.

4.2.1.1 USABC Test Method for Peak Power

The test method for USABC peak power is formulated to test the ability of discharging power of a battery for 30 s, without allowing the voltage to fall below 2/3 OCV, at different depths of discharge (DOD) [3]. The power is defined with 80% DOD, which is of great importance for power capability. However, the method cannot be used to measure the actual peak power of a battery.

Only the discharging current is mentioned in this method, without the charging current. The test high current is the maximum rated discharge current given by the manufacturer, and the smaller value of the rated peak current can be obtained from the ratio of the rated peak power at 80% DOD to 2/3 OCV. The basic discharging current can be calculated as the difference between the discharging current multiplied by12 and the testing high current divided by 35. The baseline of the discharging voltage is the maximum value between the discharging cut-off voltage supplied by the manufacturer and 2/3 OCV of 80% DOD at the beginning of the cycle.

The test steps are as follows:

1. Batteries are fully charged.
2. Discharging: First, the batteries are discharged by a basic discharging current for 30 s, then by a high current for 30 s, and, finally, by a continuous basic current until the DOD increases by 10%.
3. First, the discharging is repeated at intervals of 10% DOD from 0 to 90% DOD. Then the battery is discharged by a basic current and high test current, respectively, for 30 s at 90% DOD. Finally the residual capacity is released by a basic discharging current. The voltage of the battery must be higher than the baseline of the discharging voltage in the experiment. If this is not achievable then the experiment is repeated with a lower discharging current.
4. The peak power of the battery at a DOD can be obtained from the internal resistance of the present DOD, and is equivalent to the non-load voltage. By recording the voltage variation ΔU and the current variation ΔI of the discharging process under high current at 1 s and 30 s, the impedance and non-load voltage can be obtained at the DOD:

$$R = \frac{\Delta U}{\Delta I} \tag{4.4}$$

$$U_{\mathrm{IRFree}} = U - IR \tag{4.5}$$

The peak power at a DOD can be calculated from the minimum value via the following four equations:

$$\text{Peak Power Capability} = \frac{(-2/9)(U_{\mathrm{IRFree}})^2}{R} \tag{4.6}$$

$$\text{Peak Power Capability} = \frac{-U_{\min}\left(U_{\text{IRFree}} - U_{\min}\right)}{R} \tag{4.7}$$

$$\text{Peak Power Capability} = I_{\max}\left(U_{\text{IRFree}} + I_{\max} R\right) \tag{4.8}$$

$$\begin{aligned}\text{Peak Power Capability} &= \text{Actual Power at the end of step} \\ &\quad \text{(only if voltage or current limit occurs)}\end{aligned} \tag{4.9}$$

where $U_{\min}$ is the discharging cut-off voltage, $I_{\max}$ is the maximum rated current given by the manufacturer, the discharging power is negative value.

4.2.1.2 The HPPC Test Method

The HPPC is used to test the ability of the discharging power of the minimum voltage at the end of a 10 s discharging pulse, and the charging power of the maximum voltage at the end of a 10 s charging pulse [4]. It is applied in the test system including the charging and discharging pulses within the limits of the charging voltage of the available test platform. The method is used to verify whether the power performance of the discharging and charging pulses can satisfy the auxiliary power target established by the Freedom CAR. The test system is shown in Figure 4.2.

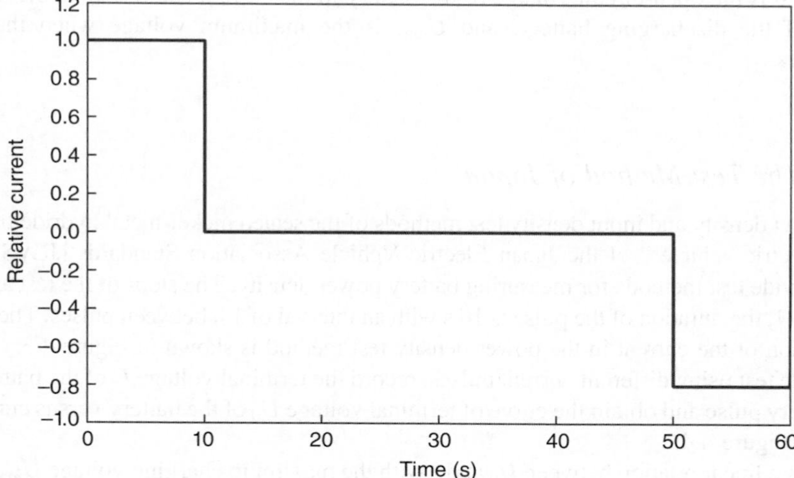

Figure 4.2 The test system for pulse power. The current is a relative current, positive values indicate a discharging current and negative values a charging current. The Freedom CAR Battery Test manual points out that there are two available currents to be selected for the pulse test. One test is performed with a low current, the peak value of the discharging current is 25% I_{MAX}, where I_{MAX} is the maximum current set by the manufacturer for a 10 s discharging pulse, at least above 5 C; the other test is with high current, the peak value of the current is 75% I_{MAX}.

Test method:

1. The battery is discharged at a rate of 1 C before HPPC testing until it is empty, then fully charged at constant current and voltage, then rested for 1 h.
2. The battery is discharged at a rate of 1 C and the DOD is adjusted to 90%, then rested for 1 h.
3. Implement the test system once and then discharge at the rate of 1 C to 10% DOD, then rest for 1 h. Then repeat the test system at 10% DOD increments until 90% DOD.
4. After implementing the test system at 90% DOD, the battery is discharged at the rate of 1 C to 10%DOD, then rested for 1 h.

Impedance performance can be obtained with the test method, and the ability of the minimum voltage discharging and maximum voltage feedback can be obtained at every DOD, as shown in Equations 4.10 and 4.11. Those powers can be used to ensure the total available SOC range and energy amplitude of the battery at the level of specific discharging and power-feedback.

$$P_{\text{discharge}} = \frac{U_{\min}(U_{\text{OCV}} - U_{\min})}{R_{\text{discharge}}} \tag{4.10}$$

$$P_{\text{charge}} = \frac{U_{\max}(U_{\max} - U_{\text{OCV}})}{R_{\text{charge}}} \tag{4.11}$$

where $R_{\text{discharge}}$ is the discharging internal resistance, R_{charge} is the charging regeneration resistance, U_{OCV} is the open-circuit voltage of the battery in the current state, $U_{\min}$ is the minimum voltage of the discharging battery, and $U_{\max}$ is the maximum voltage when the battery regenerates.

4.2.2 *The Test Method of Japan*

"The output density and input density test methods of the sealed nickel-metal hydride battery for hybrid electric vehicles" of the Japan Electric Vehicle Association Standards (JEVS) (D713-2003) provide test methods for measuring battery power density. The steps of the test are shown in Table 4.1, the duration of the pulse is 10 s with an interval of 1 h between pulses. The curve of the variation of the current in the power density test method is shown in Figure 4.3.

With the test using different current pulses, record the terminal voltage U of the battery at the end of every pulse and obtain the curve of terminal voltage U_0 of the battery versus current I, as shown in Figure 4.4.

There is a linear relation between U_0 and I with the maximum charging voltage $U_{\max}$ and the minimum discharging voltage $U_{\min}$ as constraints.

Then the OCV of the battery U_{OCV}, the maximum pulse discharge current $I_{\max}^{\text{dis}}$, and the maximum pulse charge current $I_{\max}^{\text{chr}}$ can be obtained. The peak power that can be charged or discharged can be calculated by the following equations:

$$P_{\max}^{\text{dis}} = U_{\min} I_{\max}^{\text{dis}} \tag{4.12}$$

Table 4.1 Steps of the JEVS power density test.

Single-step time (s)	Accumulative time (s)	Current rate (C)
3600	3 600	0
10	3 610	−1
3600	7 210	0
10	7 220	1
3600	10 820	0
10	10 830	−2
3600	14 430	0
10	14 440	2
3600	18 040	0
10	18 050	−3
3600	21 650	0
10	21 660	3
…	…	…

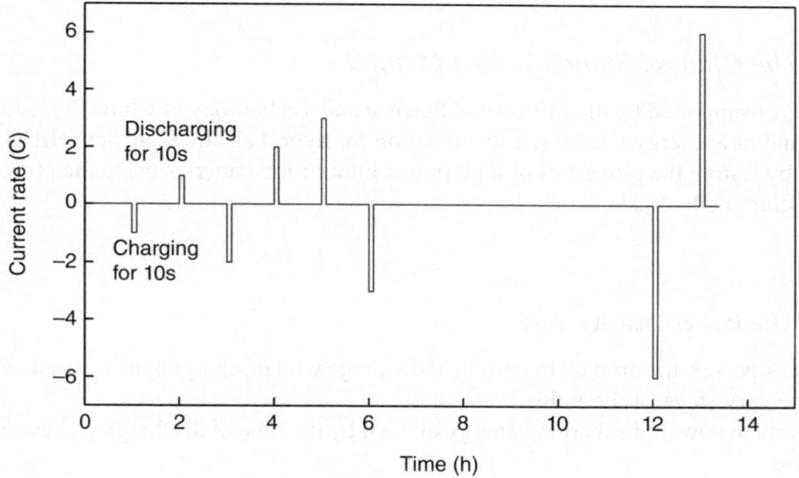

Figure 4.3 The current curve for the JEVS test.

$$P_{\max}^{\text{chr}} = U_{\max} I_{\max}^{\text{chr}} \tag{4.13}$$

In order to get the ability of charging and discharging power at the different DOD pulses, adjust batteries to the target DOD, and then repeat the test process described in Table 4.1.

Even though the JEVS test method for battery power is tested at many current rates, constant current is used in the pulse process, which is different from the constant power test. The result of the test is suitable for judging the power density of batteries.

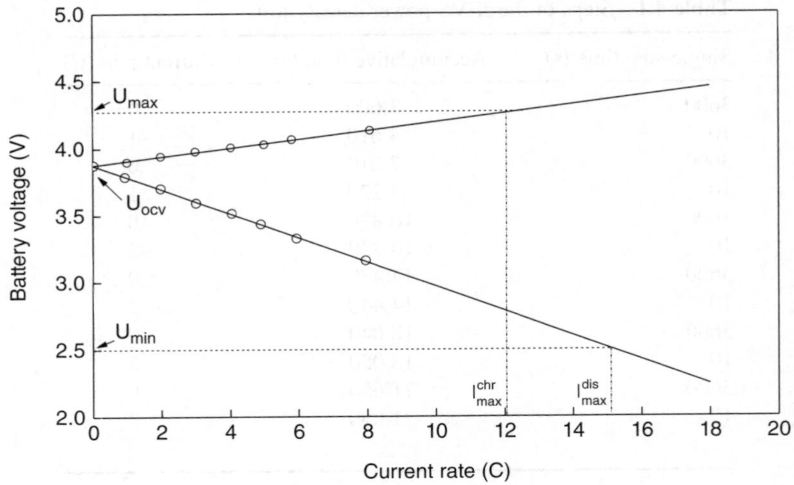

Figure 4.4 Voltage versus current for the JEVS test.

4.2.3 The Chinese Standard Test Method

In the project supported by the Ministry of Science and Technology of China (863) for energy-efficient and new energy vehicles, a specification for hybrid electric vehicles (HEV) has been proposed by testing the properties of high power lithium-ion batteries using the power characteristic testing method [5].

4.2.3.1 The Power Density Test

Battery peak power: the product of current and voltage with discharging in a closed circuit after 0.1 s at every stage of the pulse.

Battery average power: discharging energy divided by the time of discharging at every stage of the pulse.

The Test Method

First, after being fully charged at constant current and constant voltage, the battery is discharged for 6 min at a rate of 1 C at 20 ± 2 °C, then rested for 1 h.

Next, the battery is discharged for 10 s at a rate of 6 C, rested for 40 s, charged at a rate of 3 C, discharged for 330 s at a rate of 1 C and rested for 1 h. This stage is carried out six times with the SOC of the battery being 90, 80, 70, 60, 50, and 40% then discharged to a final voltage ($n \times 3.0$ V) with constant current at a rate of 1 C. The discharging is stopped when the voltage of the battery cell is lower than the discharging cut-off voltage during the stage of discharging. The power density of every stage and the DC internal resistance can be calculated.

Table 4.2 The pulse testing profile.

Capacity (Ah)	SOC (%)	Discharge rate (C)	Time (s)	SOC (%)	Discharge rate (C)	Time (s)
8–30	100	1	540	100	1	1800
	85	15	10	50	15	10
	100	1	540	100	1	1800
	85	20	10	50	20	10
30–55	100	1	540	100	1	1800
	85	10	10	50	10	10
	100	1	540	100	1	1800
	85	15	10	50	15	10

Power Test

After being fully charged, the battery is rested for 1 h at 20 ± 2 °C. The battery is then discharged according to Table 4.2. After the tests, the battery is discharged to the lowest limited voltage at a rate of 1 C. The average and the peak power of the battery are calculated at 1, 3, 5, and 10 s.

The Rated Power Test

After being fully charged, the battery is discharged with a rated power 480 W kg^{-1} at the rated temperature 20 ± 2 °C, rested for 1 h and then the final voltage of the battery is ($n \times 2.8$) V, accumulating the discharge capacity and energy. If the voltage of the battery cell is less than ($n \times 2.8$) V at the stage of discharging, then the discharging is stopped.

4.2.4 The Constant Power Test Method

In order to judge the power capability of a battery in different situations and states, based on the two power test methods of HPPC and JEVS, the power capability of the battery when charging and discharging is tested via the constant power method. The battery used in the experiment is a lithium manganese oxide battery whose rated capacity is 8 Ah and the specific parameters are shown in Table 4.3.

Because the working range of the battery discharging and charging voltage is given by the battery manufacturers, based on the protecting voltage in charging and discharging, batteries can be discharged with constant power P_1. Discharging is stopped when the terminal voltage of the battery declines to the minimum working voltage U_{min}, and the discharging time t_1 with constant power (as shown in Figure 4.5) is recorded. Adjusting the DOD of the battery to the initial state of discharging, after fully resting, the discharging power of the battery is adjusted to P_2 and it starts to discharge with constant power up to the minimum available voltage when the discharge time t_2 is recorded. The curve of discharging power P and time T can be obtained from fitting after the multiple cycle test (Figure 4.6). The peak discharging power (P_{max}, 10 s) at 10 s can be obtained from the fitting curve. The operation is repeated to give the peak power at 5 and 15 s.

Table 4.3 The performance parameters of a power lithium manganese oxide battery.

Positive and negative material	Lithium manganese oxide/graphite
Rated capacity	8 Ah
Nominal voltage	3.6 V
Maximum working voltage	4.2 V
Minimum working voltage	2.5 V
Internal resistance	<1.5 mΩ
Energy density	100 Wh kg^{-1}
Pulse output power	2500 W kg^{-1} (50% SOC, 10 s)
Pulse input power	2700 W kg^{-1} (50% SOC, 10 s)
Maximum sustainable discharge current	200 A
Temperature range	−20–55°C
Self-discharge rate	≤5% (100% SOC, 28 days)
Cycle performance at 1 C	>2000 cycles (70% rated capacity left)
Size	9×120×190 mm
Weight	290 g

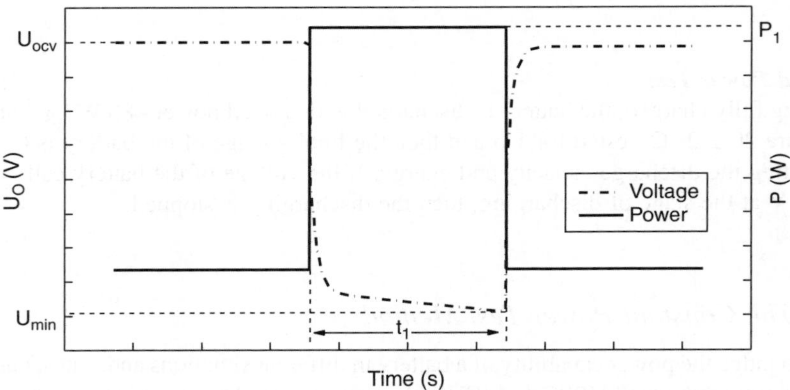

Figure 4.5 Testing curve for pulse discharging with constant power.

The charging peak power of the battery at 5, 10, or 15 s can be obtained similarly by the test method shown in Figures 4.7 and 4.8.

The peak power obtained in Figure 4.7 is a test power under a certain DOD and temperature. In order to get the charging or discharging power over the whole range of SOC, the SOC should be adjusted and the testing procedure repeated.

When the battery power is tested by the constant power method, if the initial value is selected appropriately, the test points and time will be reduced. So the selected initial test power is the peak power at normal temperatures provided by the manufacturer. The next test power is adjusted according to the last test result, and the optimum is to make the test results cover the range 5–15 s. To ensure the accuracy, the test temperature must be constant and the battery must be kept at a high or low temperature. Meanwhile, the peak power of the battery at different

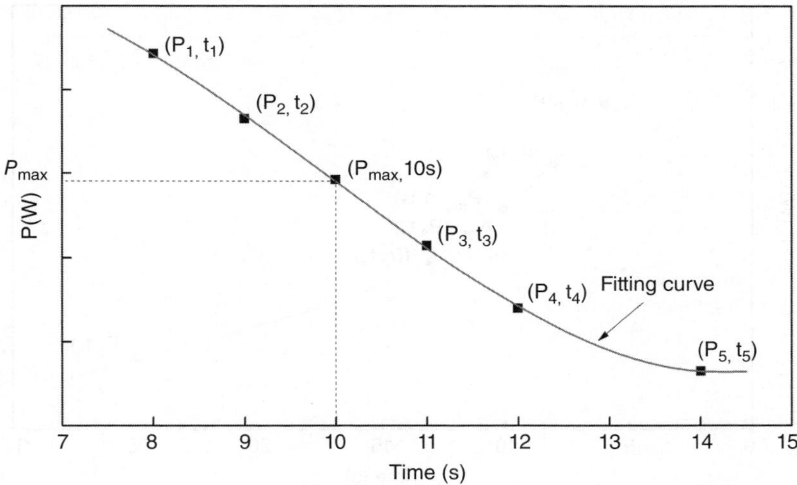

Figure 4.6 Testing curve fitting for pulse discharging with constant power.

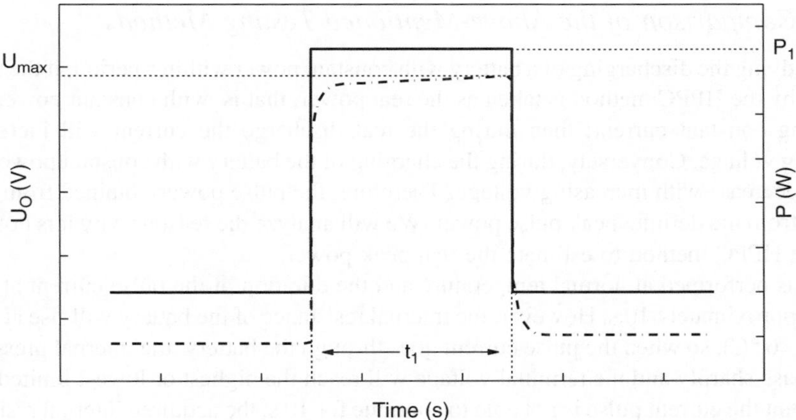

Figure 4.7 Testing curve for pulse charging with constant power.

temperatures can be tested after fully resting, just by adjusting the temperature of the environment.

During the test with constant power, the maximum charging or discharging power cannot exceed the peak power given by the manufacturer. As there are restrictions on the anode and cathode material, plate connection, electrolyte solution and diaphragm conductive rate, and restrictions on the scope of discharging and charging of the battery and power, the capacity of a battery may be sharply reduced and irreparable damage caused when using a battery outside the design limits. Similarly, the peak power obtained by the fitting method is likely to be beyond the designed maximum power. So the maximum value of the two will be selected as the real peak power of the battery under such conditions.

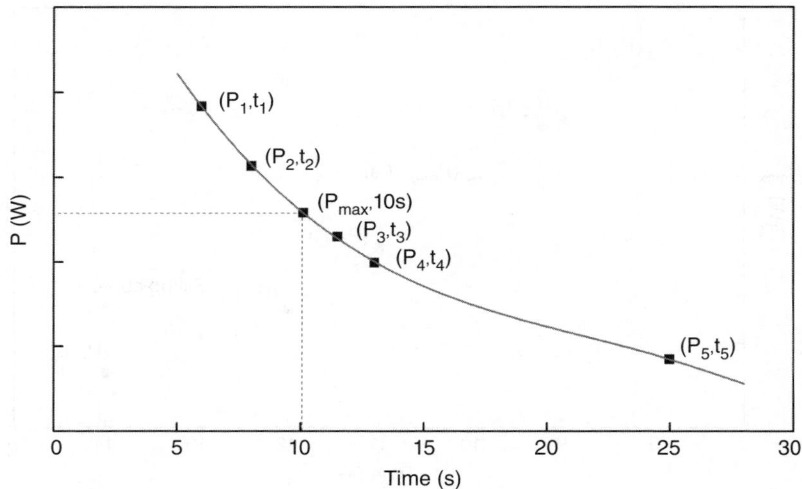

Figure 4.8 Testing curve fitting for pulse charging with constant power.

4.2.5 Comparison of the Above-Mentioned Testing Methods

When studying the discharging of a battery with constant power within a period, if the test result obtained by the HPPC method is taken as the real power, that is, with constant power discharging using constant current; then during the real discharge the current will increase with decreasing voltage. Conversely, during the charging of the battery with constant power the current will decrease with increasing voltage. Therefore, the pulse power obtained from HPPC is different from the definite peak pulse power. We will analyze the reasons why it is not possible to use the HPPC method to estimate the real peak power.

HPPC is performed at normal temperature and the duration of the pulse current at different DOD is approximately 10 s. However, the internal resistance of the battery will rise at low temperature (<0 °C), so when the pulse current goes through the battery, the internal pressure drop will increase sharply and the terminal voltage will reach the highest or lowest limited voltage. At this point the current pulse is not able to continue for 10 s, the acquired internal resistance of the HPPC method cannot be applied to the calculation of the pulse power. In testing the power of a battery, the JEVS has adopted current of various rates, it is testing with constant current that is different from the tests with constant power. The testing result can merely be used to estimate the power density but not real-time power.

4.3 Peak Power

Figure 4.9 shows the power characteristics obtained by applying a constant power method and HPPC. It can be seen that the results from the two methods are very different. To verify the agreement between the two methods and the real power, when discharging the battery according to the fitting power density, the peak power of 10 s obtained by the constant power method is quite different from the real one, and the discharging duration is over 13 s. The peak power at

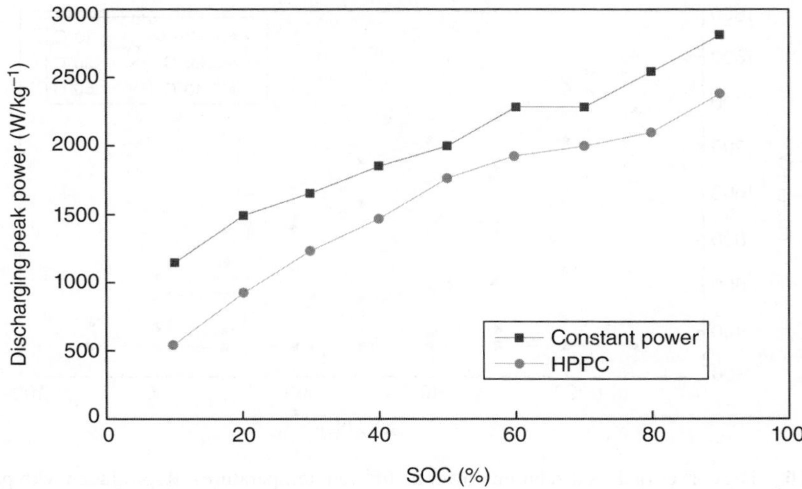

Figure 4.9 Power characteristics obtained by applying a constant power method and HPPC.

10 s obtained by the HPPC method is a little higher than that obtained by the other method and the duration is 8 s. This is because there is a lag in the PI adjustment of the equipment, which cannot keep the same power during discharging, when testing with constant power. Moreover, because the resting time is short, the test results are very different.

HPPC can also be regarded as a special case as its Ohmic internal resistance and the peak power are obtained at the same time, so the model can be researched via the statistics of peak power obtained by HPPC. Then if the peak power of a single battery is taken as an example the relation between peak power and temperature, and between SOC and Ohmic internal resistance can be analyzed by the constant power method.

4.3.1 The Relation between Peak Power and Temperature

The conductivity of an electrolyte and the activity of anode and cathode materials change with temperature, so the charging and discharging power of batteries will be affected by temperature. The reaction rate of the electrode decreases with decreasing temperature. The temperature also affects the rate of transport of ions and electrons by the electrolyte. The rate increases when the temperature rises, and *vice versa.* Moreover, the charging and discharging can also be affected. If the temperature is too high and over a specified temperature limit value, the chemical equilibrium in the battery will be destroyed, which leads to side reactions. Figure 4.10 shows the discharging power at different temperatures.

There is a major change in chemical reaction rate with temperature, the speed of reactive particles slows down with decreasing temperature and the internal resistance of the liquid phase mass transfer will increase. To study the peak output power a specified temperature and SOC are selected from the smallest changing range (20–80% SOC) of the output and input power. Within a range of temperatures, the change in power and internal resistance is observed, and the

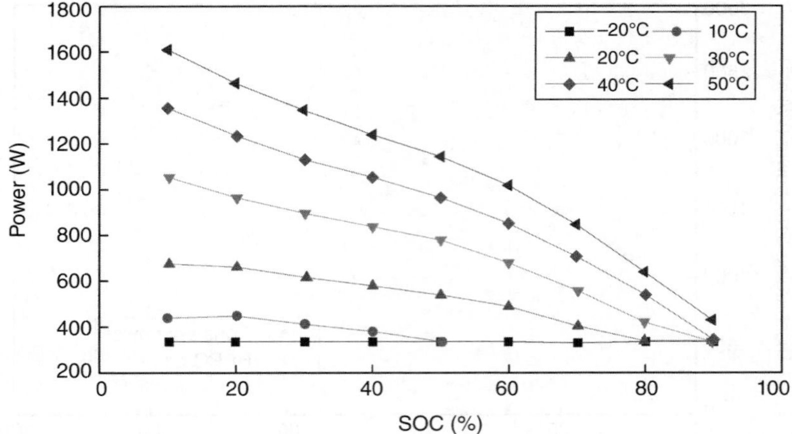

Figure 4.10 The curves of 10 s discharge power at different temperatures. Reproduced with permission from Hongyu Guo, "Research on Power Capability Prediction Method of HEV Lithium-ion Battery," Beijing Jiaotong University © 2012

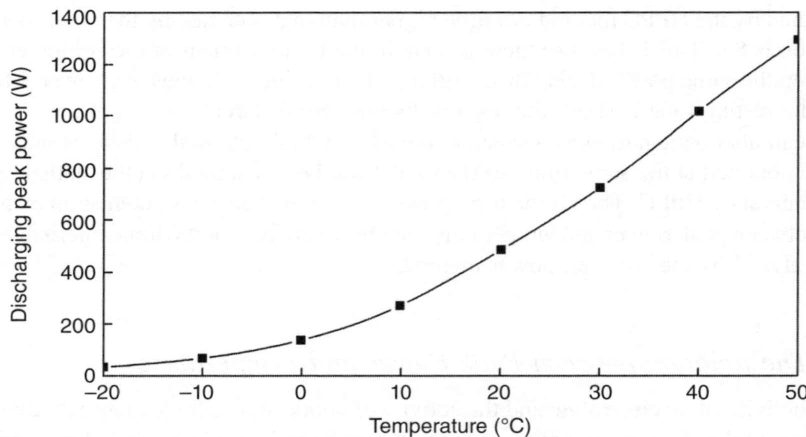

Figure 4.11 Plot of discharging peak power versus temperature with 60% SOC.

normal temperature 25 °C is selected to avoid the low and high temperatures at which the characteristics of the battery change a lot. Therefore, we choose normal temperature 25 °C and 50% SOC of the battery to investigate the output power.

When studying the relations between peak power and temperature, the thermal chamber is used to maintain the temperature. The battery must be put in the thermostat with a set temperature for at least 4 h to make sure the temperature of the battery is stable. The relation between discharging peak power (taking a SOC of 60% as an example) and temperature is shown in Figure 4.11.

From Figure 4.11, it can be seen that the peak power of the battery changes with temperature and the curve is obviously nonlinear. When the temperature is decreased, the peak power is reduced, changing slowly at low temperatures. When the temperature increases, the peak power increases but if the temperature is too high, heat dissipation of the battery will be difficult and detrimental to the service life of the battery.

4.3.2 *The Relation between Peak Power and SOC*

The main aim of HPPC is to get the power characteristics of the battery in the range of the available SOC, which are shown as a variable function of SOC. When studying the relation between peak power and SOC, accuracy of the SOC is strictly required. The capacity of the battery at different temperatures should be checked to adjust the discharging electric quantity in accordance with the temperature; moreover, the discharged capacity 10 s large pulse current has to be taken into account to obtain more accurate experimental data. The curve relating peak power and SOC at a temperature of 30 °C is shown in Figure 4.12. This demonstrates a non-linear relation. The lower the SOC, the smaller the peak power and the faster the peak power changes. For other values of SOC the peak power changes more slowly.

To test the influence of SOC on the peak power of the battery, the $LiMn_2O_4$ power battery of 8Ah is charged and discharged, and the 10 s pulse discharging ability under different DOD conditions is calculated by the constant power method.

Figure 4.13 shows the relation between discharge–charge power and the SOC of a single battery. It shows that the discharging power capacity increases as SOC increasing, while the charging power capability decreases. For example, the discharging peak power increases from 222 to 693 W when the SOC increases from 10 to 90% whereas the charging peak power falls from 675 to 300W for the same SOC range. The difference in power capability in various conditions of SOC occurs mainly because the acceptance ability of the battery to current is different. For low SOC (<30%), the amount of available Li-ion in the anode is small, so, when discharging with a large current, the cathode cannot move a large amount of Li-ions through

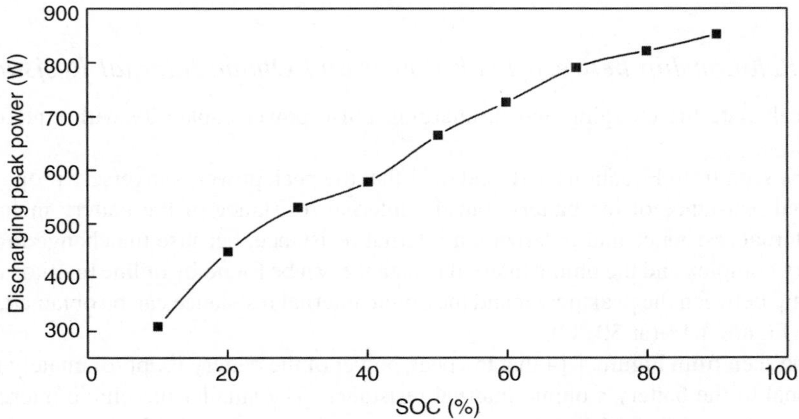

Figure 4.12 Discharging peak power versus SOC at 30 °C.

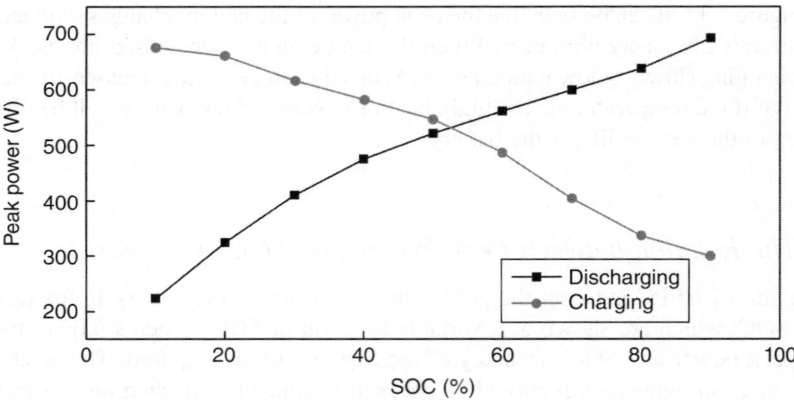

Figure 4.13 Plots of peak power versus SOC at normal temperature. Reproduced with permission from Hongyu Guo, "Research on Power Capability Prediction Method of HEV Lithium-ion Battery," Beijing Jiaotong University © 2012

the electrolyte solutions and the battery separator to the cathode, reducing the voltage between the cathode and anode sharply in a short time, which affects the discharging power. For high SOC (>80%), the density of Li-ion transferred from the cathode is small, so when charged with a large current instantly, the cathode cannot provide enough Li-ions to be transferred and imbedded in the anode material to combine with electrons, which leads to an increase in the potential of the cathode and anode. Quickly, it reaches the maximum voltage allowed by a single battery and then influences the charging peak power of the battery.

The peak power reflects the power capability of the power battery. Research on peak power at different SOCs can estimate the technical capacity of a power battery to provide data and technical support for its use in electric vehicles.

4.3.3 *Relationship between Peak Power and Ohmic Internal Resistance*

We can calculate the charging and discharging pulse power capability with equations 4.10 and 4.11.

It can be seen from Equations 4.10 and 4.11 that the peak power is inversely proportional to the internal resistance of the battery, but the internal resistance of the battery includes both ohmic internal resistance and polarization internal resistance, because the change in polarization is very complex and the ohmic internal resistance can be found by online monitoring, so the relationship between the peak power and the ohmic internal resistance can be obtained, which is shown in Figure 4.14 (at 30 °C).

It can be seen from Figure 4.14 that the peak power of the battery is approximately inversely proportional to the battery's ohmic internal resistance. The smaller the ohmic internal resistance, the greater and faster the peak power; the greater the ohmic internal resistance, the smaller and slower the peak power.

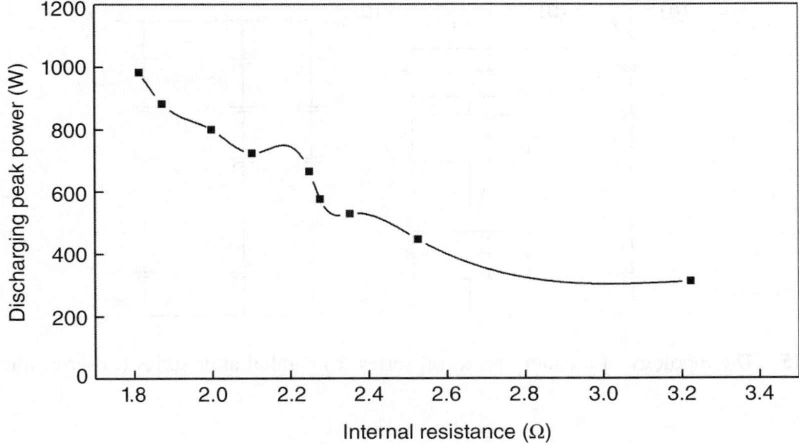

Figure 4.14 Plot of discharge peak power versus internal resistance at 30 °C.

4.4 Available Power of the Battery Pack

4.4.1 Factors Influencing Available Power

Due to the poor anti-abuse ability of lithium-ion batteries and the inconsistency among cells within a battery pack, if the current or power is beyond the limits, the fading of particular cells will be accelerated, increasing the battery pack inconsistency. Therefore, the management system of lithium-ion batteries has to monitor the single cell voltage and measure differences among single cells to estimate the group performance in real-time. It cannot improve the degree of non-uniformity by only monitoring characteristics of the battery.

In addition to differences in voltage, current and temperature, the connection method of a battery pack also has a significant impact on its available power; three common battery connection methods are shown in Figure 4.15. It can be seen that the current in Figure 4.15a which is connected in series is the same, while in the connection methods shown in Figure 4.15b and c there is current imbalance in the branch due to the presence of parallel components; at the same time, the current difference in parallel components will make the available power of the battery pack different. When the total current ratio is low, the available power of the series connection method is mainly influenced by capacity, SOC, resistance and polarization difference; when the total battery current ratio is larger, over-current may occur in the battery which has poor current acceptance ability, due to the lack of current imbalance information between the parallel branches.

In addition, although the battery management system has limited the peak current of charging and discharging based on single cells, the limit value is often considered with regard to the security of the battery, and during normal operation it generally will not exceed the range. If the battery works in the boundary conditions (low temperature, high or low SOC) or there are large differences between the battery parameters (resistance and polarization difference increases with aging), the performance of the battery pack will be affected. The single cell with the worst performance directly limits the performance of the whole battery pack, if it is not timely maintained,

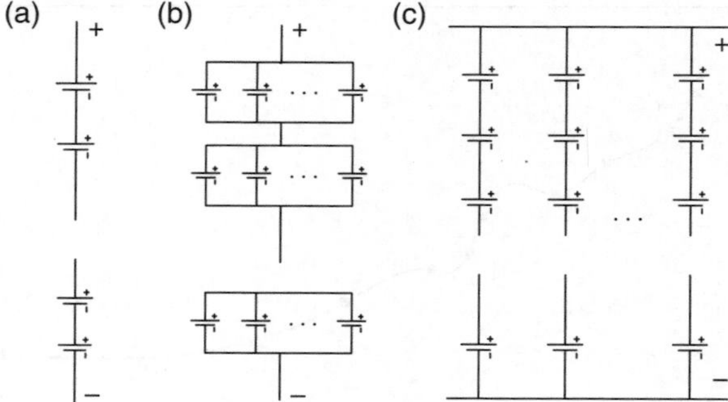

Figure 4.15 The topology of a battery pack, (a) series (b) parallel after series (c) series after parallel.

the current of a single cell with poor performance is more likely to exceed the normal range than other cells and so declines faster, at the same time the battery with better performance cannot use its power advantage, which makes the performance of the whole battery pack decline.

First we assume that the peak power during the discharging process of each cell within 10 s in the battery pack at a certain time is $P_i(i = 1, 2, \ldots, n)$, where n is the series number in the battery pack. Suppose the minimum of $P_i(i = 1, 2, \ldots, n)$ is $P_{\min}$, then

$$P_{\min} \leq P_i(i=1,2,\ldots,n) \tag{4.14}$$

If the peak power of the battery pack is P_{pack}, then

$$n \cdot P_{\min} \leq P_{\text{pack}} \leq \sum_{i=1}^{n} P_i(i=1,2,\ldots,n) \tag{4.15}$$

and

$$\Delta P = \sum_{i=1}^{n} P_i - n \cdot P_{\min} \tag{4.16}$$

At the same time, ΔP can also be expressed as

$$\Delta P = \sum_{i=1}^{n} (P_i - P_{\min}) \tag{4.17}$$

When the difference between the peak power of each battery and that of the minimum cell is a minimum, the whole battery pack power P_{pack} is a maximum.

Suppose that at a certain time, the voltage of the battery cell is V_i, then the power of a single cell is

$$P_i = U_i \cdot I \tag{4.18}$$

Similarly the power of the battery pack is

$$P_{\text{pack}} = U_{\text{pack}} \cdot I = \sum_{i=1}^{n} U_i \cdot I = \sum_{i=1}^{n} (U_i \cdot I) \tag{4.19}$$

If the power of the k single cell is the poorest, its peak power meets the condition that $P_k \le P_i (i = 1, 2, \ldots, n)$, and at the same time if the current I of the whole battery pack meets the condition that the power of the single cell is constantly P_k, then the power of the single cell can be expressed as

$$P_i = U_i \cdot I = U_i \cdot \frac{P_k}{U_k} = P_k \cdot \frac{U_i}{U_k} (i = 1, 2, \ldots, n) \tag{4.20}$$

If

$$\eta_i = \frac{U_i}{U_k} \quad (i = 1, 2, \ldots, n) \tag{4.21}$$

Then the power of the single cell is

$$P_i = \eta_i \cdot P_k \quad (i = 1, 2, \ldots, n) \tag{4.22}$$

So the power of the battery pack is

$$P_{\text{pack}} = (\eta_1 + \eta_2 + \cdots + \eta_i) \cdot P_k \quad (i = 1, 2, \ldots, n) \tag{4.23}$$

4.4.2 The Optimized Method of Available Power

From the consistency analysis of the battery pack, it can be learned that the main factors affecting the characteristics of the battery power are voltage, resistance and inconsistency of the SOC, and from the point of view of external characteristics, the main difference between the single cells of a battery pack in series is reflected in the inconsistency of the terminal voltage. When the current is large, the ohmic voltage drops across the internal resistance and the polarization will further increase the differences in the terminal voltage.

In order to improve the consistency of the whole battery pack, various hardware circuits are normally used to manage the battery and make it balanced. There are a variety of classifications for the balanced approaches. By the mode of battery energy dissipation, they can be classified into the charging balance, the discharging balance, and a combined balance of charging and discharging. It can be seen from the SOC–OCV curve of a manganese acid lithium battery (Figure 4.16) that there is a clear correspondence between its OCV and the battery SOC. Moreover, the OCV of the charging and discharging states overlap substantially, which shows that the SOC difference of the battery pack can be reflected by the battery terminal voltage on standing. The SOC difference between each single cell can be obtained by

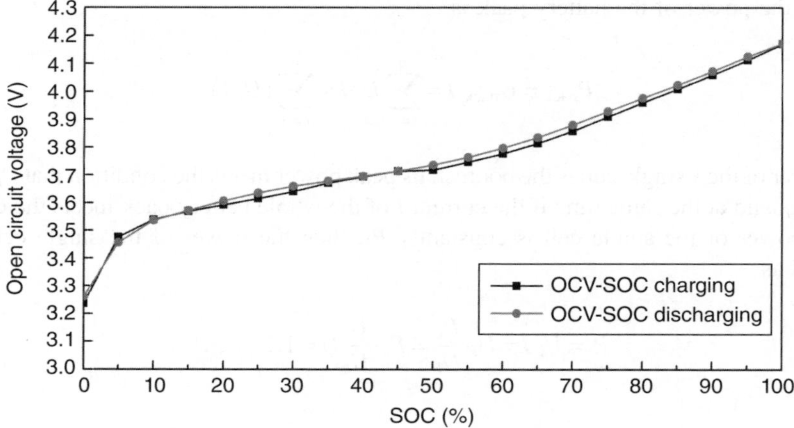

Figure 4.16 Plot of OCV versus SOC for a power lithium manganese battery. Reproduced with permission from Hongyu Guo, "Research on Power Capability Prediction Method of HEV Lithium-ion Battery," Beijing Jiaotong University © 2012

checking the SOC–OCV table, which could also help us to determine the balanced energy needed by each battery.

The available capacity inconsistency of single cells will become more obvious with aging of the battery. With the same current, the cell with the minimum available capacity will limit the efficiency of the battery pack. Because the difference between the SOC–OCV curve of aged batteries and that of new batteries is very small, a balanced SOC goal can be achieved by controlling the consistency of the cell voltage in the standing state.

The peak power of the battery will show a significant decline if the battery pack operates at low temperatures; the temperature difference among cells will be enlarged due to increased battery resistance and polarization at low temperatures. When batteries operate at high or low SOC, 10% differences in SOC may result in more than double deviations in peak power as the charging curve for peak power is relatively steep, thus the SOC differences have a severe influence on the battery power of the whole group, especially when the disparities in actual capacity of each battery become wider, the SOC differences of each battery will be much more distinct at high or low SOC. Consider a battery group with n single batteries, the nominal capacity of these batteries is Q_{nom}, the actual capacity of each battery is denoted as Q_1, Q_2, …, Q_n, and, at a certain moment, the SOC of each battery is $SOC_1, SOC_2, \ldots, SOC_n$, the OCV points are U_1, U_2, …, U_n, the temperature of each battery is T_1, T_2, …, T_n, the charging power and discharging power of each battery are $P_1^{\text{chr}}, P_2^{\text{chr}}, \ldots, P_n^{\text{chr}}$ and $P_1^{\text{dch}}, P_2^{\text{dch}}, \ldots, P_n^{\text{dch}}$. respectively. It can be seen from the multiple factors affecting the available battery power that the main task is to improve the working state of the weakest battery to increase the available power of the battery pack, especially at the temperature and SOC of the weakest battery. From the relationship between battery peak power and SOC, shown in Figure 4.17, it is apparent the peak power of charging and SOC have a monotonically decreasing trend whereas the peak power of discharging rises while the SOC is increasing. The comprehensive power curve can be obtained by combining the charging power and the discharging power of the battery pack.

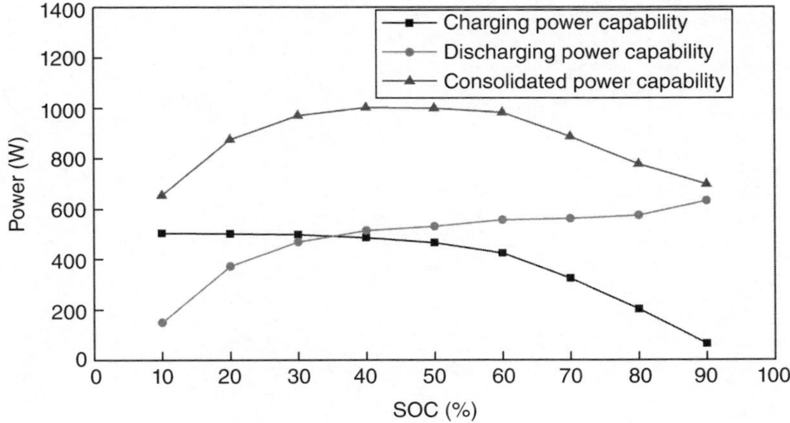

Figure 4.17 The relationship between the power capacity and the SOC of a battery. Reproduced with permission from Hongyu Guo, "Research on Power Capability Prediction Method of HEV Lithium-ion Battery," Beijing Jiaotong University © 2012

If the SOC of the battery group is fully consistent, the peak power of the whole group is the sum of that of the single batteries, however, this is an extremely ideal state; if the SOC of the batteries has issues of inconsistency, it can be seen that the peak power of the battery pack will be limited according to the curve of peak power and SOC. At any moment, the battery with the highest SOC will directly limit the peak power of charging of the battery group while the battery with the lowest SOC will limit the peak power of discharging of the group. Therefore, it is required to fully ensure battery consistency to improve the comprehensive charging and discharging capacity of the battery group.

Therefore, it is important to carry out further research on the battery capacity fading characteristics of components in series and parallel, and to propose a connection method for group optimization of the battery pack based on the battery management strategy and the balance strategy, which will not only improve the comprehensive ability of charging and discharging of the battery pack, but also improve the cycle performance of the battery pack.

References

[1] Hongyu, G. (2012) *Research on power capability prediction method of HEV lithium-ion battery*, Beijing Jiaotong University, Doctoral dissertation.

[2] van Bree, P.J., Veltman, A., Hendrix, W.H.A. *et al.* (2009) Prediction of battery behavior subject to high-rate partial state of charge. *IEEE Transactions on Vehicular Technology*, **58**(2), 588–595.

[3] US Advanced Battery Consortium, USABC Electric Vehicle Battery Test Procedures Manual. 1996. DOE/ID-10479, pp 9–10.

[4] Hunt, G. (2003) Freedom CAR Power Assist Battery Test Manual. Idaho National Engineering & Environmental Laboratory. Idaho Falls, DOE/ID-11069.

[5] 863 Modern Traffic Technology Office of Ministry of Science and Technology (2008) Annual Performance Test Specification by HEV with High Power Lithium Ion Power Battery.

5

Charging Control Technologies for Lithium-ion Batteries

5.1 Literature Review on Lithium-ion Battery Charging Technologies

The demand for lithium-ion batteries (LIBs) leads to a rapid growth in the related technologies. Lithium-ion batteries have a large capacity and have their own characteristics. The development of charging technologies is of great significance to relevant industries. The development of charging modes occurred in three stages: the charger-alone control mode, the BMS (battery management system)-cooperation control mode and the BMS-dominant control mode. There are also a variety of charging methods available now, including the current control method, the voltage control method, the Mas Law charging method and the modern intelligent charging method.

5.1.1 The Academic Significance of Charging Technologies of Lithium-ion Batteries

The development of the EV industry and the construction of charging infrastructure not only promote large-scale application of LIBs, but also pose higher requirements for the development of LIBs. Unlike lead-acid batteries, LIBs have strong technical advantages in terms of energy density, power density and cycle life. However, like lead-acid batteries, their level of resistance to abuse is currently undesirable, and their charging–discharging performance and cycle life can be greatly affected by the environment in which they are used. Thus, in order to improve the properties and manufacturing techniques of a battery, it is important to control and manage its charging and discharging process [1–4], so that the battery will perform well in an optimum environment and within its allowable stress range.

The application technologies for LIBs mainly include technologies for charging and discharging management. For EVs, the charging of their batteries is the only way to providing energy, but there are many problems regarding the charging process of LIBs [5]. First, overcharging can frequently be observed during the charging process. Overcharged batteries will trigger side

Fundamentals and Applications of Lithium-ion Batteries in Electric Drive Vehicles, First Edition.
Jiuchun Jiang and Caiping Zhang.

reactions, which result in internal thermal runaway. Meanwhile, overcharging causes the precipitation of Li^+ from the cathode material, which will damage the cathode crystal structure and may lead to oxygen evolution. The precipitated Li^+ deposited on the anode electrode may pierce the battery films, leading to cathode to anode short circuit. Related surveys show that 80% of the commercial LIB safety accidents occurred in the charging process. In the early stage of EV development, due to the lack of knowledge and experience of using LIBs, the charging process was controlled by using the traditional charging methods for lead-acid batteries, and over-heating or even burning of the batteries happened frequently due to overcharging. Therefore, good management of the charging process is key to guaranteeing the safety issues of LIBs. Secondly, from the view point of electrochemistry, the charging of a LIB involves substantially the insertion of Li^+ into the anode; in other words, the chemical reaction rate (i.e., the current flowing through the battery) has a direct relation to the efficiency of Li^+ insertion. If the reaction rate is too low, the efficiency of Li^+ insertion will accordingly be low, and the LIB cannot achieve its best performance. If the reaction rate is too high, some Li^+ from the cathode will remain on the electrode surface, the concentration difference between the two electrodes will rise, and hence a higher polarization will occur. In addition, the precipitated Li^+ on the anode will be permanently sedimentary, which results in a decline of the battery cycle life and an increase in the cost of LIB usage. Also, charging stations are the most important infrastructure for the industrialization of EVs. In the battery-swapping and battery-leasing combined mode for developing EVs, when the departure frequency and the number of vehicles are fixed, the shorter the battery charging time, the smaller will be the number of spare batteries in the station, and the lower the cost of the charging station construction. Therefore, the charging process requires not only security, but also rapidity. That is to say, we should minimize the battery charging time while ensuring its safety and cycle life. In summary, the charging technology is a critical issue in the LIB industry and even in the new energy industry, and it is also related to the spread of LIBs into other disciplines.

5.1.2 Development of Charging Technologies for Lithium-ion Batteries

Both LIBs and traditional batteries have many advantages, but the research on the charging technology for lithium-ion batteries is flourishing. Currently, the study of battery charging primarily involves charging methods and the battery's life. Charging methods are in line with the charging characteristics of the battery (charging capacity, charging time, etc.). Existing studies on charging methods mainly focus on charging modes [6–10] and the control of the battery's external parameters. Charging modes include charger-alone control mode, BMS-cooperation control mode, and BMS-dominant control mode; and charging methods include the current control method, the voltage control method, the Mas law control method and the intelligent charging method. The battery life is the basis for optimizing charging methods, and also an important standard for assessing charging methods.

5.1.2.1 Development of Charging Modes [8, 9]

Charger-Alone Control Mode

Early EVs commonly used lead-acid batteries as the power source, and lead-acid batteries have a good performance in sustainable overcharging and are not much of a security risk at high or

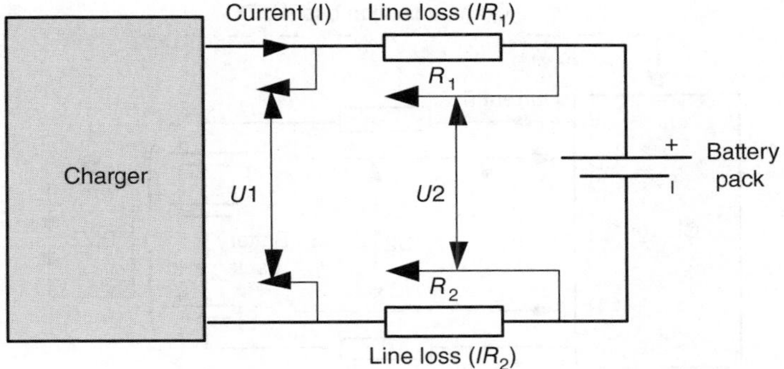

Figure 5.1 Schematic diagram of the charger-alone control mode.

low temperatures. Therefore, the charger-alone control mode was normally used ($U1$ control mode) (Figure 5.1) for lead-acid batteries. When the charger's output voltage $U1$ is lower than the clamping voltage (pre-set), the battery pack will be charged at a constant current (pre-set, generally low rate), and the $U1$ of the charger will increase gradually. Once $U1$ reaches the clamping voltage, the charging is complete. In this mode, the charger only needs connection of the two power lines to both poles of the battery pack, which is easy to assemble. However, since the power lines are too long, when the battery pack is charged at current I, the line impedance ($R_1 + R_2$) will generate a line voltage drop ($I\,(R_1 + R_2)$). Thus, there will be a difference between the actual voltage of the battery pack ($U2$) and the output voltage of the charger ($U1$), which is: $U1 = U2 + I\,(R_1 + R_2)$. Therefore, when $U1$, which is the feedback variable for controlling the charging voltage and current, reaches the upper voltage limit, the voltage of the battery pack will be smaller than the upper voltage limit since $U1 > U2$, leading to a decrease in charging capacity.

Based on its characteristics mentioned above, researchers made some improvements to the charger-alone control mode ($U2$ control mode). The actual voltage $U2$ of the battery pack is measured through two added signal testing wires, and a constant voltage control loop will be added when $U2$ reaches the upper voltage limit ($U2$ remains constant at the upper voltage limit and the charging current tapers until it declines to the pre-set cut-off value); thus, the control of constant battery pack voltage is realized and the charging capacity also increases. However, with the use of LIBs instead of lead-acid batteries in EVs, this charger-alone control mode has great risk of causing potential security problems. Lithium-ion batteries can easily cause accidents in cases of overcharging and over-heating. Though the $U2$ control mode can effectively control the battery pack voltage, it cannot control the voltage and temperature of each cell in the battery pack.

The BMS-Cooperation Control Mode

Because of its shortcomings, the charger-alone control mode cannot provide all the information on each cell in the battery pack. Therefore, the researchers put forward a BMS-cooperation control mode (see Figure 5.2). In this mode, the charger plays a dominant role, and the BMS equipment cooperates with the charger in completing the process of charging. The

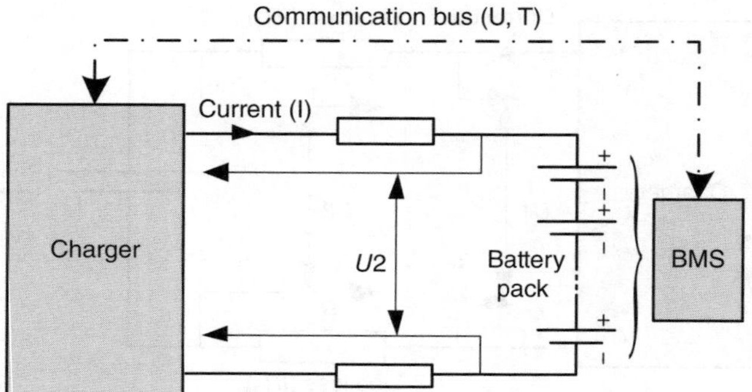

Figure 5.2 Schematic diagram of the BMS-cooperated control mode.

BMS is used to collect information about the voltage, temperature and current of each cell, and to deliver these data to the charger through the communication bus. While the charger measures the terminal voltage of the battery pack, it can obtain each cell's data from the BMS and combine the data with the terminal voltage as the feedback variables for the closed-loop control system to adjust the charging current. This mode can effectively monitor the voltage and temperature of each cell and prevent abnormal conditions from occurring, and it improves the charging security of LIBs.

This charging mode can achieve data interaction through communication, the data of each cell is collected by the BMS, the charging logic is controlled by the charger. Different from the charger-alone control mode, it simply needs two more communication wires, in order to avoid cumbersome connections and unreliability problems caused by the large quantity of cells and the uncertainty concerning their layout. This mode also solves security issues about the charging of the LIB.

The BMS-Dominant Control Mode

The basic hardware structure of the BMS-dominant control mode is shown in Figure 5.2, but the control strategies are unique. The BMS can get the battery data directly, and record the key information of the corresponding battery pack at the same time, such as battery type, configuration, parameters, control thresholds, and so on. The BMS also undertakes important tasks of estimating battery model parameters and controlling the temperatures. So when there is much data from the battery and a large amount of calculation for the model, which are then communicated to the charger, and the calculation of the charging current is done by the charger, a delay occurs and will affect the estimated speed. Therefore, in the BMS-dominant control mode, the output current of the charger is given, and the battery data and model parameters are taken as feedbacks to optimize the control of the battery charging process. The charger simply accepts the charging current and voltage and adds the most suitable protection thresholds, which can not only ensure the synchronization of data acquisition of the battery pack, battery state estimation and current control, but also can enhance the safety and efficiency of the charging process.

Since the hardware of BMSs is gradually improving, the BMS-dominant control mode increasingly displays its rationality and efficiency when charging the batteries, and becomes an advanced charging control mode.

5.1.2.2 The Present Charging Methods

Current Control Method

The constant-current (CC) charging method is used for charging batteries at low rates, which prevents the thermal runaway and overvoltage caused by the sharp increase in the batteries' internal potential and temperature at the end of charging that occurs when batteries are charged at high rates [10–16]. However, both the charging capacity and the battery life should be taken into consideration when the charging current is selected, which is overly dependent on the experience of the operator. Moreover, the rate of charging current is always small, leading to long charging times. Thus, the practicability of this method is far from acceptable.

The multi-step CC charging method is an enhanced version of the CC charging method. The amplitude and duration of the CC are adjusted according to the state of the battery during the charging process. That is, first the battery is charged at a higher CC for an initial period, then the current amplitude is lowered when the cell polarization becomes more serious. This method takes the polarization during the charging process into account and consciously changes the current amplitude to reduce the polarization, that is, it ensures high rate fast charging when the battery can benefit from a larger current, and also minimizes side reactions and gas precipitation when the polarization gets serious. On the basis of the multi-step CC charging method, some research on an approach of stopping the charging of the battery for some time under different current amplitudes was conducted, this is called the intermittent alternating current charging method.

Therefore, the key to using the current control method is to determine the magnitude and duration of the charging current. With more reasonable control, the charging effect will be better; otherwise, accidents of over-temperature, over-voltage, and so on, will happen during the charging process.

The Voltage Control Method

The constant voltage (CV) charging method keeps the battery charging voltage constant, the charging current is larger at the beginning than at the end of the charging process due to the decrease in battery voltage [10–16]. With the appearance of polarization during charging, the charging current gradually decreases, and the charging stops when the current decreases to a certain value. The advantage of this method is that the relevant charger is relatively simple, and the current will be adjusted automatically to avoid polarization during charging. The disadvantage is that the current is too large at the beginning of charging, which might damage the battery.

Based on the CV control method, a charging method that uses CC first and then CV is proposed. This method is an improved version aiming at solving problems with the CV charging method, such as battery damage. By this improved method, the battery is charged at a given CC and then when the battery voltage reaches a certain value, the battery is charged at a CV. This improved method ensures fast charging with CC when the battery is able to accept big currents; and it also ensures CV charging when the voltage is high and polarization is serious, thereby causing the charging current to decrease gradually. However, the values of the CC and the CV are critical, since both can affect the safety and service life of the battery.

Mas Law Charging Method

Joseph A. Mas, an American scholar, announced three laws of the charging process at the second session of the annual meeting of the World Electric Vehicle Association in 1971, which were later called the "Mas Three Laws". They are (i) the charging receptivity is proportional to the square root of the discharging capacity; (ii) the charging receptivity is proportional to the logarithm of the discharging current; (iii) the charging receptivity after several different discharging rates is equal to the total charging receptivity after each rate [17–24]. According to Mas's third law, using positive and negative pulses for charging can greatly improve the charging receptivity. Based on the above principles, researchers put forward a method of pulse charging for lead-acid batteries. The Mas laws are the empirical summary of methods which aim at determining the gas evolution rate of a lead-acid battery electrode, which can objectively describe the optimal charging curve of lead-acid batteries. However, in terms of the electrode materials, electrolyte, manufacture and electrical properties, LIBs are different from lead-acid batteries, so the amplitude and duty ratio of the current pulses for lead-acid batteries cannot be quantified for LIBs. The charging curve characteristics of the batteries will also change with different degrees of aging and different charging conditions, even when charged under the same condition. Therefore, simply obeying the Mas laws cannot be arbitrarily applied to the charging of LIBs.

Modern Intelligent Charging Method

Much theory and practice have proved that the charging and discharging of batteries is an extremely complex electrochemical process which has the following characteristics: (i) Multi-variables: there are many factors affecting the charging effect, such as electrolyte concentration, the concentration of the electrode active materials and ambient temperature [25–29]. All will lead to big differences in the charging effect. (ii) Nonlinearity: the acceptable battery charging current declines nonlinearly over time, and the charging current should track the maximum acceptable charging current curve of the battery. (iii) Discreteness: due to the differences in discharging status and aging states, even two batteries from the same manufacturer have different charging characteristics. For such a multi-variable, nonlinear, and strong coupling control object, if we take a general mathematical model, it will be difficult to design a controller whose reliability and dynamic performance meet the requirements. Such controllers have been designed by means of modern intelligent charging technologies [25, 26, 28, 30–34]. The references [28, 33, 34], respectively, used evolutionary computing approaches, such as genetic algorithms, an ant colony algorithm and particle swarm optimization, to obtain the optimal charging parameters. But the designers faced difficulties and challenges when choosing the appropriate parameters, such as the number of particles and the learning factor in evolutionary computing approaches. In reference [34] the LIB is considered as a Gray system, replacing the conventional CV charging curve with one estimated by a Gray prediction algorithm to improve the charging characteristics of the batteries. It is reported that this charging method for LIBs increases the charging time by 23% and the charging efficiency by 1.6%. However, the above method mainly focuses on the optimization of charging time and charging efficiency, and rarely covers the effect of the charging mechanism on battery life. Its implementation requires much related data, and new tests will be needed for researching into battery aging. Since the amount of data is too large, requiring a complex processing mechanism, and this method has not formulated optimized systematic charging standards, it is not universally accepted.

5.1.2.3 Deficiencies of the Lithium-ion Battery Charging Methods in Practical Application

Ignoring Differences in Battery Type

Professor Mas obtained the optimal charging curve of the lead-acid battery-attenuation model through years of experiment in controlling the gassing rate on the lead-acid battery electrode. When a LIB is charged, the electrode reaction is different from that for lead-acid batteries, and the current capability of LIBs is much larger than that of lead-acid batteries. Just following the parameters and forms of the optimal charging curve from the Mas three laws to control the charging will reduce the charging efficiency and does not maximize the currently accepted advantage of LIBs.

Ignoring Changes in Battery Status

In the battery charging process, the charging characteristics and current capability change with the remaining capacity of the battery. On the premise that voltage or current is taken as the target to control, most of the charging parameters in the charging process are constant and empirical, such as the charging voltage and charging current. If the degree of aging, the ambient temperature and other factors change, the charging characteristics of the battery will change accordingly. Therefore, simply controlling the charging process according to the experience of the operator cannot bring the battery characteristics into full play and will have an unfavorable impact on battery life.

Lacking Basic Theoretical Support

Since power batteries, especially LIBs, do not have a long history the relevant test data are scarce and the research is superficial. There is also a lack of systematic analysis on the charging characteristics and their influencing factors, and also a lack of measuring standards and theoretical foundations for optimizing charging methods of LIBs. Therefore, the charging control methods are mostly derived from experimental data and field experience, and are not perfect in guiding relative research.

Existing charging methods for LIBs do not take account of the relationship between charging performance and battery life, which may lead to reduction in battery charging performance or charging life, and have an impact on the economic efficiency of practical applications. Developing new charging methods should start from the charging characteristics and charging requirements of LIBs, combine experimental testing with theoretical analysis, and make systematic research into optimization theory and the optimal charging method of power LIBs, in order to achieve fast, efficient and lossless charging.

5.2 Key Indicators for Measuring Charging Characteristics

From the viewpoint of electrical engineering, the LIB circuit model describes the voltage distribution and variation under the excitation of charging current and all other kinds of charging characteristics. Therefore, this chapter mainly focuses on the basic circuit model and then combines this with experimental data analysis to explore the impact of various model parameters on key indicators of the charging characteristics, such as charging capacity, charging efficiency, charging time and charging life, in the hope of providing a theoretical basis for optimizing the battery charging characteristics.

5.2.1 Charge Capacity

5.2.1.1 Calculation of Charging Capacity under Constant Current

Charging capacity refers to the integral value of the charging current in a charging process. However, since the charging cut-off condition is determined by the battery terminal voltage, the difference between the battery terminal voltage and the battery internal potential E (i.e., OCV after a sufficient standing time) results in differences in the charging capacities.

Charging Capacity Calculation

Suppose that a battery is charged with a constant current I_s at a certain temperature. Throughout the charging process, define the set M as [0%, 100%],

When SOC $\in M$, there must be:

$$E(\mathrm{SOC}) + U_{\mathrm{OE}}(I_s, \mathrm{SOC}) < U_s \tag{5.1}$$

And only when SOC = 100% will

$$E(100\%) + U_{\mathrm{OE}}(I_s, 100\%) \geq U_s \tag{5.2}$$

E (SOC) represents the battery internal potential when the remaining capacity of a battery is SOC, $U_{\mathrm{OE}}(I_s, \mathrm{SOC})$ represents the overpotential under current I_s. When the battery's state-of-charge is SOC, U_s represents the upper limit of the battery charging voltage. All current values that satisfy the above conditions are in the set $\{I_{s1}, I_{s2}, \ldots\}$. The maximum value among all the values is the critical charging current:

$$I_{\mathrm{smax}} = \max\{I_{s1}, I_{s2}, \ldots\} \tag{5.3}$$

When the charging current $I_{\mathrm{cha}} < I_{\mathrm{smax}}$, the overpotential generated at any moment during the charging process will not cause the battery terminal voltage to exceed the maximum charging voltage U_s, and the battery can reach its maximum usable charging capacity C_N, that is:

$$\int_0^{T_2} I_{\mathrm{cha}}(\mathrm{t})\mathrm{d}t = C_N \tag{5.4}$$

When the charging current $I_{\mathrm{cha}} > I_{\mathrm{smax}}$, the overpotential generated during the charging process will cause the battery terminal voltage to exceed the maximum charging voltage V_s

at a certain moment, so the charging capacity will be smaller than the maximum usable charging capacity of the battery, that is:

$$\int_{0}^{T_2} I_{\text{Cha}}(t)\mathrm{d}t < C_{\text{N}} \tag{5.5}$$

In other words, when the battery charging current reaches the maximum charging voltage under current I for the first time, if SOC = x, then:

$$E(x) + U_{\text{OE}}(I, x) = U_{\text{s}} \tag{5.6}$$

You can get:

$$x = E^{-1}(U_{\text{s}} - U_{\text{OE max}}) \tag{5.7}$$

If $0 < x < 100\%$, the charging capacity of the battery will be $C_{\text{N}} \times E^{-1}(U_{\text{s}} - U_{\text{OE}}(I, N))$.

Given $0 \le E^{-1}(U_{\text{s}} - U_{\text{OE}}(I, N)) \le 1$, the value of the charging capacity will be:

$$0 \le C_{\text{N}} \times E^{-1}(U_{\text{s}} - U_{\text{OE}}(I, N)) \le C_{\text{N}} \tag{5.8}$$

When $0 < x < 100\%$, the charging capacity is smaller than the battery rated capacity.

When and only when $x = 100\%$, the charging capacity is equal to the battery rated capacity.

For the 90 Ah $LiMn_2O_4$ battery, CC charging capacity tests were carried out with different current rates and different numbers of cycles. The experimental data are shown in Figure 5.3.

As found from the comparative experiments of the charging capacity under different conditions, when the charging rate increases, the overpotential will increase and the charging capacity will decrease. When the current is constant, the battery life shortens, the overpotential increases and the charging capacity decreases. As the charging stress increases or charging conditions change, the critical value of the charging current also changes, resulting in change in the charging capacity. Thus, under the condition that the maximum charging voltage is the cut-off condition, the key to improve the charging capacity of batteries is to control the overpotential in a way to make sure that the timing of the maximum charging capacity is ahead of the time when the terminal voltage reaches the maximum charging voltage. However, for the same battery, current, temperature and aging state will affect the value of the overpotential [9], and will also affect the value of the critical charging current. Therefore, in order to maximize the charging capacity, the constant charging current should be adjusted along with the change in the charging state.

5.2.1.2 Improvements in Charging Capacity in Variable-Current Charging

Charging capacity is a key factor in determining the driving range of an electrical vehicle. The analysis of the previous section shows that among all the factors which affect the charging capacity, the change in charging conditions (ambient temperature, aging state) is variable. Therefore, the key to improving battery charging capacity is to control the overpotential

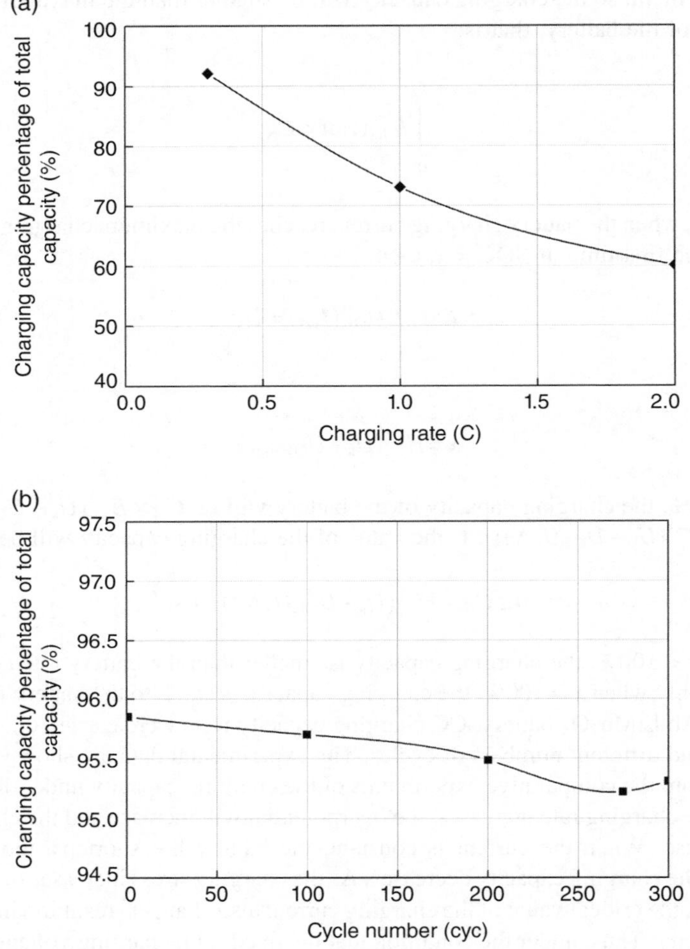

Figure 5.3 CC charging capacity (a) with different charging rates and (b) for different numbers of cycles.

through regulation of the charging current, to ensure that the maximum terminal voltage is reached before the maximum battery capacity.

CC Charging

Derived from the analysis of CC charging capacity, when the charging current I is less than the critical charging current I_{smax}, the charging capacity will reach its maximum value, that is, it can achieve the rated capacity C_N of the battery. When the charging current I is larger than the critical current I_{smax}, the charging capacity will be smaller than the rated capacity of the battery C_N. During CC charging, in order to achieve the maximum charging capacity, the charging current must be lower than the critical charging current I_{smax}. However, since the critical charging current changes with changing charging conditions, the value is difficult to determine. The charging current cannot be unlimitedly reduced in order to ensure a certain length of charging time, so, in actual applications, the maximum charging capacity of the battery is hard to monitor to realize the CC charging mode.

Variable-Current Charging

When the charging current I_1 is larger than the critical current I_{smax}, the charging capacity of a battery is inevitably smaller than the maximum available capacity of the battery. Therefore, when the battery reaches the charging upper limit voltage U_s of the battery for the first time, the actual charging capacity is only $C_N E^{-1}(U_s - U_{OE}(I_1, \text{SOC}_1))$. In this case, by changing the charging current I_1 to a current I_2 (generally $I_{i+1} < I_i$, $i = 0, 1 \ldots$), because $I_2 < I_1$, the internal resistance voltage decreases with the decline in current. The polarization voltage is reduced, $U_{OE}(I_2, \text{SOC}_1) < U_{OE}(I_1, \text{SOC}_1)$, thus the overpotential decreases and the terminal voltage is reduced, which satisfies $E(\text{SOC}_1) + U_{OE}(I_2, \text{SOC}_2) < U_s$. This means that the charging can be continued. When the battery voltage reaches the charging upper limit voltage again, the battery capacity is $C_N E^{-1}(U_S - U_{OE}(I_2, \text{SOC}_2))$; the charging capacity continues during the period from the starting time to the moment when I_2 reaches the charging upper limit voltage again. The charging current is reduced from I_2 to I_3 and, in the same way, when the current of the battery changes after N times, the total charging capacity is:

$$C_{\text{all}} = C_{I_1} + C_{I_2} + C_{I_3} + \cdots = \sum_{k=1}^{N} C_{I_k} \tag{5.9}$$

When $N \to \infty$, and $I \to 0$, the total charging capacity gets infinitely close to the maximum usable capacity of the battery. So the variable current charging process solves problems in the constant current charging process such as difficulty in monitoring the current and a too low charging efficiency. By changing the charging current step by step, the overpotential is reduced while the charging capacity increases.

As analyzed above, when the current declines from I_k to I_{k+1}, the internal resistance voltage drops to $(I_{k+1} - I_k)R_\Omega$, and the polarization voltage drops to $(I_{k+1} - I_k)R_P\tau(t)$. The delay function is $\tau(t)$, $\lim_{t\to 0} \tau(t) = 0$, $\lim_{t\to\infty} \tau(t) = 1$.

To meet the requirement that the continuous charging time of the current I_{k+1} is longer than Δt_{k+1}, that is, that the change in charging capacity is $I_{k+1}\Delta t_{k+1}$ (corresponding inner potential change is $\Delta E(I_{k+1}\Delta t_{k+1})$).

The condition in Equation 5.10 must be met:

$$\Delta E(I_{k+1}\Delta t_{k+1}) \le (I_{k+1} - I_k)R_\Omega + (I_{k+1} - I_k)R_P\tau(t) \tag{5.10}$$

However, the difference in distribution ratio between the polarization voltage and the internal resistance voltage causes changes in the dependent factors of the charging capacity.

When the current $(I_{k+1} - I_k)R_\Omega \gg (I_{k+1} - I_k)R_P$ (i.e., the current change is large), the Equation 5.10 can be simplified as:

$$\Delta E(I_{k+1}\Delta t_{k+1}) \le (I_{k+1} - I_k)R_\Omega \tag{5.11}$$

When the current change is large, the corresponding inner potential increments and the duration Δt_{k+1} of the current I_{k+1} are determined by the voltage drop of the internal resistance. The hysteretic nature of the voltage decline with current is reducing at this time.

When the current $(I_{k+1} - I_k)R_\Omega \ll (I_{k+1} - I_k)R_P$ (i.e., at the end of charging when the charging current is very small), Equation 5.10 can be simplified as:

$$\Delta E(I_{k+1}\Delta t_{k+1}) \leq (I_{k+1} - I_k)R_P\tau(t) \tag{5.12}$$

Due to the strong polarization voltage hysteresis effect, the duration Δt_{k+1} of I_{k+1} is determined by the value of the polarization delay function $\tau(\Delta t_{k+1})$.

If $\tau(\Delta t_{k+1}) \to 0$, then $E(I_{k+1}\Delta t_{k+1}) \to 0$, $\Delta t_{k+1} \to 0$ can be obtained, indicating that the current is reduced from I_k to I_{k+1}, and the polarization voltage becomes lower with a lag, the duration of the charging current I_{k+1} is very short. In a similar line of reasoning, along with the increasing of the hysteresis time of the polarization voltage, the charging current drops quickly to zero. Therefore, at the end of charging

$$U_{OE} = U_s - E(SOC) \tag{5.13}$$

At the time when the charging current $U_{OE} = U_s - E(SOC)$, the terminal voltage must be larger than the OCV, and the difference between the battery voltage and the OCV voltage is the polarization voltage.

If $U_{OE} \geq U_s - E(100\ \%)$, the polarization amplitude and hysteresis effect will cause the current to drop to zero and the charging capacity will not reach the maximum;

If $U_{OE} \leq U_s - E(100\ \%)$, the polarization voltage amplitude will be small and the charging capacity will reach the maximum.

Thus, the magnitude and the hysteresis effect of the polarization voltage are key factors in determining battery capacity. The key to improving charging capacity is to effectively control the overpotential amplitude and the balance of the hysteresis effect polarization voltage and the hysteresis effect at the end of the charging.

5.2.1.3 Charging Capacity at Low Temperatures

According to the variable-current charging process, charging capacity can be increased by adjusting the charging current. However, when the battery works in a low temperature environment, the overpotential characteristics and the charging capacity will change. Tests of charging and discharging capacities were done on an 8 Ah $LiMn_2O_4$ battery at different temperatures, experimental data are shown in Figure 5.4.

According to the analysis in the previous chapter, the main factors which affect the charging capacity are the potential magnitude and the current hysteresis effect in a low temperature environment, and the lithium-ion power battery's overpotential amplitude and seriously enhanced hysteresis quality greatly reduce the charging capacity. As shown in Figure 5.4, the variable current charging capacity at 0 °C is only 91% of that at 20 °C (charging is strictly forbidden below 0 °C).

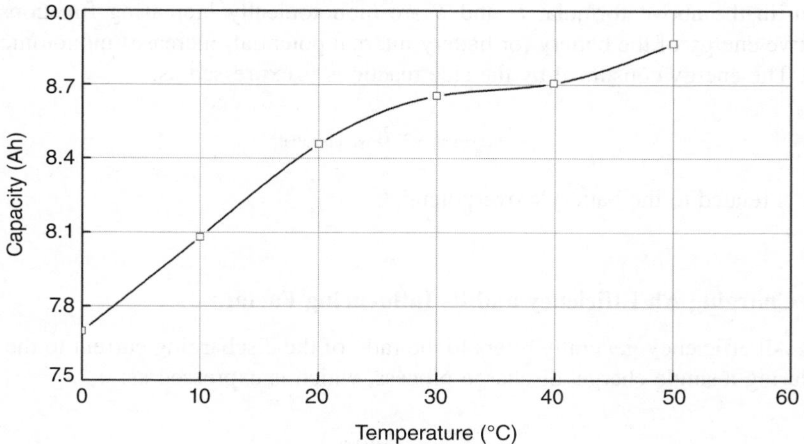

Figure 5.4 Variable-current charging capacity at different temperatures.

Thus, the charging capacity is the primary variable to measure the quality of a charging method. To improve the charging capacity during the charging process, it is necessary to adapt automatically to the current hysteresis effects of overpotential and have effective control of the overpotential amplitude. Under low temperature conditions, due to the enhancements of overpotential amplitudes and hysteresis effects, even regulating the charging current cannot ensure the charging capacity. Therefore, the charging capacity at low temperatures is bound to be lower.

5.2.2 Charging Efficiency

Charging efficiency generally refers to the battery energy ratio between discharging and charging during a single charge–discharge cycle. For lithium batteries there is battery charging efficiency and energy efficiency. Battery charging efficiency refers to the proportion of the charging current which is effectively changed into the battery's output capacity; the energy efficiency of the battery is the proportion of the charging energy which is effectively converted to the battery output energy. It can be concluded from the above analysis that the input energy of the battery is divided into two parts: the effective chemical energy and the energy consumed by the side reaction:

$$E_{\text{input}} = E_{\text{effective}} + E_{\text{side reaction}} \tag{5.14}$$

According to the definition of the battery circuit model, the increase in the effective energy of a battery is manifested as the increase in the internal potential (or the SOC), expressed as:

$$E_{\text{effective}} = E(\text{SOC}) = \theta(\text{OCV}) \tag{5.15}$$

As shown in the above formula, E and θ are monotonically increasing functions, that is, the effective energy of the battery (or battery internal potential) increases monotonically with the SOC. The energy consumed by the side reactions is expressed as:

$$E_{\text{side reaction}} = E_{\text{overpotential}} \tag{5.16}$$

that is, it is related to the battery's overpotential.

5.2.2.1 Charging Ah Efficiency and its Influencing Factors

Charging Ah efficiency generally refers to the ratio of the discharging current to the charging current during a single charge–discharge process, which is expressed as:

$$\eta_{\text{Ah}} = \frac{\text{Ah}_{\text{Dch}}}{\text{Ah}_{\text{Cha}}} \tag{5.17}$$

If the total charging time is T_2, the total discharging time T_1, the charging current I_{Dch} and the discharging current I_{Cha}, then the charging Ah efficiency will be:

$$\eta_{\text{Ah}} = \frac{\int_0^{T_1} I_{\text{Dch}}(t)\text{d}t}{\int_0^{T_2} I_{\text{Cha}}(\text{t})\text{d}t} \tag{5.18}$$

From the definition, Equation 5.18, of the charging efficiency, it can be concluded that the battery charging Ah efficiency is related to the charging and discharging capacity.

Influence of the Charging/Discharging Conditions

According to the analysis, different charging conditions have a big influence on the charging and discharging capacity. When the charging conditions remain the same, if the discharging conditions change, the charging capacity also changes accordingly. As for the charging Ah efficiency, in order to avoid negative effects under different charging and discharging conditions, defining the charging capacity should take into account the charging conditions.

Influence of the Permanent Loss of Capacity

In a single cycle of the charging process, the main factors that affect the charging Ah efficiency are the battery efficiency loss of capacity and the permanent loss of capacity, expressed as follows:

$$\eta_{\text{Ah}} = \frac{C_{\text{Cha}} - Q_{\text{loss}} - Q_{\eta}}{C_{\text{Cha}}} \tag{5.19}$$

Table 5.1 The continuous capacity decline with increasing number of cycles.

Cycle index (cyc)	Charging capacity (Ah)	Discharging capacity (Ah)	Ah efficiency (%)	Charge–discharge capacity difference (Ah)	Single capacity decline (charging differential Ah)	Single capacity decline (discharging difference Ah)
1	79.06	78.93	99.8356	0.13		
2	79.02	78.89	99.8355	0.13	0.04	0.04
3	78.89	78.75	99.8225	0.14	0.13	0.14
4	78.86	78.74	99.8478	0.12	0.03	0.01
5	78.82	78.7	99.8478	0.12	0.04	0.04
6	78.76	78.62	99.8222	0.14	0.06	0.08
7	78.65	78.53	99.8474	0.12	0.11	0.09
8	78.56	78.43	99.8345	0.13	0.09	0.1
9	78.48	78.38	99.8726	0.1	0.08	0.05
10	78.45	78.38	99.9108	0.07	0.03	0
11	78.46	78.34	99.8471	0.12	−0.01	0.04
12	78.45	78.32	99.8343	0.13	0.01	0.02
13	78.39	78.26	99.8342	0.13	0.06	0.06
14	78.36	78.21	99.8086	0.15	0.03	0.05
15	78.23	78.07	99.7955	0.16	0.13	0.14
16	78.12	78	99.8464	0.12	0.11	0.07
17	78.04	77.9	99.8206	0.14	0.08	0.1
18	77.96	77.83	99.8332	0.13	0.08	0.07
19	77.92	77.81	99.8588	0.11	0.04	0.02
20	77.82	77.66	99.7944	0.16	0.1	0.15
21	77.6	77.49	99.8582	0.11	0.22	0.17
22	77.52	77.41	99.8581	0.11	0.08	0.08
23	77.5	77.39	99.8581	0.11	0.02	0.02
Average value				0.12	0.07	0.07

Q_η indicates the efficiency loss of battery charging Ah capacity, and Q_{loss} indicates the decline of single battery charging and discharging capacity loss, which is permanent capacity loss. Equation 5.19 indicates that the charging Ah efficiency value reflects two variables: the rate of decline of the capacity and the Ah conversion ratio. Based on the life test, tests of LIBs are executed by using constant current and constant voltage in different ratios, and the results are shown in Table 5.1.

We can infer from the table that the difference between two adjacent charging capacities or discharging capacities is the same. The declining rate reflects the battery single capacity. According to the calculated result, the efficiency of single charging and discharging is about 99.8%, in the transformation of the single charging Ah capacity, and the permanent loss capacity accounts for about 58% of the total Ah transformations, so single charging Ah transformation efficiency is found to be around 99.9%.

For lithium-ion power batteries, the single charging and discharging Ah efficiency is about 99.8%, and the transformation efficiency of Ah is high. Therefore, the efficiency of charging

Ah actually reflects the rate of decline of the battery single capacity. So the formula of Ah efficiency can be simplified as:

$$\eta_{\mathrm{Ah}} \approx \frac{C_{\mathrm{Cha}} - Q_{\mathrm{loss}}}{C_{\mathrm{Cha}}} \tag{5.20}$$

In the practical process of charging, because the single charging and discharging capacity decline rate is very small (generally lower than 0.05%), and there are some errors in measurement, it is difficult to measure the effectiveness of a charging method by using the efficiency of charging Ah.

5.2.2.2 Structure of Charging Efficiency (Wh) and its Influencing Factors

Charging efficiency (Wh) is the battery's ratio of releasing energy to filling energy in a single charging and discharging process, formulated as:

$$\eta_{\mathrm{Wh}} = \frac{\mathrm{Wh}_{\mathrm{Dch}}}{\mathrm{Wh}_{\mathrm{Cha}}} \times 100\% \tag{5.21}$$

If the total charging time is T_2, the total discharging time T_1, the charging current I_{Cha}, the discharging current I_{Dch}, then the charging efficiency is:

$$\eta_{\mathrm{Wh}} = \frac{\int_0^{T_1} U_{\mathrm{Dch}} I_{\mathrm{Dch}} \mathrm{d}t}{\int_0^{T_2} U_{\mathrm{Cha}} I_{\mathrm{Cha}} \mathrm{d}t} \times 100\% \tag{5.22}$$

Incorporating some variables in the battery model, the above formula can be converted to:

$$\eta_{\mathrm{Wh}} = \frac{\int_0^{T_1} U_{\mathrm{ODch}} I_{\mathrm{Dch}} \mathrm{d}t}{\int_0^{T_2} U_{\mathrm{OCha}} I_{\mathrm{Cha}} \mathrm{d}t} = \frac{\int_0^{T_1} (E(t) - I_{\mathrm{Dch}}(t) R_{\Omega}(t) - I_{\mathrm{Dch}}(t) R_{\mathrm{P}}(t)) I_{\mathrm{Dch}}(t) \mathrm{d}t}{\int_0^{T_2} (E(t) + I_{\mathrm{Cha}}(t) R_{\Omega}(t) + I_{\mathrm{Cha}}(t) R_{\mathrm{P}}(t)) I_{\mathrm{Cha}}(t) \mathrm{d}t} \tag{5.23}$$

If the minimum unit time is Δt, the total charging time $M\ \Delta t$ and the total discharging time is $N\ \Delta t$ (M,N = 1,2,3…), then we improve the above equation and get:

$$\eta_{\mathrm{WH}} = \frac{\sum_{k=1}^{N} \left\{ (E(k\Delta t) I_{\mathrm{Dch}}(k\Delta t) - [I_{\mathrm{Dch}}(k\Delta t)]^2 R_{\Omega}(k\Delta t) - [I_{\mathrm{Dch}}(k\Delta t)]^2 R_{\mathrm{P}}(k\Delta t) \right\} \Delta t}{\sum_{l=1}^{M} \left\{ (E(l\Delta t) I_{\mathrm{Cha}}(l\Delta t) + [I_{\mathrm{Cha}}(l\Delta t)]^2 R_{\Omega}(l\Delta t) + [I_{\mathrm{Cha}}(l\Delta t)]^2 R_{\mathrm{P}}(l\Delta t) \right\} \Delta t} \tag{5.24}$$

Definition of Charging Efficiency (Wh)

Similar to that of the charging efficiency (Ah), the accurate definition of charging efficiency (Wh) is also influenced by the charging and discharging conditions. When calculating the charging efficiency, it should keep the consistency of the charging and discharging conditions, in Equation 5.20.

If

$$\sum_{l=1}^{M}(E(l\Delta t)\times I_{\mathrm{Cha}}(l\Delta t)\times \Delta t = W_{\mathrm{che}}$$

then

$$\sum_{k=1}^{N}E(k\Delta t)I_{\mathrm{Dch}}(k\Delta t)=\lambda W_{\mathrm{che}}(\lambda\le 1),$$

Thus Equation 5.22 can be rewritten as:

$$\eta_{\mathrm{Wh}}=\frac{\lambda W_{\mathrm{che}}-\sum_{k=1}^{N}[I_{\mathrm{Dch}}(k\Delta t)]^2R_{\Omega}(k\Delta t)\Delta t-[I_{\mathrm{Dch}}(k\Delta t)]^2R_{\mathrm{P}}(k\Delta t)\Delta t}{W_{\mathrm{che}}+\sum_{l=1}^{M}[I_{\mathrm{Cha}}(l\Delta t)]^2R_{\Omega}(l\Delta t)\Delta t+[I_{\mathrm{Cha}}(l\Delta t)]^2R_{\mathrm{P}}(l\Delta t)\Delta t} \tag{5.25}$$

From Equation 5.25 we can see that the overpotential energy consumption makes the charging efficiency (Wh) constant and less than 1, and the magnitude of the overpotential directly determines the charging efficiency (Wh).

Factors Influencing Charging Efficiency (Wh)

We can infer from Equation 5.25 that charging energy efficiency (Wh) depends on the overpotential of the battery during the charging process or current increasing.

If we formulate with a monotonically increasing function L:

$$\eta_{\mathrm{Wh}}=L(R_{\mathrm{P}},R_{\Omega},I_{\mathrm{cha}}) \tag{5.26}$$

1. The influence of charging conditions. We know from the influence factors [9] of DC internal resistance and polarization resistance that when the ambient temperature is low or cell aging occurs, even if $I_{\mathrm{cha2}}=I_{\mathrm{cha1}}$, with $R_{\mathrm{P2}}>R_{\mathrm{P1}}$ or $R_{\Omega 2}>R_{\Omega 1}$, the charging efficiency of the battery $\eta_{\mathrm{Wh1}}>\eta_{\mathrm{Wh2}}$. That is to say, the same charging current effect produces different charging efficiencies (Wh).
2. The influence of current stress. Under the same battery charging condition, $R_{\mathrm{P2}}=R_{\mathrm{P1}}$ and $R_{\Omega 2}=R_{\Omega 1}$, but along with increasing current, set $I_{\mathrm{cha2}}>I_{\mathrm{cha1}}$, before and after the changing of state, the charging efficiency of the battery will be $\eta_{\mathrm{Wh1}}>\eta_{\mathrm{Wh2}}$.

The Charging Efficiency Reflects the Current Acceptance Characteristic of the Battery Itself

From the above analysis, we can infer that the charging efficiency of a battery depends on the charging conditions and the charging stress. In other words, the higher the charging efficiency, the better the charging conditions and the smaller the charging and discharging stress. Conversely, the smaller the charging stress, the worse the charging conditions will get. Hence, the charging efficiency (Wh) reflects the acceptance characteristic of the charging current under current conditions. With charging current conditions, as above, the higher the charging efficiency (Wh), the better the charging conditions or the more active the battery will be. Likewise, if a battery has higher charging efficiency (Wh), that means it has a higher transformation efficiency of current.

Taking an 8 Ah manganese lithium-ion battery as an example, a series of experiments were performed in which the battery was charged with the CCCV mode and various charging

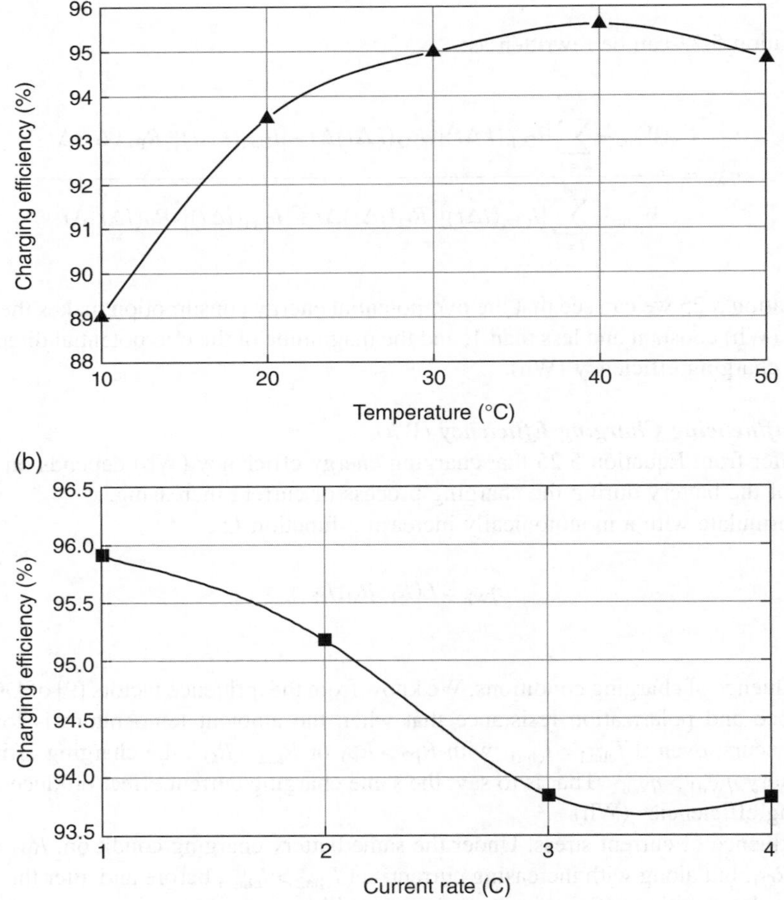

Figure 5.5 Charging efficiency (Wh) (a) at different temperatures and (b) with different charging rates.

rates were used to test the battery charging performance. We extract the battery data shown in Figure 5.5.We can see that when analyzing the efficiency (Wh) of a battery, two types of influence should be comprehensively taken into account, namely charging conditions (temperature, etc.) and charging stress (current). The lower the current ratio, the more appropriate the charging temperature, and the higher the charging efficiency (Wh) will be. Consequently, for the purpose of improving the charging efficiency (Wh) of the battery, we should: (i) Provide reasonable temperature conditions for the charging of the battery, trying to avoid low-temperature charging, and thus reduce the influence of the charging conditions on the charging efficiency. (ii) Avoid getting into the situation where the charging efficiency (Wh) is too low. The battery polarization state should be considered when increasing the current, and the current can be increased at times of low overpotential.

5.2.3 Charging Time

Charging time is the time taken to charge the battery from the initial capacity to full capacity (SOC = 100%). The longest charging time occurs when the initial capacity is 0, in accordance with the charging capacity analysis in Section 5.2.1. Based on different charging processes (constant current or variable current), the charging time is analyzed as follows.

5.2.3.1 Analysis of Constant Current Charging Time

Based on the previous analysis of the charging capacity, when the charging current I is smaller than the critical current $I_{s\,max}$, the charging capacity reaches a maximum and can achieve the rated capacity of the battery C_N. If the constant current is I, then the charging time is

$$T = \frac{C_N}{I} \tag{5.27}$$

From Equation 5.27, the charging time of the constant current is only related to the rated capacity of the battery and the charging current. Therefore, the larger the charging current ratio, the shorter the charging time. Similarly, after decline in the battery capacity, the charging time gets shorter and the charging capacity is less. The calculation of constant current charging time is simple. However, from the above analysis we know that the critical charging current is usually much smaller than $C_N/10$. When the battery ages or the temperature falls, the critical charging current becomes smaller and hence the charging time increases. So it is difficult to meet the requirement of rapid charging.

5.2.3.2 Analysis of Variable Current Charging Time

Variable current charging is the process in which if the charging current I is larger than the critical current $I_{s\,max}$, the initial current of the process is not restricted by the critical charging current and the charging current can also be appropriately adjusted according to the requirements. Given the analysis of charging capacity, in this charging process, the external voltage reaches the upper limit of the charging voltage before the battery charging capacity reaches 100%. In order to ensure the security of the battery, the battery's terminal voltage cannot continue to rise. Therefore, it is necessary to make appropriate adjustments to the charging current of the battery. In this way the charging process can be continued, until the charging process is completed.

For this reason, the total charging capacity is the sum of the values at the various stages of the charging capacity:

$$C_{\text{total}} = C_{I_1} + C_{I_2} + C_{I_3} + \cdots = \sum_{k=1}^{N} C_{I_k} \tag{5.28}$$

And the total charging time is the sum of the charging times at each stage:

$$T_{\text{total}} = \frac{C_{I_1}}{I_1} + \frac{C_{I_2}}{I_2} + \frac{C_{I_3}}{I_3} + \cdots = \sum_{k=1}^{N} \frac{C_{I_k}}{I_k} \tag{5.29}$$

When the battery is charged in a variable current mode, the total charging time is decided by the various segments of the charging time at the constant current stages. However, each charging phase of the turning point is decided by the relationship between overpotential and the maximum charging voltage. Expressed as follows: set the series of constant currents of the battery as: $I_1, I_2, I_3, \ldots$, and for every stage, the cell polarization voltage is $V_{\text{OE}}(I_1, \text{SOC}_1)$, $V_{\text{OE}}(I_2, \text{SOC}_2)$, $V_{\text{OE}}(I_3, \text{SOC}_3)$, …, then set the current of the battery at the first stage as I_1, then the charged capacity for the first stage is:

$$C_{\text{N}} E^{-1}(U_{\text{S}} - U_{\text{OE1}}(I_1, \text{SOC}_1)) \tag{5.30}$$

The charging time t_1 of the first stage will be:

$$t_1 = \frac{C_{I_1}}{I_1} = \frac{C_{\text{N}} \times E^{-1}(U_{\text{S}} - U_{\text{OE1}}(I_1, \text{SOC}_1))}{I_1} \tag{5.31}$$

Set the current of the battery at the second stage as I_2, then the charged capacity of the second stage is:

$$C_{\text{N}} \times E^{-1}(U_{\text{S}} - U_{\text{OE2}}(I_2, \text{SOC}_2)) - C_{\text{N}} \times E^{-1}(U_{\text{S}} - U_{\text{OE1}}(I_1, \text{SOC}_1)) \tag{5.32}$$

The charging time t_2 for the second stage will be:

$$t_2 = \frac{C_{I_2}}{I_2} = \frac{C_{\text{N}} \times E^{-1}(U_{\text{S}} - U_{\text{OE2}}(I_2, \text{SOC}_2)) - C_{\text{N}} \times E^{-1}(U_{\text{S}} - U_{\text{OE1}}(I_1, \text{SOC}_1))}{I_2} \tag{5.33}$$

Likewise, the total charging time of the battery is:

$$\begin{aligned} t_{\text{all}} &= \frac{C_{\text{N}} \times E^{-1}(U_{\text{S}} - U_{\text{OE}1}(I_1, \text{SOC}_1))}{I_1} \\ &+ \frac{C_{\text{N}} \times E^{-1}(U_{\text{S}} - U_{\text{OE}2}(I_2, \text{SOC}_2)) - C_{\text{N}} \times E^{-1}(U_{\text{S}} - U_{\text{OE}1}(I_1, \text{SOC}_1))}{I_2} \\ &+ \cdots\cdots \\ &= \sum_{k=1}^{N} \frac{C_{\text{N}} \times E^{-1}(U_{\text{S}} - U_{\text{OE}k}(I_k, \text{SOC}_k)) - C_{\text{N}} \times E^{-1}\left(U_{\text{S}} - U_{\text{OE}(k-1)}\left(I_{(k-1)}, \text{SOC}_{(k-1)}\right)\right)}{I_{\text{k}}} \end{aligned} \tag{5.34}$$

From Equation 5.34, we know that, on the premise that the maximum charging voltage is limited, the size of the overpotential and the variation directly determine the charging current and the reaction time for each stage. Their value plays a key role in measuring the total charging time.

5.2.3.3 Analysis of Average Charging Time

As shown in Figure 5.6, based on the basic waveform of the charging current, the charging process is divided into N different constant current phases, and the sum of the corresponding charging capacity of each current phase (Ah_N) constitutes the total charging capacity. The battery's charging time for the whole process is:

$$T = T_1 + T_2 + \cdots + T_N \tag{5.35}$$

The total capacity of the charging process is:

$$C_N = \text{Ah}_1 + \text{Ah}_2 + \cdots + \text{Ah}_N \tag{5.36}$$

In the charging process, the charging overpotential of the battery increases gradually, the phase current decreases and the time taken to fill the same capacity increases.

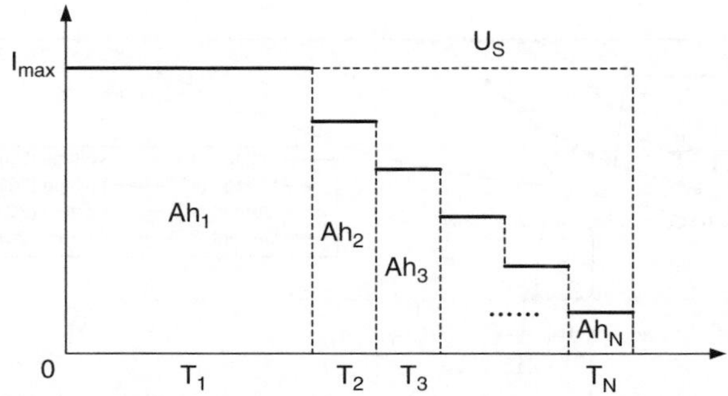

Figure 5.6 The basic waveform of a variable charging current.

We know, through analysis of the characteristics of the overpotential energy charging, that increasing the initial charging current excessively will reduce the cell's service life and increase the initial polarization voltage. As a consequence, when the battery is charged in the second, third and later stages of a longer charging time, the charging curve is concave in shape (Figure 5.7a). So, while the charging currents in the first three stages are large, in the constant current stage the average charging current of the battery is relatively low. Hereby, the charging time will not be dynamic. During a normal current charging, since the initial charging current to establish the polarization voltage remains small, the charging curve is convex in shape (Figure 5.7b) and the average current of the whole charging process is fairly large. So the charging time and the high-current charging time are roughly equal. Figure 5.8 shows the data obtained using different current ratios CCCV in the same cell sample; when the charging current increases from 1 to 2 C, the charging time is reduced by about 2100 s, when it increases

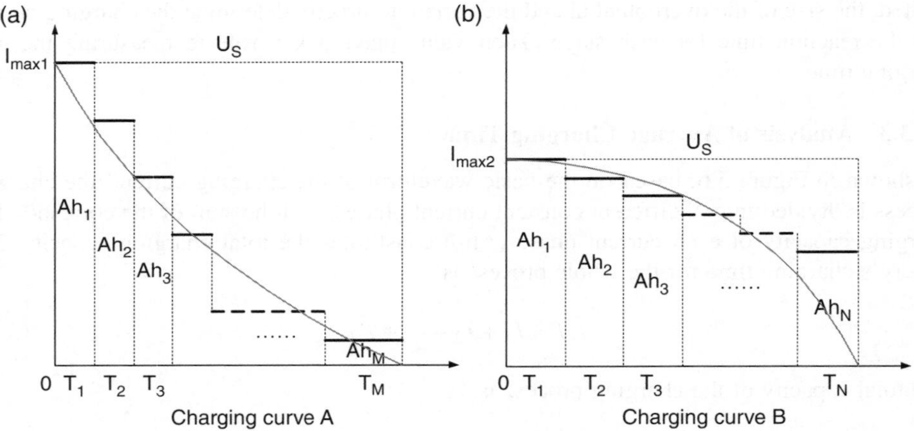

Figure 5.7 Charging curves with (a) high initial current and (b) with low initial current. Reproduced with permission from Jiangpeng Wen, "Studies of Lithium- ion Power Battery optimization charging theory for Pure Electric Vehicle," Beijing Jiaotong University, @2011

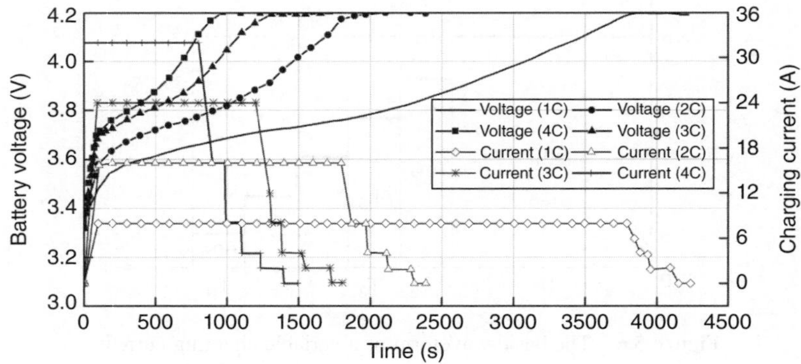

Figure 5.8 Comparison of CCCV charging times for different currents.

from 2 to 3 C, the charging time is reduced by about 500 s , and when it increases from 3 to 4 C, the charging time will be reduced by about 250 s. So we can conclude that increasing the charging current cannot achieve the purpose of exponentially reducing the charging time. On the contrary, if the charging current is too large, it will increase the potential loss, which will usually cause a rise in the charging temperature, thus posing a potential threat to the charging safety.

From analysis of the charging curve waveform, we can consider the phase current and its reaction time when reducing the charging time. According to Equations 5.35 and 5.36:

$$T = \frac{\mathrm{Ah}_1}{I_1} + \frac{\mathrm{Ah}_2}{I_2} + \frac{\mathrm{Ah}_3}{I_3} + \cdots + \frac{\mathrm{Ah}_N}{I_N} \tag{5.37}$$

Assume that the first stage Ah_1 accounts for 10% of the total capacity, for the second to N stages the average charging current is $I_{2\mathrm{a}}$, so the battery's charging time is

$$T_1 = \frac{0.01C_N}{I_1} + \frac{0.09C_N}{I_{2a}} \tag{5.38}$$

If the charging current waveform makes the first stage Ah_1 account for 20% of the total capacity, for the second to N stages the average charging current is $I_{2\mathrm{a}}$, so the battery's charging time is

$$T_2 = \frac{0.02C_N}{I_1} + \frac{0.08C_N}{I_{2a}} \tag{5.39}$$

Thereby, increasing the capacity of the first charging stage will reduce the charging time. Compared with conventional charging, if the charging current for the first k phases is larger than the conventional charging current, and the current for the later N to k phases is less than the conventional charging current, so at the battery's former k phase the charging time gets shorter, and at the later N to k phase it gets longer, the total charging time becomes:

$$\Delta T_{\text{total}} = \Delta T_1 + \Delta T_2 + \cdots + \Delta T_k - \Delta T_{k-1} \cdots - \Delta T_N \tag{5.40}$$

If $\Delta T_{\text{total}} > 0$, the total charging time gets shorter and fast charging will be achieved. On the contrary, if $\Delta T_{\text{total}} < 0$, fast charging will not be achieved. In other words, the average current determines the charging rate and the charging time of the charging curve. As shown in Figure 5.8, the average charging current is:

$$\bar{I}_n = \frac{I_{11} \times t_{11} + I_{12} \times t_{12} + \cdots + I_{1(N-1)} \times t_{1(N-1)} + I_{1N} \times t_{1N}}{\left(t_1 + t_2 + \cdots + t_{1(N-1)} + t_{1N}\right)} \tag{5.41}$$

Transform to:

$$\begin{aligned} \bar{I}_n = {} & \frac{t_{11}}{\left(t_1 + t_2 + \cdots + t_{1(N-1)} + t_{1N}\right)} \times I_{11} + \frac{t_{12}}{\left(t_1 + t_2 + \cdots + t_{1(N-1)} + t_{1N}\right)} \times I_{12} + \cdots \\ & \frac{t_{1(N-1)}}{\left(t_1 + t_2 + \cdots + t_{1(N-1)} + t_{1N}\right)} \times I_{1(N-1)} + \frac{t_{1N}}{\left(t_1 + t_2 + \cdots + t_{1(N-1)} + t_{1N}\right)} \times I_{1N} \end{aligned} \tag{5.42}$$

Then set $\alpha(t)_{Ni} = \dfrac{t_{Ni}}{(t_{N1} + t_{N2} + \cdots + t_{N(N-1)} + t_{NN})}$, the proportion that current number i accounts for in the N charging curves, so Equation 5.42 can be reduced to:

$$\bar{I}_n = \alpha(t)_{N1} \times I_{N1} + \alpha(t)_{N2} \times I_{N2} + \cdots + \alpha(t)_{N(N-1)} \times I_{N(N-1)} + \alpha(t)_{NN} \times I_{NN} \tag{5.43}$$

$$I_{N1} \neq I_{N2} \neq \cdots \neq I_{N(N-1)} \neq I_{NN}$$

We can see from the curve that max (I_{Ni}) and $\alpha(t)_{Ni}(1 < i < N)$ determine the average charging current. Therefore, the choice of the initial current of the battery and the charging time for each stage directly decides the shape of the charging curve. In order to reduce the charging time and to increase the proportion of large current, it is necessary to adjust the weight of the charging time for each stage while improving the initial current.

5.3 Charging External Characteristic Parameters of the Lithium-ion Battery

The three basic physical parameters for battery charging and discharging are current, voltage and temperature. They are all real-time measurable and belong to the battery external characteristic parameters. A brief analysis of the influence they have on the battery during its work is described below.

5.3.1 Current

Battery charging and discharging at high currents will be discussed from two aspects. On the one hand, due to the internal resistance effects, a considerable part of the electrical energy is transformed into thermal energy and dissipates. If the current is higher than a certain value, the battery internal temperature will become fairly high; thus the battery thermal stability will obviously decrease and its working performance will fall. On the other hand, the intercalating-extracting ability of lithium ions is limited during the chemical reaction on the electrodes while charging and discharging, and this determines the battery's maximum allowable charging and discharging current. When the working current is larger than the limit value, the cell polarization effect will inevitably be more obvious, and the increasing polarization voltage makes the battery reach its cut-off charge–discharge voltage more quickly. Thus the charging and discharging capacity and the percentage of used battery energy will decrease. There will also be a security risk if the lithium ions are deposited on a massive scale.

5.3.2 Voltage

When the voltage value is higher than the allowable maximum voltage 4.3 V in the charging process (overcharging), side reactions occur in the battery and a large amount of heat is generated which affects the battery stability. For example, when the decomposition potential of the solvent reaches 4.5 V, a redox reaction between the electrolyte and the electrode active materials will occur with loss of active materials by production, for example, of Mn_2O_3. When the graphite anode is not able to contain excess lithium ions from the cathode, they will be deposited on the anode surface. The decrease in available lithium ions leads to a smaller charging capacity. The lithium ions depositing on the anode can also cause an internal short circuit, leading to serious safety accidents.

In the discharging process, if the voltage is lower than the lowest voltage limit 2.8 V (over-discharging), lithium ions will be completely extracted from the anode. In order to maintain the discharging current, the oxidation of low potential substances at the anode or the reduction of lithium ions at the cathode should continue. With the decrease in lithium ions at the anode, the extraction ability decreases and the polarization voltage increases. When the discharging voltage exceeds the copper dissolution voltage, copper will be oxidized and dissolve, so the active materials at the anode will fall off. When the battery is charged again, the dissolved metal will deposit in the vicinity of the anode, causing a battery internal short circuit when the metal reaches the cathode; this is very dangerous.

5.3.3 Temperature

Temperature is one of the most important factors affecting battery performance. The battery is active and the energy can be effectively used at high temperatures, but if the temperature is too high the stability of the battery cathode crystal lattice will be lost and the safety of the battery is reduced. On the contrary, at low temperatures, the battery activity decreases significantly, and both the internal resistance and the polarization voltage increase (low temperature reduces the diffusion speed of lithium ions, and the lithium ions on the anode electrode reach the graphite very late, which causes a higher ion concentration on the graphite surface and produces a relatively high concentration polarization in other parts of the cell). Therefore, the actual usable capacity of the battery gets smaller and the discharging capability declines. Too low temperatures are more likely to cause lithium metal deposition, resulting in security risks.

5.4 Analysis of Charging Polarization Voltage Characteristics

The polarization voltage is one of the characteristic parameters of a battery, involving many factors and complex variations. It is one of the most important factors affecting battery performance. Here, we elaborate on the generation mechanism and methods for calculating the polarization voltage.

5.4.1 Calculation of the Polarization Voltage

5.4.1.1 Resting Identification Method

Figure 5.9 shows the principle of this method. When charging the battery at a constant current, stop suddenly and test the instantaneous voltage drop and its recovery process.

By taking the parameter identification precision, resting time and the 24 h resting curve of the lithium-ion battery into consideration, we believe that the battery has fully rested when the voltage recovery speed is less than 10 mV/180 s. The following formulae can be obtained:

$$R_1 = (U_0 - U_1)/I \tag{5.44}$$

$$R_2 = (U_1 - U_2)/I \tag{5.45}$$

$$C = \frac{(t_2 - t_1)}{3} R_2 \tag{5.46}$$

The DC resistance R_1, the polarization impedance R_2, and C are identified from the voltage drop curve which is acquired by using a pulse current. The resting identification method is the most effective and accurate method for battery model parameter estimation. The DC resistance R_1 is obtained from the current drop, and the calculation time is short. However, because the battery needs to be fully rested, the required time ($t_2 - t_1$) to get the final voltage V_2 is long. Consequently, R_2 and C cannot be identified online.

5.4.1.2 Parameter Fitting Method

Measure the battery voltage under a current curve, and then use the curve fitting to estimate the model parameters. Apply a short current pulse to the battery, and then rest it. The battery voltage response curve is shown in Figure 5.10b. Two parallel resistance–capacitor circuits are series connected, as in Figure 5.10a. Its zero input response is $U_{csum} = U_{01}e^{-t/\tau_1} + U_{02}e^{-t/\tau_2}$.

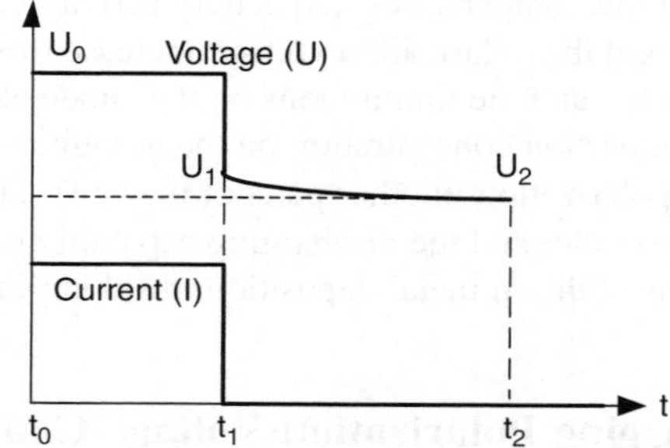

Figure 5.9 Diagram of the principle of the resting identification method.

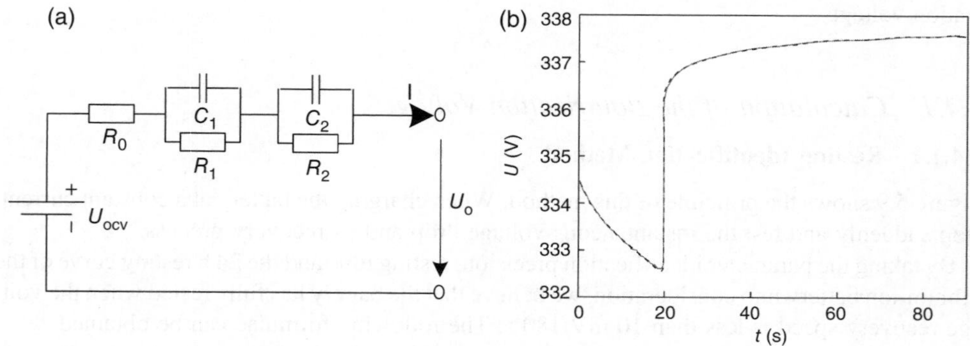

Figure 5.10 The principle of parameter fitting. (a) Circuit diagram. (b) Battery voltage response curve.

This corresponds to the voltage rebound curve. All the coefficients can be determined by least-squares curve fitting. Reference [9] proposes another least-squares curve fitting method to identify the parameters, in which a very short resting time is required due to its fast depolarization. But these methods require data points in a certain time as the fitting basis, the amount of computation is huge, and the real-time capability is also poor.

5.4.1.3 Tracking Calculating Method

The tracking calculating method is a commonly used algorithm based on the following premises.

Continuous Change of the Current

The changing of SOC is a cumulative calculation result, but polarization is an instantaneous effect. When the current changes sharply, it may cause the loss of synchronism of the polarization voltage sampling and the SOC cumulative calculation, and directly affect the estimation accuracy. For pure electric vehicles, the charging current changes constantly and slowly, which provides the premise for polarization voltage estimation.

Known SOC

SOC is the core known quantity for the polarization estimation algorithm and is the base for solving the polarization voltage function. Therefore, knowledge of the SOC is a pre-condition for estimation of the charging polarization.

OCV–SOC Curve of the Battery

Figure 5.11 shows the OCV–SOC curve for the sample battery type I at ambient temperature (25 °C). The battery is allowed to stand for a sufficient length of time at every 5% SOC interval

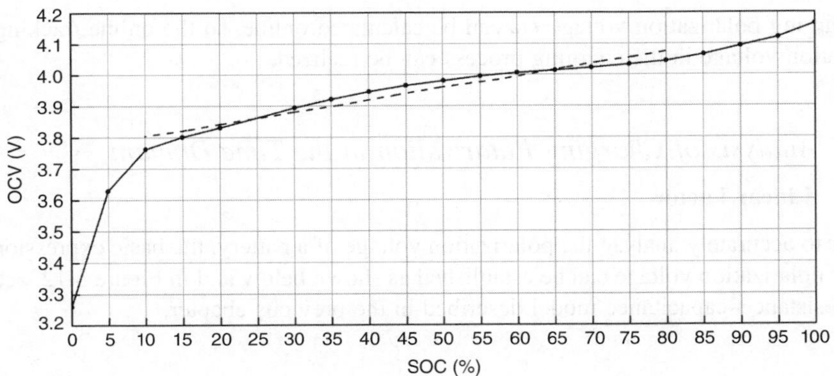

Figure 5.11 The open circuit voltage (the battery is fully rested during charging) versus capacity curve (OCV–SOC).

in the range from 0 to 100%. Between every two adjacent points a curve fitting is applied, and the battery OCV–SOC curve function is represented as follows:

$$\mathrm{OCV}=f(\mathrm{SOC})=\begin{cases} H(0)\mathrm{soc}+B(0); & 0<\mathrm{soc}\le 5 \\ H(1)\mathrm{soc}+B(1); & 5<\mathrm{soc}\le 10 \\ \vdots & \vdots \\ H(18)\mathrm{soc}+B(18); & 90<\mathrm{soc}\le 95 \\ H(19)\mathrm{soc}+B(19); & 95<\mathrm{soc}\le 100 \end{cases} \tag{5.47}$$

OCV is described as a piecewise function of SOC, and is piecewise linear, which simplifies the computational complexity and is easy for engineering realization. $H(i)$ and $B(i)$, $i = 0,1,2, \ldots, 19$ are the slopes and intercepts of each curve, respectively.

Calculation of Battery DC Resistance

According to the principle of the resting identification method, when the CC charging battery stops charging suddenly, the instantaneous voltage drop and the battery recovery process are tested. By using the changes in charging current, we have:

$$R_{\mathrm{d}}=\frac{U_{\mathrm{O2c}}-U_{\mathrm{O1c}}}{I_{\mathrm{2c}}-I_{\mathrm{1c}}} \tag{5.48}$$

Where, U_{O1c} and I_{1c} are the battery voltage and charging current at previous points of time before the change in the current, U_{O2c} and I_{2c} are the instant voltage and current after the change.

By the above method and according to the charging polarization formula:

$$U_{\mathrm{P}}=U_{\mathrm{O}}-\mathrm{OCV}(\mathrm{SOC})-I\times R_{\mathrm{d}}$$

The charging polarization voltage U_{P} can be calculated online, so the online tracking of the polarization voltage in the charging process can be realized.

5.4.2 Analysis of Charging Polarization in the Time Domain

5.4.2.1 Linear Factor

In order to accurately analyze the polarization voltage of a battery, the basic expression of the *N*-order polarization voltage can be established as shown below and in Figure 5.12, according to the resistance–capacitance model described in the previous chapter.

$$u_{\mathrm{P}}(t)=\sum_{i=1}^{n}\left(U_{\mathrm{O}i}\mathrm{e}^{-\frac{t}{R_{\mathrm{P}i}C_{\mathrm{P}i}}}+i(t)R_{\mathrm{P}i}\left(1-\mathrm{e}^{-\frac{t}{R_{\mathrm{P}i}C_{\mathrm{P}i}}}\right)\right) \tag{5.49}$$

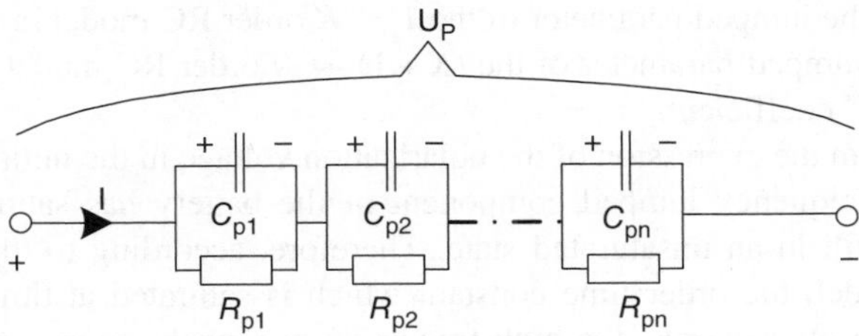

Figure 5.12 *N*-order model of polarization voltage.

When $t = 0$, the cell polarization voltage is:

$$U_{\mathrm{P}}(0) = \sum_{j=1}^{n} U_{0j} \tag{5.50}$$

That is to say, the current battery voltage status depends on the previous polarization state, and when $t \rightarrow \infty$, the cell polarization voltage is a steady-state value:

$$U_{\mathrm{P}}(\infty) = \sum_{j=1}^{n} i(t) R_{\mathrm{P}j} \tag{5.51}$$

Start at any time T_1 in the charging process, charge the battery at current I for a duration time T, so the expression of the polarization voltage change during the charging process can be formulated:

$$\begin{aligned} U_{\mathrm{P}} = U_{\mathrm{P}}(0) + I \times R_{\mathrm{P}1}(1 - \exp(-t/R_{\mathrm{P}1}C_{\mathrm{P}1})) + I \times R_{\mathrm{P}1}(1 - \exp(-t/R_{\mathrm{P}2}C_{\mathrm{P}2})) \\ \cdots + I \times R_{\mathrm{P}n}(1 - \exp(-t/R_{\mathrm{P}n}C_{\mathrm{P}n})) \end{aligned} \tag{5.52}$$

Assume that during the charging period, the RC saturation order of the battery is K, so the cell polarization voltage at $(T_1 + T)$ is:

$$\begin{aligned} U_{\mathrm{P}} = U_{\mathrm{P}}(0) + I \times R_{\mathrm{P}1} + I \times R_{\mathrm{P}2} + \ldots I \times R_{\mathrm{P}k} \\ + I \times R_{\mathrm{P}k}(1 - \exp(-t/R_{\mathrm{P}k}C_{\mathrm{P}k}))\ldots + I \times R_{\mathrm{P}n}(1 - \exp(-t/R_{\mathrm{P}n}C_{\mathrm{P}n}) \end{aligned} \tag{5.53}$$

Further simplification of the above expression gives:

$$\begin{aligned} U_{\mathrm{P}} &= U_{\mathrm{P}}(0) + I \times (R_{\mathrm{P}1} + R_{\mathrm{P}2} + \cdots R_{\mathrm{P}k}) \\ &\quad + I \times (R_{\mathrm{P}k}(1 - \exp(-t/R_{\mathrm{P}k}C_{\mathrm{P}k})) \cdots + R_{\mathrm{P}n}(1 - \exp(-t/R_{\mathrm{P}n}C_{\mathrm{P}n}))) \\ &= U_{\mathrm{P}}(0) + I \times \sum_{i=1}^{K} R_{\mathrm{P}k} + I \times \sum_{j=k+1}^{N} \left[R_{\mathrm{P}j}(1 - \exp(-t/R_{\mathrm{P}j}C_{\mathrm{P}j})) \right] \\ &= U_{\mathrm{P}}(0) + I \times R_{\mathrm{total1} \sim K} + K_{\mathrm{RC}} \times I \times R_{\mathrm{total}(K+1) \sim N} \end{aligned} \tag{5.54}$$

Where, $R_{total1\sim K}$ is the lumped parameter of the $1 \sim K$ order RC model in the saturation state, $R_{total(K+1)\sim N}$ is the lumped parameter of the $(K+1) \sim N$ order RC model in the unsaturation state, K_{RC} is the RC coefficient.

It can be seen from the expression of the polarization voltage in the initial charging stage, at (T_1+T), the high-frequency lumped component of the battery has saturated, but the low-frequency part is still in an unsaturated state. Therefore, according to the characteristics of the battery RC model, the order time constant which is saturated at time T can be written as: $3R_{Pk}C_{Pk} < T$. In other words, the high-frequency part of the time constant $\tau = R_{Pk}C_{Pk} < (T/3)$ is saturated within the period T, and the polarization voltage remains constant.

Derived from the above equation, the resistance–capacitance constant of the battery polarization voltage has a strong time effect. The polarization voltage at time T is:

$$\begin{aligned} U_P &= U_P(0) + I \times R_{P1} + I \times R_{P2} + \cdots I \times R_{Pk} \\ &\quad + I \times R_{Pk}(1-\exp(-T/R_{Pk}C_{Pk})) \cdots + I \times R_{Pn}(1-\exp(-T/R_{Pn}C_{Pn})) \end{aligned} \tag{5.55}$$

Set:

$$A_K = \begin{cases} R_{Pk} & 0<k<K \\ R_{Pk}(1-\exp(-T/R_{Pk}C_{Pk})) & K<k<N \end{cases} \tag{5.56}$$

A_0 is the initial state factor which is independent of the current, and $A_1 \cdots A_N$ are the linear influence factors of the polarization voltage, referring to the effects produced by the same current in different order models. The polarization impedance of the charging current I in the period T is divided into the saturated segment and the unsaturated segment. The amplitude of the polarization voltage, decided by the saturated polarization impedance, depends only on the resistance. However, for the unsaturated order, the polarization voltage established in the period of time T is determined by both the polarization impedance parameters and the time constant. The fading frequency is exponential to the ratio of the time T to the time constant $R_{Pn}C_{Pn}$. The basic linear relationship of the cell polarization voltage to the charging current is:

$$U_P = U_P(0) + (A_1 + A_2 + \cdots + A_N) \times I \tag{5.57}$$

$A_1, A_2 \cdots A_N$ are the RC coefficients of different order polarization voltages, and they are the linear influence factors of the polarization voltage on the charging current.

5.4.2.2 Nonlinear Factors

Distortion Factor of Current Rate

Test the polarization voltage of a 90 Ah lithium manganese battery at different initial currents. By extracting the charging time T (about 900 s), the established polarization voltages at charging time T are found and listed in Table 5.2.

The polarization voltage changes with the charging current rates with a monotonically increasing relationship. If all the battery polarization impedance parameters are decided by multi-order resistance–capacitance parameters, the polarization voltage increment is linear to the charging current increment, which means if $I_2 = 2I_1$, then $V_{P2} = 2V_{P1}$ must also hold.

Table 5.2 Polarization voltage established with different initial currents.

Charging current I(A)	Polarization voltage at time T U_P(V)	Polarization voltage ΔU_P/ current ΔI	Current distortion factor K_I
12	0.064	—	—
24	0.030	−0.0028	−4.8
36	0.039	0.00075	1.3
48	0.046	0.00058	1
72	0.076	0.00125	2.1

Table 5.3 Polarization voltage established at different initial SOC.

Initial SOC (%)	Polarization voltage U_P(V)at time T	Distortion factor of SOC K_{SOC}
10	0.012	1
20	0.006	0.5
30	0.004	0.3
40	0.028	2.3
50	0.017	1.4
60	0.014	1.2
70	0.014	1.2
80	0.02	1.7
90	0.039	3.3

However, Table 5.2 shows that the corresponding polarization voltage change of a current unit is not consistent under different currents, which means the calculated ratios of the polarization voltage to the charging current are different: $(U_{P2} - U_{P1})/(I_2 - I_1) \neq (U_{P3} - U_{P2})/(I_3 - I_2)$. Thus, in spite of the fact that there is a linear relationship between the polarization voltage and charging current, this linear relationship has a certain distortion coefficient K_I at different currents (the polarization voltage impact factor K_I is shown in Table 5.2. It is calculated from the corresponding polarization voltage caused by the unit current at 48 A.

Therefore, the N order polarization RC model, which is influenced by the current distortion factor, describes the relationship between the polarization voltage and the charging current:

$$U_P = U_P(0) + K_I \times I \times (A_1 + A_2 + \cdots + A_N) \tag{5.58}$$

If using a small current, the polarization distortion factor built by the unit current is large. With increasing current, the factor gradually becomes stable, then with further increase in current it becomes large again.

Distortion Factor of the Initial SOC

Test the polarization voltages of 90 Ah lithium manganese batteries at different initial SOCs. The polarization voltages are established under different SOCs but the same current is calculated by the charging polarization voltage tracking algorithm. The test data at the time T (about 900 s) are listed in Table 5.3.

The polarization voltage established at SOC = 10% is taken as the base for calculating the initial-SOC distortion factor K_{SOC}. With increasing SOC, the polarization voltage amplitude established by the same current rises gradually. Thus, the linear expression of SOC impact factors for the polarization voltage is corrected by:

$$U_P = U_P(0) + K_{SOC} \times I \times (A_1 + A_2 + \cdots + A_N) \tag{5.59}$$

Initial Polarization Distortion Factor

Test the established polarization voltage of the sampled 90 Ah lithium manganese batteries for different initial resting times. The polarization voltages under various situations established by the same current are obtained by using the charging polarization voltage tracking algorithm, by which we can extract the charging polarization voltage data at the time T (about 900 s), shown in Table 5.4.

From the table it is seen that the shorter the initial resting time, the smaller the polarization voltage, and vice versa. When the resting time ranges between 0 and 1 h, the change in the established polarization voltage is large but when the resting time varies between 1 and 5 h, the established polarization voltage basically stays at the same level. Therefore, by considering the influence on the initial polarization state distortion factor B_{PO^-}, the polarization voltage linear expression can be defined as:

$$U_P = U_P(0) + I \times (A_1 + A_2 + \cdots + A_N) + B_{PO^-} \tag{5.60}$$

The initial resting time reflects the battery's initial polarization state and, according to the direction of current, it can be divided into three sorts:

$$B_{PO^-} = \begin{cases} B \\ 0 \\ -B \end{cases}$$

where for B the previous time is in the charging state, for 0 it is in the fully resting state, and for –B it is in the discharging state.

Table 5.4 Polarization voltage established with different initial resting times.

Resting time before charging (h)	Charging current (A)	Polarization voltage at time T (mV)
0	24	12
0.5	24	15
1	24	25
2	24	26
5	24	25

Due to the hysteresis effects of the polarization voltage and the differences in initial resting time, the reduction of the polarization voltage at the last moment is also different, which affects the magnitude of the charging polarization voltage.

Aging Distortion Factor of the Battery

Test the established polarization voltage of the sampled 90 Ah lithium manganese batteries at different aging degrees. The polarization voltage values are obtained by using the charging polarization voltage tracking algorithm, and the charging polarization voltage data curve is extracted (Figure 5.13). When the battery capacity declines, its polarization voltage increases. Therefore, the impact of the SOH aging distortion factor must be taken into account and the polarization voltage linear formulation corrected accordingly:

$$U_P = U_P(0) + I \times (A_1 + A_2 + \cdots + A_N) + B_{SOH} \tag{5.61}$$

In conclusion, the nonlinear influence factors of the polarization voltage are studied under the action of every single factor at the same time. The amplitude of the established polarization voltage is not only linearly related to the inherent resistance and the capacitor parameters, but also influences the following four nonlinear factors, namely charging rate, initial SOC, initial resting state and the battery's aging state. Hence, the comprehensive expression of the charging polarization voltage and its influence factors in the time domain is proposed as:

$$\begin{aligned} U_P &= f_{V_P-t}(I, t, K_{SOC}, K_I, K_{SOH}, B_{P0^-}) \\ &= U_P(0) + K_{SOC} \times K_I \times I \times (A_1 + A_2 + \cdots + A_N) + B_{P0^-} + B_{SOH} \end{aligned} \tag{5.62}$$

where $A_1 \sim A_N$: that is, $R_{P1}(1-\exp(-T/R_{P1}C_{P1})) \sim R_{PN}(1-\exp(-T/R_{PN}C_{PN}))$, are the polarization RC coefficients; K_I the distortion factor of the current rate; K_{SOC} the distortion factor of the initial SOC; B_{SOH} the distortion factor of the initial SOH; and B_{P0^-} the distortion factor of the initial polarization state.

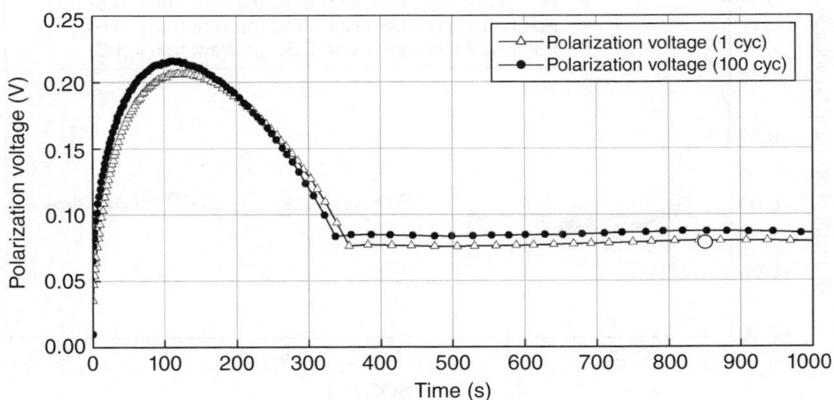

Figure 5.13 The change in polarization voltage with different aging states. Reproduced with permission form Jiapeng Wen, "Studies of Lithium -ion Power B attery optimization charging theory for Pure Electric Vehicle," Beijing Jiaotong University, @2011

The time-domain comprehensive expression of the polarization voltage shows that there are many factors impacting the polarization voltage; which leads to strong nonlinear characteristics. So we can see the variation of the polarization voltage in the time domain is fairly complex.

5.4.3 Characteristic Analysis of the Charging Polarization in the SOC Domain

5.4.3.1 Gradient Characteristics of the Charging Polarization Voltage in the SOC Domain

The gradient of the charging polarization voltage in the SOC domain is defined as: $L = \dfrac{\partial U_P}{\partial \mathrm{soc}}$; and for simplicity referred to as SOC gradient in this book.

SOC Gradient Inflection Point for Different Initial Currents

The values of the polarization voltage are tested at different initial currents, and the curves of polarization voltage gradient in the SOC domain are shown in Figure 5.14.

In the time domain, the polarization voltage amplitude is often different at different initial currents, even varying at the stable moments. However, despite the fact that the SOC gradient of the polarization voltage varies in the initial charging, with gradual decreasing of the gradient, the inflection points of the SOC gradient at different initial currents almost coincide at the same SOC value (about 3%). After the charging capacity rises to 5%, the SOC gradient of the polarization voltage gradually becomes steady and the values are minimal. The above analysis shows that if the other conditions (initial SOC, initial polarization state and degree of aging) are fixed, the SOC gradient inflection points of the charging polarization voltage will not change with the initial current.

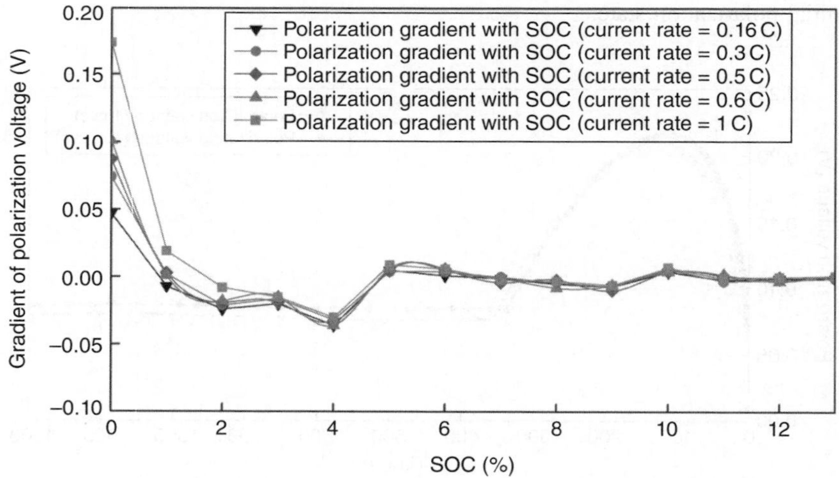

Figure 5.14 SOC gradients of the polarization voltage at different initial currents. Reproduced with permission from Jiapeng Wen, "Studies of Lithium -ion Power B attery optimization charging theory for Pure Electric Vehicle," Beijing Jiaotong University, @2011

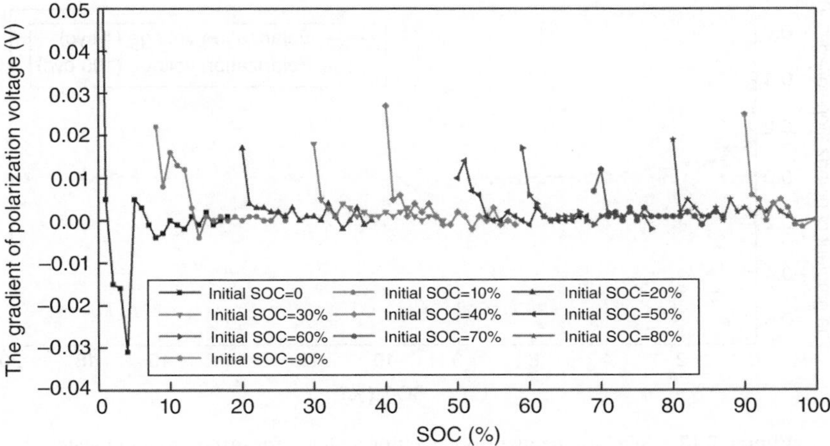

Figure 5.15 SOC gradients of the polarization voltage for different initial available capacities.

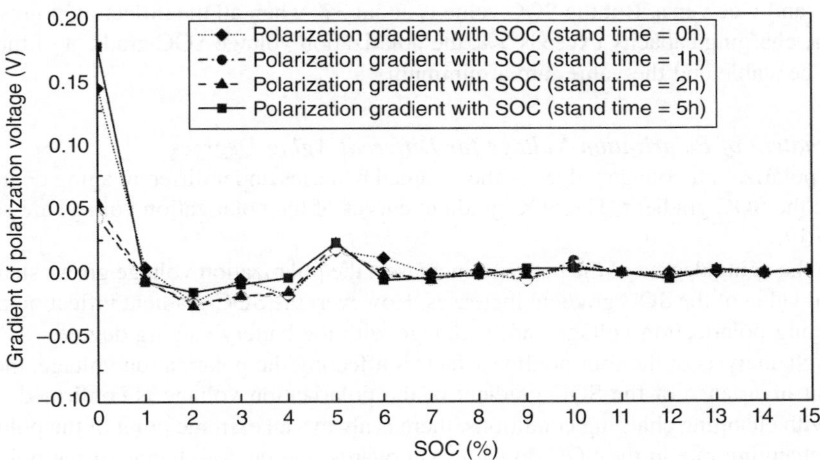

Figure 5.16 SOC gradients of the polarization voltage for different initial standing states.

SOC Gradient Inflection Point for Different Initial Available Capacities

The established polarization voltage values are tested under different initial SOC, and the calculated SOC gradient of the polarization voltage is shown in Figure 5.15.

Likewise, if the initial charging capacity varies, the SOC gradient inflection point characteristics of the polarization voltage remain consistent (<3%), the gradient stays stable and the values are a minimum when the charging capacity is around 5%.

Polarization Voltage Gradient Inflection Point for Different Initial Polarization States

Test the polarization voltage values under different periods of initial resting time. The calculated SOC gradient curves of the polarization voltage are displayed in Figure 5.16.

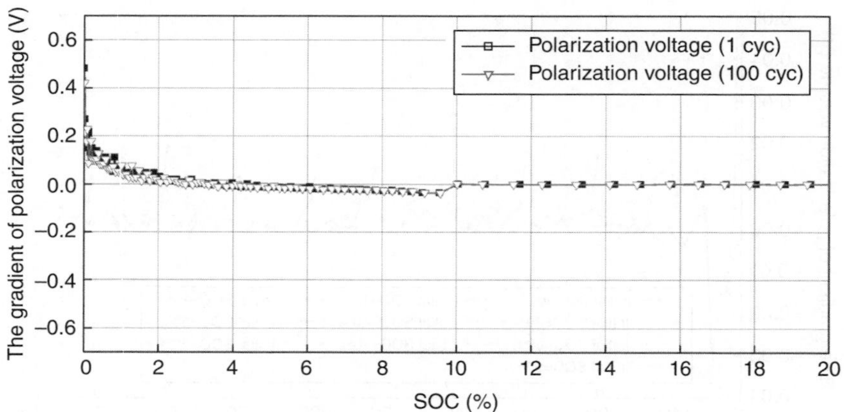

Figure 5.17 SOC gradients of polarization voltage for different aging states.

As the figure 5.16 shows, the shorter the resting time, the smaller the initial value of the SOC gradient, and vice versa. But the SOC value is about 3% when all the inflection points appear. When the charging capacity exceeds 5%, the polarization voltage SOC gradient of the battery tends to be stable and the values are a minimum.

SOC Gradient of Polarization Voltage for Different Aging Degrees

Test the polarization voltage values of the sampled batteries under different aging degrees, and calculate the SOC gradient. The SOC gradient curves of the polarization voltage are shown in Figure 5.17.

When the aging degree of the batteries increases, the polarization voltage grows sharply and the initial value of the SOC gradient increases. However, the SOC gradient inflection points of the charging polarization voltage hardly change with the battery's aging degree.

Through analysis of the four nonlinear factors affecting the polarization voltage, the inflection-point invariance of the SOC gradient of the polarization voltage is confirmed. In other words, with changing charging conditions, there is always an extreme point of the polarization voltage changing rate in the SOC domain. Afterwards, the rate of change of the polarization voltage gradually becomes stable and its value becomes quite tiny.

5.4.3.2 Amplitude Characteristics of the Polarization Voltage in the Inflection Point

Polarization Amplitude and the Current Rate

Based on the SOC gradient characteristics of the polarization voltage, different initial currents are used to establish the polarization voltage, and its values U_{PD} can be obtained when the battery was charged to 3% capacity (i.e., the inflection point of $\partial U_P/\partial soc$ occurred). The U_{PD} values are shown in Figure 5.18.

Under different current rates, when $\partial U_P/\partial soc$ reaches the inflection point, the amplitude of the polarization voltage is linear to the charging current. Fitting the data by $U_{PD} = f(I)$, we have:

$$U_{PD} = 0.0005 \times I + 0.0169 \tag{5.63}$$

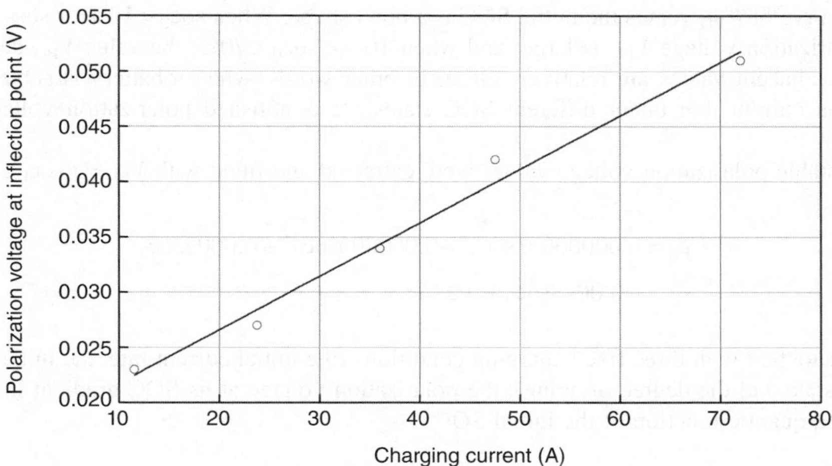

Figure 5.18 Fitting curve of the polarization voltage U_{PD} versus current *I*.

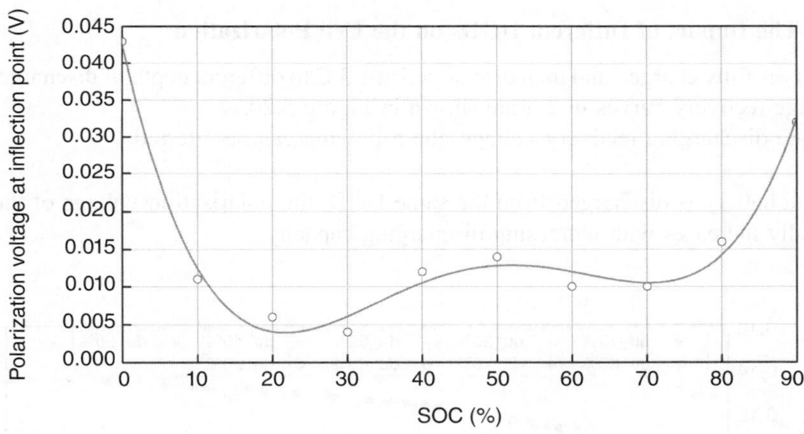

Figure 5.19 Fitting curve of the polarization voltage U_{PD} versus the initial SOC.

This shows that when the SOC gradient of the polarization voltage reaches its inflection point, the polarization voltage amplitude exhibits linear changing characteristics, and the relationship between the polarization voltage and the charging current can be described by a linear function.

Polarization Amplitude and the SOC

Based on the SOC gradient characteristic of the polarization voltage, the values of polarization voltage U_{PD} established under different initial SOCs were extracted when the battery was charged to 3% capacity (i.e., $\partial U_P/\partial \text{soc}$ stable). The polarization voltage values U_{PD} are shown in Figure 5.19.

The curve of U_{PD} versus the initial SOC is a bowl shape. When $soc_0 < 10\%$ or $soc_0 > 80\%$, the polarization voltage V_{PD} is large; and when $10\% < soc_0 < 70\%$, the value V_{PD} decreases and the adjacent values are relatively close. In other words, when a battery is subjected to the same current, but under different SOC states, its established polarization voltages are different.

The stable polarization voltage values were extracted and fitted with $V_{PD} = f(soc_0)$, thus:

$$U_{PD} = 0.00000001soc_0{}^4 - 0.000003soc_0{}^3 + 0.0002soc_0{}^2 - 0.0048soc_0 + 0.04 \quad (5.64)$$

This is satisfied with three fixed charging conditions (the initial current rate, the initial polarization state and the degree of aging), the polarization voltage at its SOC gradient inflection point is a quantic function of the initial SOC.

5.4.4 The Impact of Different SOCs and DODs on the Battery Polarization

5.4.4.1 The Impact of Different DODs on the Cell Polarization

The battery is fully charged and then discharged at 1/3 C to different depth of discharge (DOD). The voltage recovery curves in 2 h are shown in Figure 5.20.

From the discharging recovery voltage, the following can be inferred:

1. When a battery is discharged from the same DOD, the polarization voltage of the battery gradually increases with increasing discharging capacity.

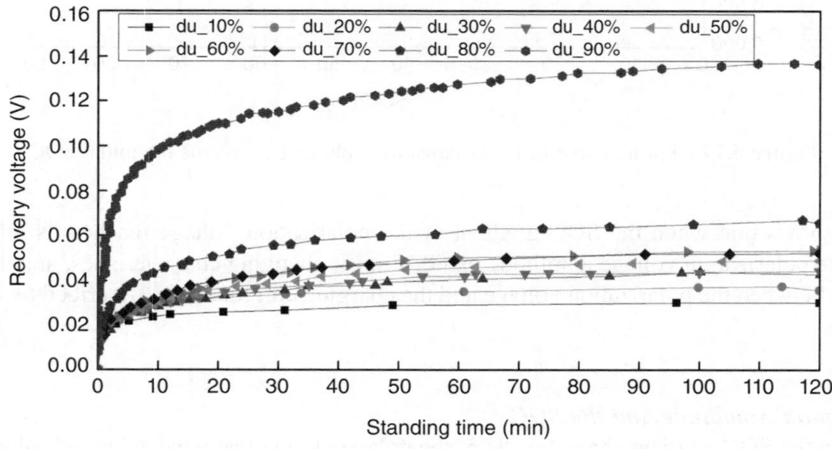

Figure 5.20 Discharging polarization voltage recovery curves for different DODs.

2. The relationship between the battery polarization voltage and the battery discharged capacity is nonlinear.
3. When the battery is discharged to 10% DOD, the battery polarization voltage is not fully established and the battery's depth of polarization stays less than for other DODs, thus the voltage recovers faster.
4. When reaching 20–70% DOD of the battery, the battery polarization voltage increases gradually, but the change is small.
5. When the battery is discharged to 80% DOD, the battery's polarization voltage begins to increase, with as much as a threefold increase at 90% DOD.

It can be concluded that, even at the same SOC, there is a huge difference between the charging and discharging polarization voltages. When the SOC is too low or too high, the battery's polarization voltages vary a lot in amplitude and recovery time. So, when modeling the battery, the model parameters of the battery need to be adjusted accordingly. The battery charging polarization at a high SOC level (SOC > 80%) increases rapidly, and this shows that the battery charging current capability declines at this point. But at the same time, the battery's discharging polarization is not that large, thus its discharging polarization voltage is small. At a high DOD level, where the battery is about to be completely discharged, the battery's discharging ability decreases and its polarization voltage increases.

5.4.4.2 The Influence of SOC on the Battery Polarization during a Charging Process

The relationship between polarization voltages and the SOC during the charging process is an important basis for studying the charging methods. A 90 Ah lithium-ion manganese battery is performed by the CCCV test. Figure 5.21 is obtained from the experimental data (by using the

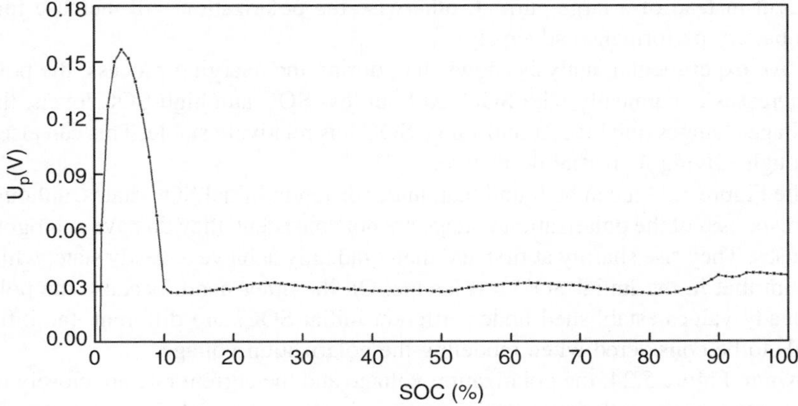

Figure 5.21 The relationship between polarization voltage and SOC when the battery is CC charged.

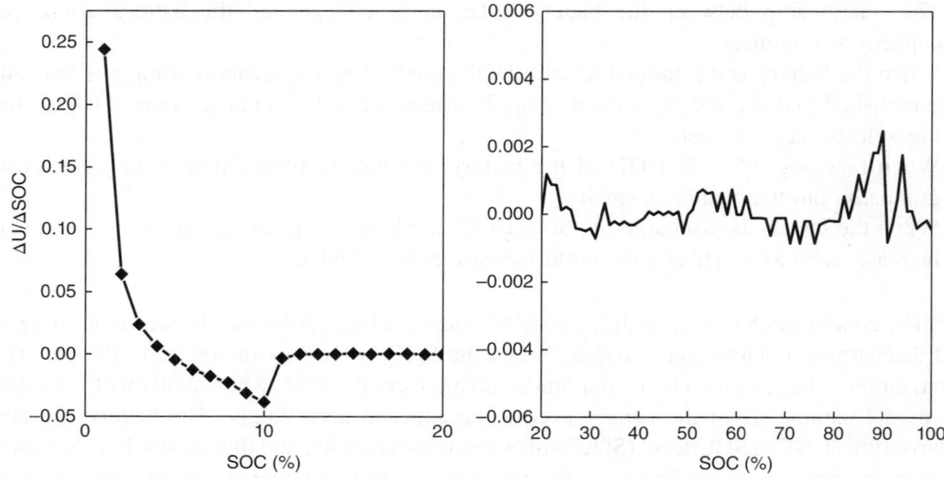

Figure 5.22 The SOC gradient of polarization voltage versus SOC.

tracking method in Section 5.4.1.2). As can be seen, during the charging process, the polarization voltage reaches its maximum value that shows as a peak at around SOC = 5%, and also arrives at a relatively stable stage when SOC = 10%. Figure 5.22 displays the relationship between the SOC gradient of polarization voltage and the SOC. It can be seen that the SOC gradient of polarization voltage in the 0–10% SOC range is much greater than that in the 10–80% SOC range. Between 80 and 90% SOC, the polarization voltage increases again, indicating that the polarization tends to spread. After reaching 90% SOC, the polarization voltage gradually reduces, since it has entered the constant-voltage stage where the charging current is gradually decreasing. The above phenomena indicate that the polarization voltage is at a high level when the SOC is below 10 or above 80% and at this time the battery should be charged at a small current instead of a large current, otherwise the polarization will increase further and affect the battery performance adversely.

The above experimental analysis shows that, during the charging process, the polarization voltage increases dynamically with SOC. At both low-SOC and high-SOC levels, the polarization voltage changes quickly. At mid-range SOC it is relatively stable. This can also be illustrated through solving its partial derivative.

From the Figure 5.23 it can be found that, under different initial SOC states, although all the dynamic responses of the polarization voltage are not consistent, they do have analogous damping responses. They rise sharply at first and then gradually achieve a steady state, which varies greatly from that in the initial SOC = 0% state. On the other hand, because the polarization voltage steady values established under different initial SOCs are different, the influence of SOC needs to be considered when modeling the polarization voltage.

As shown in Figure 5.24, the polarization voltage and the current rate are closely related. If the current rate increases, the polarization voltage is also enhanced. The processes for establishment of the polarization voltages are similar in the same initial SOC state.

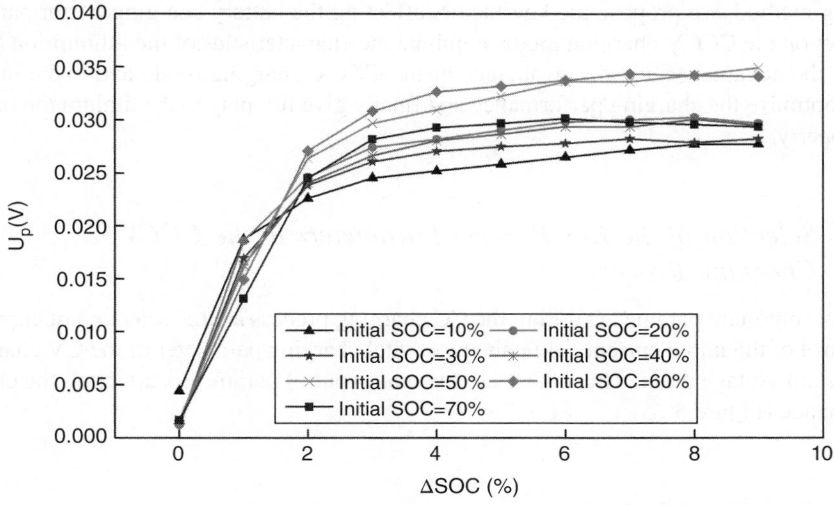

Figure 5.23 The polarization voltage versus ΔSOC for different initial SOC states.

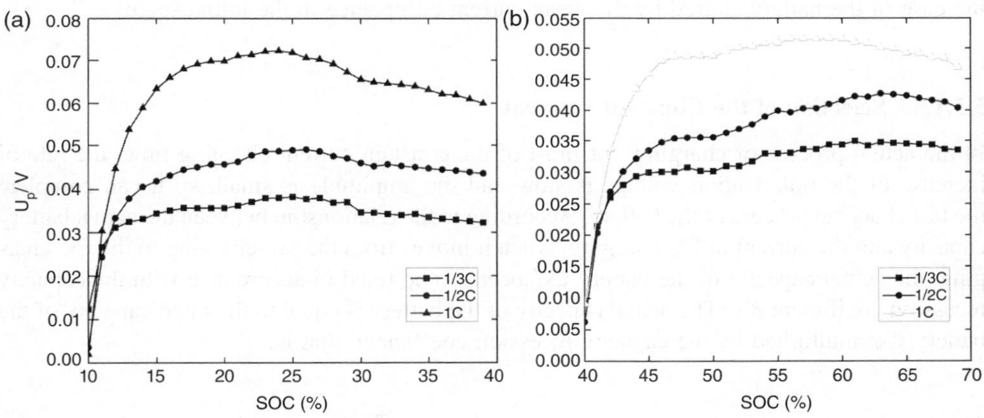

Figure 5.24 Polarization voltage versus current rate with initial SOC (a)10% and (b) 40%.

5.5 Improvement of the Constant Current and Constant Voltage Charging Method

This section is based on the conventional charging method for the lithium-ion battery (the CCCV method). Through theoretical analysis and experimental testing, we make a comprehensive analysis of the influence factors on the charging characteristics in the constant current and constant voltage charging mode by means of the basic electrical model (including capacity, internal resistance, polarization voltage, etc.). From the performance indexes for the charging method, we propose the key factors affecting the battery charging performance and the effect on the CCCV charging mode, combine the characteristics of the lithium-ion battery, analyze the advantages and disadvantages of the CCCV charging mode and make improvements, optimize the charging performance and finally give full play to the lithium ion battery's rate property.

5.5.1 Selection of the Key Process Parameters in the CCCV Charging Process

The most important parameters during the CC charging process are the selection of current and the control of the initial current, while the most vital charging parameter of the CV charging is the constant voltage point. Thus, these are the main control parameters affecting the charging performance (Figure 5.25).

5.5.1.1 Selection of the Initial Current

The initial charging current should be appropriately reduced, to slow the rate of increase of the polarization voltage, to minimize the current stress to avoid the capacity loss and polarization increase of the battery caused by the weak current acceptance in the initial stage.

5.5.1.2 Selection of the Constant Current

In the actual process of charging, for most of the constant current charging time, the rate of increase of the polarization voltage is slow and the amplitude is small, so it can complete the fast charging process of the battery. According to the relationship between the actual battery capacity and the current in CC charging, when it moves from the pre-charging to the CC charging, the actual capacity of the battery C_N' needs to be fixed in accordance with the capacity recession coefficient K_A. The actual capacity of the battery is equal to the rated capacity of the battery C_N' multiplied by the capacity recession coefficient, that is,

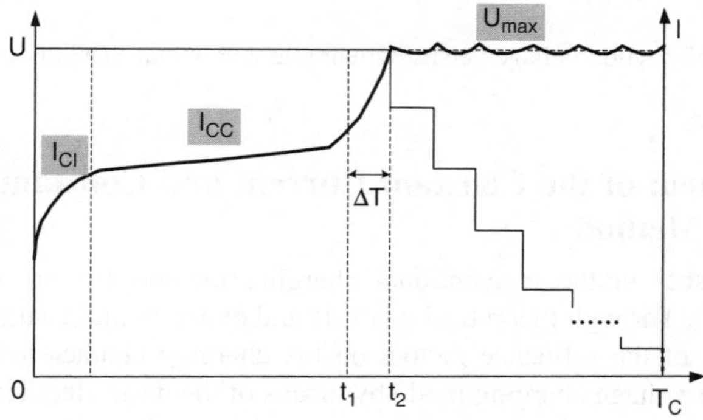

Figure 5.25 Key control parameters in the CCCV charging.

$$C'_N = K_A \times C_N \quad (5.65)$$

The constant current charging current after the capacity correction is:

$$I_{CC} = K_C \times C'_N \quad (5.66)$$

where K_C is the charging and discharging rate coefficient at room temperature, and C_N is the rated capacity of the battery.

Considering the temperature effect on the battery charging rate, we adjust the charge–discharge rate according to the ambient temperature, and the charging current is equal to the charging and discharging rate of the battery at normal temperature multiplied by the temperature coefficient of the battery:

$$K'_C = K_C \times K_T \quad (5.67)$$

The final expression of the constant charging current is

$$I_{CC} = K'_C C_A = (K_T K_C) K_A C_N \quad (5.68)$$

$0 < K_T, K_C, K_A < 1$, K'_C is the battery rate coefficient, C_A is the battery capacity coefficient.

Given the rate characteristics of the lithium-ion battery, the rate factor of the battery at normal temperature should be $K_C = (0.5 \sim 1)$.

5.5.1.3 Selection of the Constant Voltage Point

The constant voltage point is an important parameter related to both the battery life and the charging capacity. The constant voltage point value is small, so the charging capacity does not reach the available maximum capacity. If the constant voltage point increases, the battery life will be reduced. Therefore, the controlling process of the constant voltage point must be able to meet the demand of maximizing the charging capacity of the battery. In order to reduce the constant voltage values of the battery at the same time, under the premise of rapid changes in the polarization voltage, switching to the CV charging ahead of time, in the late period of the CC stage, can reduce the initial polarization voltage of the CV.

5.5.2 Optimization Strategy for the CCCV Charging

5.5.2.1 Improved CCCV Charging Method

The analysis of the critical process parameters from the CC stage of the charging shows that in order to improve the battery charging rate, the charging current in the constant current stage must be increased. The charging capacity ought to be taken into consideration, and the charging current at the end of the CV charging must be sufficiently small so that the battery capacity and the battery voltage can reach the maximum value synchronously. With regard to the main factors affecting the battery life, the pre-charging link should be added, and the CC charging current should be adjusted with the actual battery capacity and ambient temperature. Moreover, we should reduce the polarization voltage in the late charging appropriately at the same time and reduce the terminal voltage of the battery in the CV charging process. As a result, improvements in the control method for the traditional CCCV and charging are listed as follows:

1. Adding Pre-Charging: Add the pre-charging link into the CC charging stage in the CCCV charging. In the pre-charging part, the battery charging current should be kept small; when the charging capacity reaches a preset value, the battery charging current rises to a constant charging current.
2. Adjusting the Constant Current Charging Current: Improve the CC charging rate (usually 0.3–0.2 C), and update it according to the actual battery capacity and ambient temperature.
3. Pre-Advancing into the Constant Voltage Charging, and Reducing the Constant Voltage Control Value: According to the starting conditions in the CV charging process and the variation in the polarization voltage and the battery voltage, pre-advancing into the CV charging process, from the end of the CC charging to the decrease of the polarization voltage, and reduce the constant voltage value appropriately.

In the light of the related principles, the improved CCCV charging curve can be divided into three sections: the pre-charging stage, the CC fast charging stage and the optimized CV charging stage.

Compared to the original CCCV charging mode, the improved process reduces the time for the initial charging process, and the rapidly increasing polarization voltage in the late charging causes the magnitude of the charging polarization voltage to drop, thereby reducing the voltage upper limit value of the battery in the late charging and leading to a reduction in the overcharging phenomenon of the voltage stress and current stress.

5.5.2.2 Comparison with the Performance of the Traditional CCCV Charging Mode

We extract the experimental data of the 90 Ah manganese oxide lithium battery in a CCCV charging life test with different charging rate, and compare with the life test with the improved CC and CV charging (1 C) in Table 5.5.

Table 5.5 shows that when the charging rate increases (from 0.3 to 1 C), the charging time in the traditional CCCV is greatly reduced, and the charging life is also greatly reduced (Figure 5.26). While the charging time for the improved-CCCV mode is 20 min (the basic

Table 5.5 Comparison of parameters for the charging performance of the unimproved and improved CCCV charging modes.

Charging mode	Maximum current (°C)	Charge capacity (%)	Charging time (h)	Maximum temperature rise (°C)
Unimproved CCCV(1°C)	1	100	1.3	7
Improved CCCV(1°C)	1	100	1.7	6

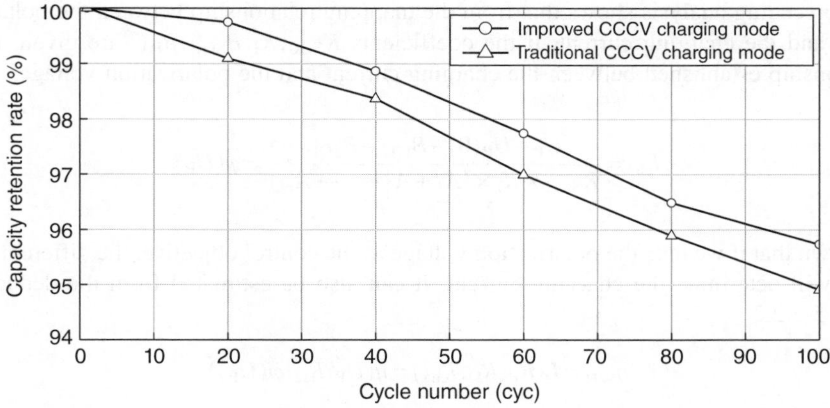

Figure 5.26 Life decay curves for traditional and improved CCCV charging modes.

pre-charging time) longer at 1 C, the charging life will be increased, boosting the charging life performance under the same conditions.

Therefore, the improved CCCV charging method can correct the charging current according to the actual capacity of the battery and increase the CC charging current appropriately, thus adding the pre-charging link to the pre-advancing into the CV charging. So the improvements can contribute to shortening the charging time and ensure the service life of the battery.

5.6 Principles and Methods of the Polarization Voltage Control Charging Method

5.6.1 Principles

The overpotential, including the internal resistance and the polarization overpotentials, which is generated during the charging process is the main factor affecting the rechargeable performance of the lithium battery. The internal resistance overpotential originates in the hindering effects on the current made by the internal battery and the connections, which present resistive properties. The polarization overpotential is directly reflected in the balance of the rates of the chemical reaction at the battery poles and in the differences in the solution concentration near the poles. Starting from the generation mechanism of the overpotential, the internal resistance overpotential is the resistive instantaneous feedback to the current density; while the polarization overpotential reflects the chemical reaction rate of the internal battery and the balance of the poles, it can be used to measure the quantitative performance of the charging efficiency and the charging acceptance force. In the charging process, the polarization voltage is the external manifestation of the effects on the battery exerted by the charging current, and is also the main influencing factor of the charging efficiency. Therefore, the polarization voltage can be considered as the indirect feedback to the chemical reaction rate from the electrochemical perspective.

The preceding analysis shows that from the mapping relationship between the polarization voltage and the charging current, if the coefficients $K_{SOC}, K_I, B_{SOH}, B_{P0^-}$ are given, there is a relationship established between the charging current and the polarization voltage.

$$I_{cha} = \frac{U_P - U_P(0) - B_{P0^-} - B_{SOH}}{K_{SOC} \times K_I \times (A_1 + A_2 + \cdots + A_N)} = \varphi(U_P) \tag{5.69}$$

It is shown that if we take the polarization voltage as the control objective, the different impact factors will determine the charging current. It can also be estimated from the deduction in Section 5.2.2.2

$$\eta_{WH} = L(R_P, R_\Omega, I_{cha}) = \eta(U_P, R_\Omega, \varphi(U_P)) \tag{5.70}$$

In terms of the charging characteristics, under the condition that the battery DC resistance is a constant value, the charging efficiency of the battery and the charging polarization voltage are closely related to each other monotonically. That is, the polarization voltage of the battery corresponds to the charging current, and it also determines the constant charging efficiency. Under the same charging conditions (such as aging, temperature), the higher the polarization voltage produced by the same current, the lower the charging efficiency, and the greater the amount of energy consumed by the batteries' side reactions, which means that the cell energy conversion efficiency under this current will be lower.

From an electrochemical perspective, by controlling the polarization voltage we can effectively control the chemical concentration of the internal battery and the positive and negative reaction rates in the charging process. From the electrical perspective, the expression based on the relationship between the charging current and the polarization voltage shows that controlling the charging current according to a certain polarization voltage can help to control the battery charging efficiency, to make the battery adaptive to changes in charging conditions such as SOC, aging and standing time, thus realizing the automatic optimization of the maximum charging process current under the current state of the battery. The constant polarization voltage controls the charging based on the above principles, which keeps the charging polarization voltage constant over the whole charging course and automatically adjusts the charging current to complete the charging process.

As shown in Figure 5.27, in the charging process, set the polarization voltage control value as a closed-loop input, and calculate the charging current value in the current state through the forward gain and output, then calculate the SOC value of the battery and the internal resistance voltage from the charging current of the battery. The actual value of the polarization voltage should be compared to the set value, further adjust the charging current to reach the dynamic equilibrium of the polarization voltage, and then realize the effective control of the polarization voltage in the process of charging.

Between the charging current and the polarization voltage there is a certain linear relationship in the charging process (resistance-capacitance coefficient), but due to the nonlinear distortion factor coefficients $(K_{SOC}, K_I, B_{SOH}, B_{P0^-})$ and the uncertainty of the initial polarization state ($U_P(0)$) which directly affect the estimation of the cell polarization voltage and the selection of the current, it is necessary to introducing a fuzzy algorithm for the purpose of controlling the current and realizing the online tracking of the polarization voltage.

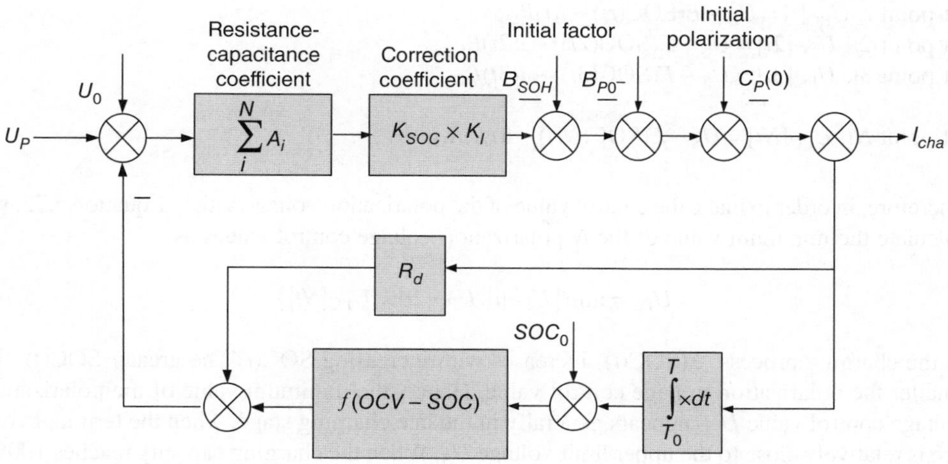

Figure 5.27 The polarization voltage control logic diagram.

5.6.2 *Implementation Methods*

The charging method based on polarization voltage control needs a certain strategy to be realized, and it can be specifically divided into the maximum voltage selection method and the maximum current selection method.

5.6.2.1 Maximum Voltage Selection Method

Under the limit of the maximum charging voltage, the relation between the charging capacity of the battery, the terminal voltage and the energy overpotential is:

$$E(\mathrm{SOC}(t)) + U_{\mathrm{OE}}(I_s, \mathrm{SOC}(t)) \le U_s \tag{5.71}$$

where U_s is the upper charging limit voltage, $E(\mathrm{SOC}(t))$ the battery potential related to the battery SOC which can be obtained through experiment, and the $U_{\mathrm{OE}}(I_s, \mathrm{SOC}(t)) = U_{\mathrm{PC}} + U_{\Omega}$.

To get the maximum charging capacity, the value of the charging polarization voltage control is:

$$E(\mathrm{SOC}(t)) + U_{\mathrm{PC}} + U_{\Omega} \le U_{\mathrm{S}} \tag{5.72}$$

Then we can get the calculation formulation:

$$U_{\mathrm{PC}} \le U_{\mathrm{S}} - E(\mathrm{SOC}(t)) - i \times R_{\Omega} \tag{5.73}$$

The whole charging time is Nt (t is the smallest unit of time in the charging process and N is the number of dividing segments for the charging time). Make the charging polarization voltage at various times reach the maximum value,

At point t, $U_{PC}[t] = U_S - E(SOC(t)) - i(t)R_\Omega$;
At point $2t$, $U_{PC}[2t] = U_S - E(SOC(2t)) - i(2t)R_\Omega$;
At point 3t, $U_{PC}[3t] = U_S - E(SOC(3t)) - i(3t)R_\Omega$;
……
At point Nt, $U_{PC}[Nt] = U_S - E(SOC(Nt)) - i(Nt)R_\Omega$

Therefore, in order to make the control value of the polarization voltage satisfy Equation 5.72, we calculate the minimum value of the N polarization voltage control values as:

$$U_{PC} = \min\{U_{PC}[t], U_{PC}[2t] \ldots U_{PC}[Nt]\} \tag{5.74}$$

In the charging process, $E(SOC(t))$ increases with increasing SOC(t).The greater SOC(t), the smaller the polarization voltage control value. Hence, the minimum value of the polarization voltage control value U_{PC} appears generally in the late charging stage, when the terminal voltage is relatively close to the upper limit voltage U_S. When the charging capacity reaches 100%, the polarization voltage of the battery satisfies:

$$U_{PC} \le U_S - E(100\%) \tag{5.75}$$

Thus the control value of the polarization voltage must satisfy:

$$0 < U_{PC} \le U_S - E(100\%) \tag{5.76}$$

Equation 5.76 shows the polarization voltage (U_{PC}) range selected by the maximum voltage selection method, which may further determine the actual control value of the polarization voltage by experiments within this range under the premise of ensuring that the charging capacity is a maximum.

The maximum voltage selection method is the constant polarization control strategy aiming at achieving the maximum charging capacity when the charging voltage is not overrun. With change in the battery charging state, the polarization voltage will change accordingly, which directly makes the charging characteristics unstable. For example, when battery aging occurs, the polarization characteristics deteriorate. When charging a battery at a fixed polarization voltage, the charging current is smaller, and the charging time will get longer.

The maximum voltage selection method can ensure the battery is fully charged if the battery voltage is within the limit, but the selected control value of the polarization voltage is experiential to a great extent and needs a timing correction with the aging of the battery. In other words, the actual operation is difficult.

5.6.2.2 Maximum Current Selection Method

In order to eliminate the effects on the polarization voltage of the initial standing time and the aging conditions, and to get steady charging curves, the maximum current selection method is presented.

The analysis of the polarization voltage gradient shows that with increasing SOC, the relationship between the values of the polarization voltage inflection point and SOC can be

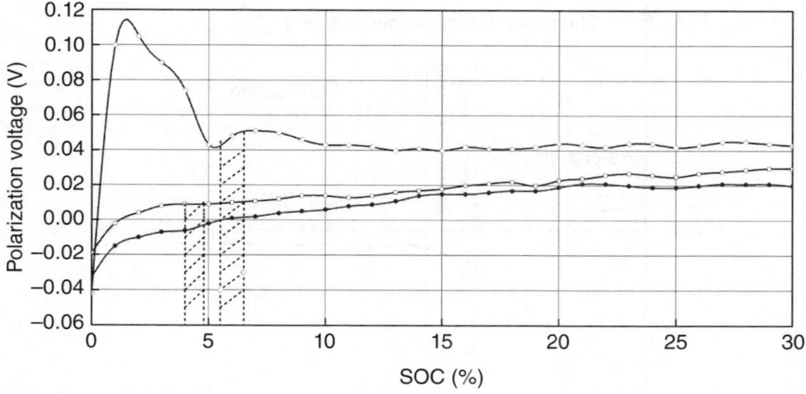

Figure 5.28 Basic characteristics of the initial charging polarization voltage.

represented by a quartic polynomial. For mid-range SOC, the lower the polarization voltage produced by the same current, the better the current accepting ability. The closer the SOC is to the front section or the end section of charging, the higher the polarization voltage produced by the same current, and the worse the current accepting ability. The battery's capability of accepting the current varies at different moments. In other words, in the battery charging process, the maximum charging current should act on the battery acceptance interval, when the polarization voltage reaches a minimum. The maximum current selection method is the method used to make the maximum charging current act on battery at suitable acceptance intervals for the purpose of controlling the polarization voltage. This method can automatically find the minimum polarization voltage at the maximum charging current according to the charging condition of the battery in the initial charging, and it has a strong self-adaptability which will promote a reasonable average current and reduce the charging time.

The changing gradient of the polarization voltage in the initial charging phase has the inflection point effect (as P1, P2 in Figure 5.28).

After completion of the initial pre-charging, we use the maximum allowable charging current to charge the battery. When the inflection point occurs in the battery, we make the minimum polarization voltage platform which is established by the maximum current calculated within area A_1 or A_2 as the constant polarization control value U_{PC}.

As shown in Figure 5.29, set the maximum charging current as I_{max} and the polarization voltage as U_P. Start from the inflection point P of the polarization voltage and take samples of the polarization voltage U_P at intervals t. Suppose that the sampling number of the polarization voltage is m in the period of A, then the average polarization voltage within {P~(Pmt)} is:

$$\overline{U}_P = \frac{\sum_{i=0}^{m} U_P(m \times N)}{m} \tag{5.77}$$

Set $U_{PC} = \overline{U}_P$, then use the U_{PC} as the polarization voltage control value for charging over the entire charging process.

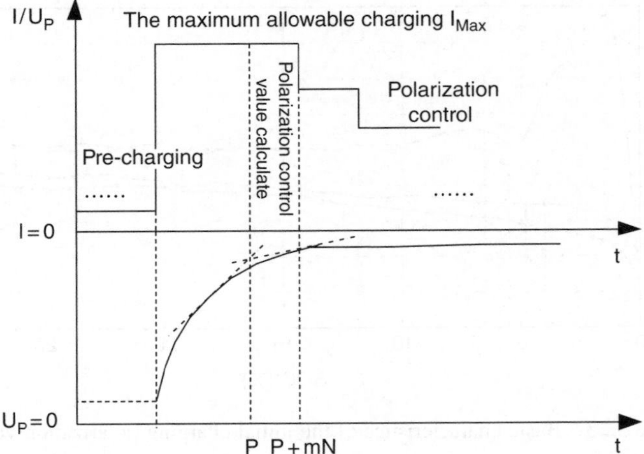

Figure 5.29 Diagram of the principle of the maximum current method to determine the polarization voltage control value. Reproduced with permission from Jiapeng Wen, "Studies of Lithium -ion Power Battery optimization charging theory for Pure Electric Vehicle," Beijing Jiaotong University, @2011.

From comparison of the two kinds of polarization voltage control values for the two selection methods, it can be inferred that the maximum voltage control method manages to take into consideration the initial standing time and the aging state which affects the charging current. However, the actual operation becomes very difficult, while the maximum current identification method can automatically select the polarization voltage control value in the current state at the maximum current of the battery, considering the influence from different SOC stages on the charging current in the charging process, but ignoring the initial state and the degree of aging.

5.6.3 Comparison of the Constant Polarization Charging Method and the Traditional Charging Method

In order to highlight the advantages of the constant polarization charging method (shorten as CVP method), the constant polarization control charging curve of 30 mV and the CCCV charging curve of 1/3 C are compared and shown in Figure 5.30.

5.6.3.1 Polarization Voltage Control

Due to the adoption of new batteries, the early charging of the CCCV polarization platform is not high (about 20 mV), but the polarization effect increases significantly due to the uncontrolled charging in the late stage. The polarization voltage rises rapidly to be twice as much as that in the mid-stage. In contrast, the CVP charging method is based on the fact that U_p, which keeps the polarization voltage within the lower range roughly constant, pulling down the constant polarization in the late charging stage.

Because the deep polarization of the battery will eventually lead to rapid decline in the capacity or cause damage to the battery, an excellent charging method must take the polarization control into account. Especially, for the battery in the late stage of charging (or discharging), the concentration of the internal charging has reached a very high level, the ion diffusion decreases and the polarization resistance increases. The accepted charging current can become much smaller, which restricts the large current charging. It needs special attention for the fast charging in the late charging of the battery to shorten the charging time of the battery and not to affect the battery life.

5.6.3.2 Charging Acceptance Characteristics

As Figure 5.30 shows, CCCV with 1/3 C constant current charging the battery to 95% SOC, begins to change after entering the constant phase and the current is gradually decreasing. On the contrary, the CVP method realizes the variable current charging over the entire process, the charging current is small in the early stage, large in the middle stage and small in the later stage. The timely changeable charging current in the middle stage is also conducive to the realization of fast charging.

The charging and discharging process of the lithium-ion battery is the deintercalation and intercalation of the lithium in the battery. When the charging and discharging current is greater than the ability of the lithium deintercalation and intercalation, it will cause side reactions, such as increase in the polarization voltage of the battery, lithium-ion deposition, loss of the recycled lithium ions, heating in the internal battery, electrolyte oxidation and decomposition. The capability of the lithium deintercalation and intercalation is limited during the charging and discharging process, and the limit is the maximum allowable charge–discharge current of the battery. However, the capacity of lithium deintercalation and intercalation is

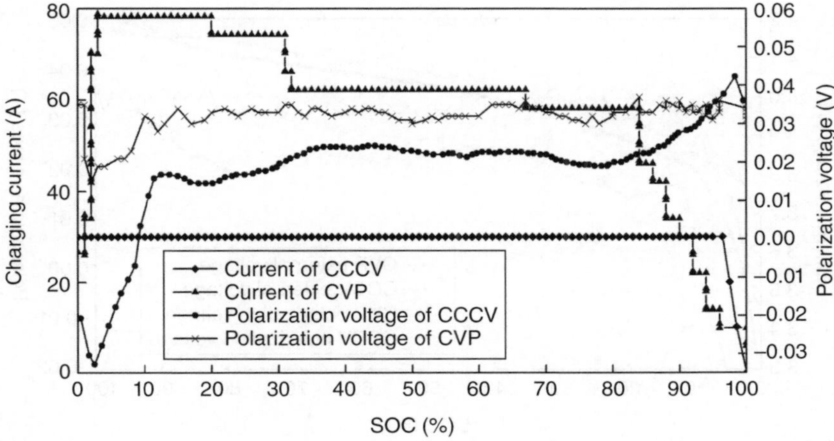

Figure 5.30 Comparison of the CCCV and CPV charging methods plotting charging current and polarization voltage versus SOC. Reproduce with permission from Jiapeng Wen, "Studies of Lithium -ion Power Battery optimization charging theory for Pure Electric Vehicle," Beijing Jiaotong University, @2011

different at different temperatures, different SOC and different degrees of aging. Therefore, the maximum allowable charge–discharge current of the battery is not constant. At the low-end and high-end SOC, the capacity of the battery to remove the lithium ion is weak, and the charging acceptance capacity of the battery is not strong; therefore, a small current charging should be used at this moment. Different from the old battery with low temperature and high SOC, the new battery with intermediate SOC at room temperature has a better charging acceptance ability; that is, the maximum allowable charging current of the battery is larger than the old one, and a higher current should be taken to charge the battery at this moment. In terms of the charging current acceptable characteristics of the battery, though the CCCV method meets the small current in the early and late stages, the small constant current in the mid-stage does not effectively follow the maximum allowable charging current, and the charging acceptance characteristics play an inadequate role in extending the charging time. The CVP method conforms to the charging acceptance characteristics, the charging current is small in the early stages, large in the middle and small in the later stages. In addition, the CVP method controls the polarization effects through the constant polarization, improves the charging acceptance of the battery, removes blockage of the reaction diffusion, achieves the maximum current under different SOC, and reaches the charging acceptance limit.

5.6.3.3 Control Point and the Upper Limit of the Cut-Off Voltage

The CCCV method is based on the external voltage control, and the upper limit of the external voltage frequently reaches 4.2 V before and after the end of the charging. In contrast, the CVP is based on the polarization voltage control, and the external voltage is only 4.17 V after charging. The contrast of external voltage and the polarization voltage is illustrated in Figure 5.31.

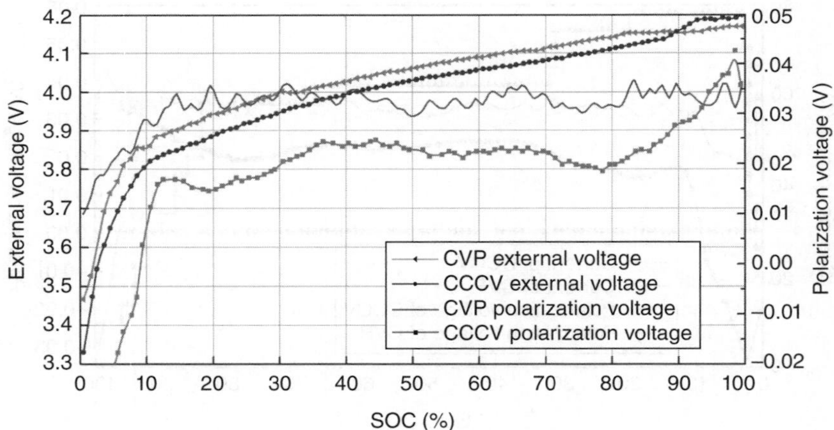

Figure 5.31 Comparison of the CCCV and CPV charging methods plotting external voltage and polarization voltage versus SOC. Reproduced with permission from Jiapeng Wen, "Studies of Lithium -ion Power Battery optimization charging theory for Pure Electric Vehicle," Beijing Jiaotong University, @2011

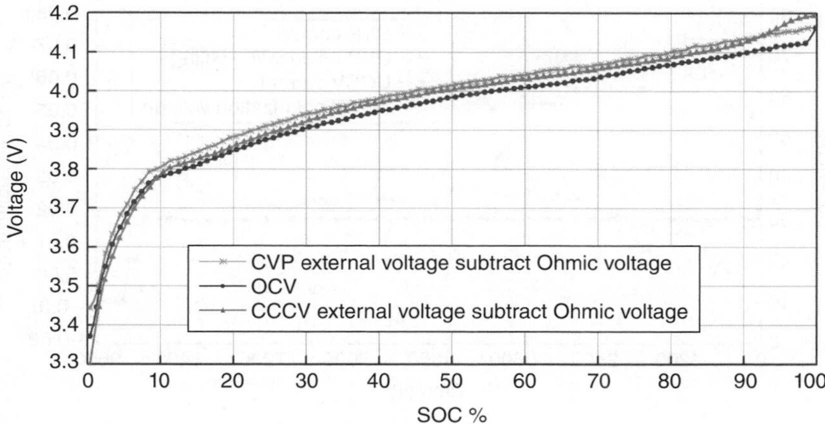

Figure 5.32 Comparisons of the SOC–OCV curve following characteristics for various charging modes.

In order to improve the capacity release of the battery in use, there is often a misunderstanding in the late stage of the CCCV charging, that is, that increasing the charging cut-off voltage can increase the charged power. Although the cut-off voltage of the battery is higher when the charging power is larger and the utilization of the single capacity is better, the battery life will decrease. Predecessors in this field carried out experiments in which, for the lithium-ion battery, when the battery voltage exceeds 4.2 V, the cycle capacity of the battery fades fast, however, the difference in the total charged capacity for the two modes is tiny [35]. From the viewpoint of the upper limit of the cut-off voltage, the control based on the external voltage in the constant voltage stage of the CCCV late stage is not reasonable, and the cut-off voltage of 4.2 V is frequently reached, leading to reduction in the battery life. The CVP method starting from the nature of the reaction mechanism, effectively curbs the late polarization control through the rising of the constant polarization voltage, thereby reducing the external battery voltage and ensuring that the battery life will be extended, as the cut-off voltage is less than 4.2 V at the end of the final charging.

The comparisons of the SOC–OCV curve following characteristics for various charging modes are shown in Figure 5.32. The curve at the bottom is the tested SOC–OCV curve, and the curves in the middle and at the top are the voltage curves when the ohmic voltages are removed from the battery terminal voltage for the CCCV mode and CVP mode, respectively. The difference between the voltage curve in CCCV charging mode and the tested SOC–OCV curve is enlarged as SOC increases, especially at the end of the SOC. The voltage curve in the CVP charging mode, however, follows the tested SOC–OCV curve well. The SOC–OCV curve characterizes the percentage of battery remaining capacity and the equilibrium potential, which represents the internal characteristics of the battery. Therefore, the better the following of the SOC–OCV curve in the charging process, the fewer the side reactions in the battery, and the longer the battery life. At the different stages, each charging method satisfies the formula: $U_{\text{external voltage}} - U_{\text{ohm voltage}} = U_{\text{P}} + U_{\text{OCV}}$, so as long as the U_{p} is controlled to be constant, the performance of the voltage curve following the SOC–OCV curve will be good. Therefore, from the perspective of the voltage curve following the open-circuit voltage characteristics, constant polarization charging has advantages over other ways of charging.

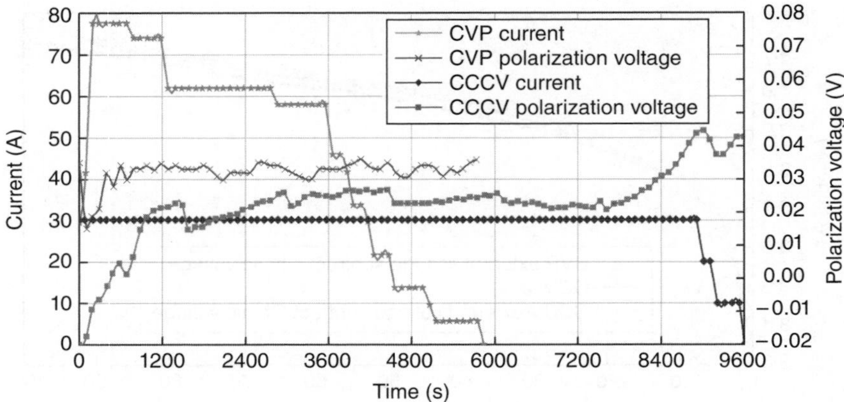

Figure 5.33 Comparison of the CCCV and CPV charging methods with respect to charging time.

5.6.3.4 Measurable External Characteristics

From Figure 5.33, the CCCV charging time is up to 3 h, but that of the CVP is only 1.6 h. Because the CVP has a greater charging current to charge at room temperature, the increase in temperature is slightly higher than with CCCV, so the different needs should be considered. When the actual capacity is 74.61 Ah, the CCCV capacity efficiency is 98.16% and the CVP capacity efficiency is 98.56%, so the cell charging capacity is increased by 0.29 Ah.

The polarization effect of the battery is divided into three processes, namely the initial establishment, the medium-term stability and late stage sudden increase. In the medium-term stable state, a linear relationship exists between the charging current and the polarization voltage. When a charging current is applied to the battery, the polarization voltage will change rapidly and become steady. This means that the current can be appropriately increased in the interim charging. The charging time can be shortened by increasing the current appropriately.

The traditional CCCV charging uses empirical control, and the current in the constant current stage is an empirical value. So as the battery ages, the rate characteristics of the battery are not good, which will generate hyper-polarization. Therefore, the current effect is not efficiently converted into chemical energy of the battery, but is consumed in the battery's heating and affects the charging efficiency of the battery seriously. The constant polarization control can reduce the energy loss to the minimum which is consumed by the polarization effect, and improve the charging efficiency of the battery.

5.6.3.5 Life Test

It was validated by 100 cycles of the battery (Figure 5.34) that the estimated life of the battery with the constant polarization charging method is the same as that by the 1/3 C rate CCCV charging method. The charging capacity is 0.4% more efficient with the constant polarization method than with the CCCV, and the charging time is also greatly reduced to 53% of the time with the CCCV. For the 1 C rate CCCV, the higher charging current platform helps

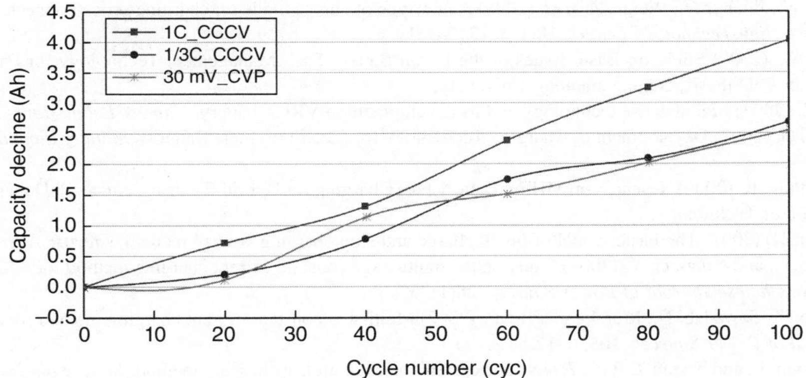

Figure 5.34 Comparison of the performance of CCCV and CPV in the100 cycle life test.

to realize fast charging, but the capacity fading rate for the constant polarization charging method is low compared to that with CCCV, leading to greatly improved battery life. Overall, under the premise of protecting or even improving the battery life, the constant polarization charging method realizes the best solution.

5.7 Summary

This chapter presents a systematic discussion on the lithium-ion rechargeable technologies, starting from the analysis of key indicators to measure the charging performance, and making further analysis of the charging characteristics of the lithium-ion battery. It focuses on the law of the charging polarization voltage, and in addition proposes improved CCCV charging method and the constant polarization charging method. The related data show that these charging methods can help to improve the performance of the batteries, establishing a foundation for further in-depth study of lithium-ion battery charging methods.

References

[1] Liying, W. (2009) The key technologies of electric vehicles and development prospects. *Auto Industry Research*, **8**, 12–15.

[2] Haibo, H., Han, W., and Yanping, Z. (2009) The research on key technologies of electric vehicle power systems. *Highways and Automotive Applications*, **5**, 4–6.

[3] Lin, L. (2008) Development of battery management design. *Electronic Test*, **12**, 7–11.

[4] Jiuchun, J. (2007) The general situation of battery management system and development trend. *Advanced Materials Industry*, **8**, 40–43.

[5] Japan Electronics and Information Technology Industries Association and Battery Association of Japan. *A Guide to the Safe Use of Secondary Lithium Ion Batteries in Notebook-type Personal Computers*. JEITA 2007.

[6] Yuejiao, D. and Wankui, Z. (2008) Power batteries for electric automobiles and their recharging techniques. *Journal of Hunan Institute of Science and Technology·Natural Sciences*, **21**(3), 59–61.

[7] Gang, W., Rong, Z., and Weigao, Q. (2009) The research on charging technologies for electric vehicle. *Agricultural Equipment and Vehicle Engineering.*, **6**, 7–9.

[8] Feng, W., Jiuchun, J., Weige, Z. *et al.* (2008) A new type of charge mode for charging series-connected batteries. *Chinese High Technology Letters*, **18**(12), 1310–1314.
[9] Feng, W. (2009) Study on Basic Issues of the Li-ion Battery Pack Management Technology for Pure Electric Vehicles.PhD thesis, Beijing Jiaotong University.
[10] Jun, Z. (2003) Maintenance technology and its development in VRLA battery. *Chinese Labat Man*, **4**, 183–186.
[11] Liang, G. (2002) Development of Charging Technology for Electric Vehicle Batteries. *Battery Bimonthly*, **32**(4), 245–246.
[12] Daozhong, H. (2006) Research on MH/NI Battery Fast Charging for Hybrid Electric Vehicle. PhD thesis, Beijing Institute of Technology.
[13] Dayou, H. (2001) The basic model of quick charge and stop charging control method. *Electric Age*, **7**, 16–17.
[14] Jingjin, C. and Ningmei, Y. (2004) Study on the multi-stage constant current charging method for VRLA battery system. *Chinese Journal of Power Sources*, **28**(1), 32–33.
[15] Ikeyaa, T., Sawadab, N., Murakami, J. *et al.* (2002) Multi-step constant-current charging method for an electric. *Journal of Power Sources*, **105**, 6–12.
[16] Qinglong, L. and Tixian, C. (1997) Variable current intermittent fast charging method. *Battery Bimonthly*, **27**(6), 266–281.
[17] Yichun, M. (2006) Application of slow pulse fast charge in capacity restoration of VRLA battery. *Battery Bimonthly*, **36**(2), 146–147.
[18] Rui, C.L. (2007) A design of an optimal battery pulse charge system by frequency-varied technique. *IEEE Transactions on Industrial Electronics*, **54**(1), 398–405.
[19] Cheung, T.K., Cheng, K.W.E., Chan, H.L., *et al.* (2006) Maintenance techniques for rechargeable battery using pulse charging. Power Electronics Systems and Applications Conference, 12–14 November 2006. Hong Kong.
[20] Kirchev, A., Mattera, F., Lemaire, E. *et al.* (2009) Studies of the pulse charge of lead-acid batteries for photovoltaic applications, Part I. Factors influencing the mechanism of the pulse charge of the positive plate. *Journal of Power Sources*, **191**, 82–90.
[21] Kirchev, A., Mattera, F., Lemaire, E. *et al.* (2009) Studies of the pulse charge of lead-acid batteries for photovoltaic applications, Part IV. Pulse charge of the negative plate. *Journal of Power Sources*, **191**, 82–90.
[22] Wang, J.B. and Chuang, C.Y. (2005) A multiphase battery charger with pulse charging scheme. *Industrial Electronics Society*, **1**, 1248–1253.
[23] Chengning Z., Jinghong Z., Wang Z. (2006) Quick charge system of lead acid battery for electric vehical. *Chinese Journal of Mechanical Engineering*, **42**(B05): 103–105, 110.
[24] Bei, L. and Liping, H. (2008) Study on decaying exponential of battery's charging current based on mass theory. *Journal of Hunan University (Natural Sciences).*, **35**(10), 26–30.
[25] Jingzhao, L. (2003). The optimal charging lead-acid battery technology by using the neural network prediction and variable structure fuzzy control. Dissertation. Hefei University of Technology.
[26] Liu Danwei, Y. and Haizhou, H.S. (2001) Reasearch of intelligent charge technique with fuzzy control. *Electrotechnical Journal*, **12**, 1–3.
[27] Lyn, C.E., Rahim, N.A., and Mekhilef, S. (2002) DSP-based fuzzy logic controller for a battery charger. Proceedings of IEEE TENCON, 28–31 October 2002, Beijing, China 1512–1515.
[28] Liu, Y., Teng, J., and Lin, Y. (2005) Search for an optimal rapid charging pattern for lithium–ion batteries using ant colony system algorithm. *IEEE Transactions on Industrial Electronics*, **52**(5), 1328–1336.
[29] Yugang, S., Weijiong, D., Qiang, C. *et al.* (2008) Rapid and intelligent charging technology for series lithium-ion battery. *Journal of Chongqing Institute of Technology*, **22**(1), 89–93.
[30] Hsieh, G., Chen, L., and Huang, K. (2001) Fuzzy-controlled Li–ion battery charge system with active state-of-charge controller. *IEEE Transactions on Industrial Electronics*, **48**(3), 585–593.
[31] Lyn, C., Rahim, N.A., and Mekhilef, S. (2002) DSP-based fuzzy logic controller for a battery charger. Proceedings of IEEE TENCON, 28–31 October2002, Beijing, China, 1512–1515.
[32] Hasanien, H.M., Abd-Rabou, A.S., and Sakr, S.M. (2010) Design optimization of transverse flux linear motor for weight reduction and performance improvement using response surface methodology and genetic algorithms. *IEEE Transactions on Energy Conversion*, **25**(3), 598–605.
[33] Wang, L. and Singh, C. (2009) Multicriteria design of hybrid power generation systems based on a modified particle swarm optimization algorithm. *IEEE Transactions on Energy Conversion*, **24**(1), 163–172.
[34] Chen, L., Chaoming Hsu, R., and Liu, C.-S. (2008) A design of a Grey-predicted Li-ion battery charge system. *IEEE Transactions on Industrial Electronics*, **55**(10), 3692–3701.
[35] Barsoukov E., Kim J., Yoon C., Lee H., (1999) Universal battery parameterization to yield a non-linear equivalent circuit valid for battery simulation at arbitrary load, *Journal of Power Source*, **83**(1/2), pp 61–70.

6

Evaluation and Equalization of Battery Consistency

The evaluation and equalization of battery consistency is the core of the battery group application technology, directly affecting the batteries' safety and efficiency. There is much literature that discusses battery consistency evaluation methods, and a variety of equalization circuits have been proposed. All are based on the use of the voltage difference of the batteries to evaluate the battery consistency and implement the equalization controls, but the previous evaluation methods cannot effectively detect inherent differences in the batteries, which may cause degradation of equilibrium effects. Therefore, it is very difficult to control such factors as the capacity, volume, cost, and heat dissipation of the battery online equalizers. Offline equalization increases the maintenance workload of the battery and eventually hinders the large-scale popularization and application of electric vehicles. Based on the battery modeling, parameter identification and state estimation mentioned above, we will discuss the mechanism leading to battery inconsistency and its influence on the battery performance, consistency evaluation index selection, consistency quantitative evaluation methods, and equilibrium problems, then propose a more reasonable equalization control strategy which can effectively improve the utilization efficiency of the equalizer, reduce the equalizer's design capacity, provide a theoretical basis and data support for battery efficient equilibrium, and make it possible to apply the online equalization technology to large capacity batteries for pure electric vehicles (PEVs).

6.1 Analysis of Battery Consistency

No matter what type of battery is considered, a single battery's voltage and capacity are unable to meet the demands of an electric vehicle. In order to achieve a certain level of voltage, power and energy, groups of batteries assembled in series or parallel are needed to provide energy for electric vehicles. However, even the battery group has been rigorously screened in practice, because in use its capacity utilization, safety, service life, and other aspects of its performance are much less than a single battery due to battery inconsistency. Battery inconsistency is the weakest link in the application of the battery group and has great impact on battery life and

Fundamentals and Applications of Lithium-ion Batteries in Electric Drive Vehicles, First Edition.
Jiuchun Jiang and Caiping Zhang.

reliability. This section starts with the actual application characteristics of PEVs and analyzes and discusses the reasons behind battery inconsistency and its influence on the batteries' performance.

6.1.1 Causes of Batteries Inconsistency

Battery inconsistency refers to the inconsistency among the cells within the battery group, which were of the same specification and same type. The performance differences exist in group application, and are mainly reflected by the inconsistency of the cell performance parameters, such as the battery voltage, polarization voltage, DC resistance, capacity, state of charge (SOC), and so on. The consistency problem is the main cause of performance degradation in the application of battery groups.

Battery inconsistency is caused by two factors: inconsistency in the initial state of the battery and inconsistency in battery performance degradation speed. Because it is impossible to keep complete consistency among the cells within the battery group in the process of producing and using them, battery consistency is relative, and inconsistency is absolute [1].

6.1.1.1 Inconsistency in the Production Process

In the production process, such factors as the techniques, coating, ingredients, and unevenness of the impure contents of the battery will cause differences in the battery initial performance (initial capacity, DC resistance, charging efficiency, self-discharge, etc.). To date, batteries, especially large capacity traction batteries, have not yet been fully produced on an automatic assembly line, so the consistency of their initial performance cannot be effectively guaranteed during the manufacturing process. When the requirements of the battery consistency are too high, the cell yield will drop greatly, which makes the price of the batteries rise significantly, thus pushing up the cost of the electrical equipment.

6.1.1.2 Inconsistency in Use

Differences in temperature, operation conditions, self-discharge degree and the charging and discharging process of each battery cause differences in the self-discharge rate and rate of internal side reactions of each battery. The main causes of inconsistency in use are:

1. Differences in the battery initial performance will be gradually reflected in the actual capacity, internal resistance and SOC. In the manufacturing process, the differences in internal impurity content, resistance and charging efficiency can be reflected in the battery self-discharge characteristic, the SOC, internal heating, and speed of heat degradation. Taking the charging efficiency as an example, when the initial capacities of two batteries are 100 Ah, charging efficiency 99.9 and 99.95%, respectively, after 100 cycles, the capacities are: $100 \times (99.9\,\%)^{100} = 90$ Ah and $100 \times (99.95\,\%)^{100} = 95$ Ah, respectively; after 200 cycles, they are 81 and 90 Ah; after 300 cycles, 74 and 86 Ah. Suffice to say, even a very small difference in charging efficiency (0.05%) may cause a tremendous capacity difference after a certain number of charging cycles.

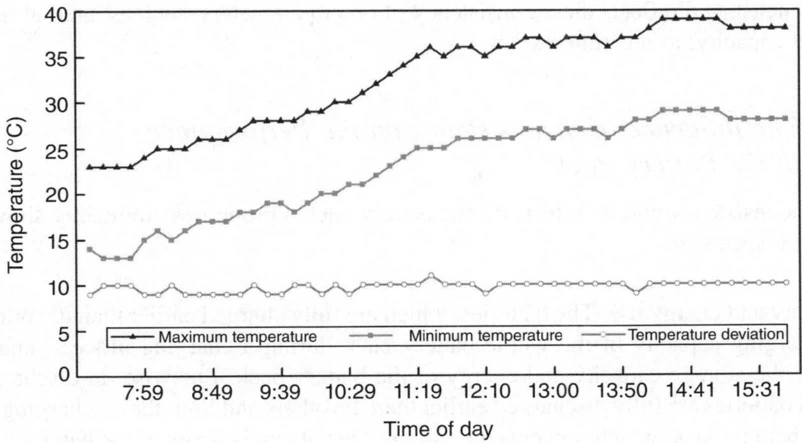

Figure 6.1 Difference between maximum temperature and minimum temperature of a vehicle battery.

2. When the PEV battery is used in a narrow space, the effectiveness of the battery thermal distribution and thermal management will directly affect the battery temperature. Figure 6.1 shows the variation of temperature, at the maximum and minimum temperature sampling points obtained from the PEV's battery pack during operation over a day. We can see that the difference between the maximum and minimum temperature is around 10 °C, which will ultimately lead to decline in cell performance at different rates, and eventually will increase battery inconsistency [2].
3. The battery that has small capacity, large internal resistance and a high self-discharge rate is easier to become over-charged, over-discharged, and to over-heat in the process of charging and discharging. This accelerates the decline in battery performance, demonstrating "positive feedback", which amplifies the battery consistency problem quickly. If the initial capacity is different and the current capacity of the cell with small capacity is poor, then when the charging and discharging current is the same, the deterioration rate of the cell with small capacity will be greater than that of the cell with large capacity, and this will lead to a vicious spiral and enlarge battery inconsistency.
4. Due to the accumulation of battery initial performance differences and the difference in the battery's working environment, the differences in the cell polarization phenomenon and polarization depth are gradually enlarged. Polarization overpotential reflects the reaction rate and conditions of the internal battery chemistry to some extent. On the one hand, the manufacturing process of the battery cannot ensure that the internal battery active material and ingredients are exactly the same, which leads to differences in the internal chemical reaction inside the battery in use. On the other hand, with the accumulation and amplification of the initial performance differences in use, there are also differences in the battery's internal chemical reactions at different aging stages, and these differences also trigger differences in the battery voltage.

In conclusion, the inconsistent parameters of the cells within the battery pack are not independent of each other. The battery voltage inconsistency is the most direct manifestation of all the inconsistencies, and reflects the inconsistency of other parameters, such as internal resistance, SOC, and capacity, to a certain extent.

6.1.2 The Influence of Inconsistency on the Performance of the Battery Pack

Battery inconsistency and its effects on the battery pack's properties are mainly shown in the following respects:

1. Capacity and energy use. The batteries, which are fully charged earlier than the others, limit the charging capacity of the whole battery pack during a charging process, and thereby reduce the storage capacity and energy of the battery pack. Likewise, in discharging, the partial batteries are fully discharged earlier than the others and limit the discharging capacity of the battery pack, which prevents the storage capacity and energy of the battery pack from being fully utilized. Thus, it is very difficult to balance safety and the efficiency of the battery pack in use due to the battery pack inconsistency. From the above analysis, in the charging and discharging process, when the first fully charged or discharged battery is not the same one, the maximum usable capacity of the battery pack may be smaller than that of the minimum charged or discharged cell in the group, which makes the effect of inconsistency more prominent.
2. Power output. When the battery is approaching the state of being fully charged or fully discharged, the maximum allowable current of charging and discharging will decrease. When different batteries are series linked in the battery group, in the process of charging, the battery with high SOC will limit the charging current of the whole battery group, so that the charging time will be increased. Similarly in the discharging process, the battery with low SOC will limit the power output capability of the whole battery group, and reduce the control of dynamic performance. In addition, when the battery is close to the state of being fully charged or discharged, its polarization voltage increases obviously, and the energy utilization efficiency decreases.
3. Estimation of battery state. The differences in capacity of the cells and the SOC make it hard to estimate the series battery state and the optimized energy use. We know from the above discussion, when the maximum usable capacity and SOC are inconsistent, the degree of difficulty in estimating the SOC and SOE of the battery group and the computational complexity will increase significantly. In order to meet the requirements of actual operation and on-line estimation, recognition of the battery status is often simplified, even the whole battery pack will be dealt with as a "large battery" for state recognition, which greatly increases the probability of abuse and inappropriate use of the battery.

In conclusion, compared with single cells, battery inconsistency makes it more complex to put them into practical use after being connected in series circuits. The simple way used for single cells, which is based on the state estimation of terminal voltage and the charge and discharge control mode, cannot effectively ensure the safety of the battery pack in use; and, when applying the method of voltage control for single cells to the voltage control of a battery pack, the battery capacity will not be utilized effectively and its high efficiency cannot be effectively guaranteed in use. So, setting up an evaluation system to test battery consistency and balancing battery groups according to this evaluation system can effectively increase the battery pack's capacity and energy use efficiency, and ensure the efficiency of the whole battery pack.

6.2 Evaluation Indexes of Battery Consistency

There are five common ways to construct battery groups: series, parallel, series after parallel, parallel after series, and mixing connection. The different types of battery groups have significant influence on the battery consistency. First, when the batteries are connected in parallel within the group, the battery voltage is consistent, but it is difficult to assess the charge and discharge properties and the aging mechanism of each parallel branch. Based on the analysis of the characteristics of batteries in parallel, this section examines how connection in parallel influences the battery's charging and discharging performance, the battery capacity, and the rate of energy utilization. We can evaluate the suitability of different ways of battery connection by estimating the degree of imbalance of parallel branch currents. This provides the basis for reasonable capacity design and improvement of the performance of the battery system. Secondly, when the batteries are connected in series, the method of evaluating the battery consistency and implementing the equalization control based on the batteries voltage differences cannot effectively indicate the batteries' internal differences, which cause degradation of the equilibrium effect. As far as the series battery pack is concerned, this section analyzes how the differences in internal resistance, polarization, capacity, and SOC influence the battery voltage, and attempts to find the internal reasons causing the differences in battery voltage. Based on the analysis of series parallel group characteristics, the comprehensive evaluation indexes of the battery consistency are proposed, using five parameters: the degree of current imbalance, internal resistance, polarization, capacity, and SOC.

6.2.1 The Natural Parameters Influencing Parallel Connected Battery Characteristics

Based on the analysis of the causes of battery inconsistency, this section researches the parallel characteristics, and analyzes the common mechanism of the action of internal resistance, capacity and other differences in the $LiFePO_4$ battery in a parallel circuit.

6.2.1.1 $LiFePO_4$ Cell Characteristics

When the $LiFePO_4$ battery is connected in parallel, the main factors to affect the charging and discharging current are the battery actual capacity Q, the ohm internal resistance R, the internal voltage U_{OCV}, the polarization voltage U_p. The difference between parallel battery voltage U_O and U_{OCV} is expressed as U_Δ, which reflects the difference in the cell voltage and changes with the changes in ohm voltage drop U_R and the polarization voltage U_p, which are induced by the parallel branches of unbalanced current.

Figure 6.2 shows, for C/3 current, the polarization voltage change during charging of the $LiFePO_4$ cell with different initial SOC. It can be seen that the polarization at either end (low or high initial SOC) is serious, and the rate of change of the polarizing voltage is greater at the end of charging, so the parallel battery U_Δ also changes much more at this point. Figure 6.3 shows

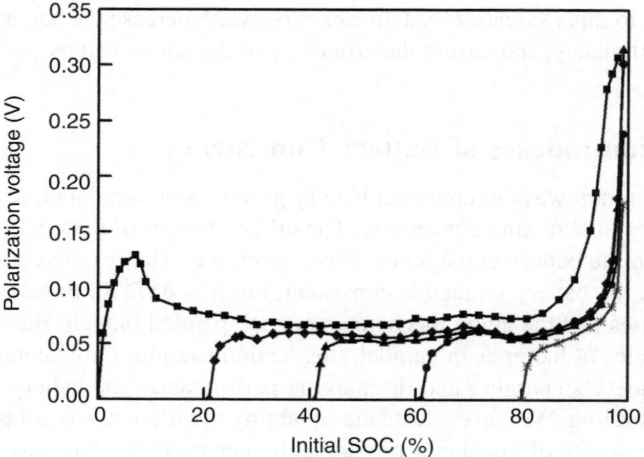

Figure 6.2 The polarization voltage of the $LiFePO_4$ battery with different initial SOC.

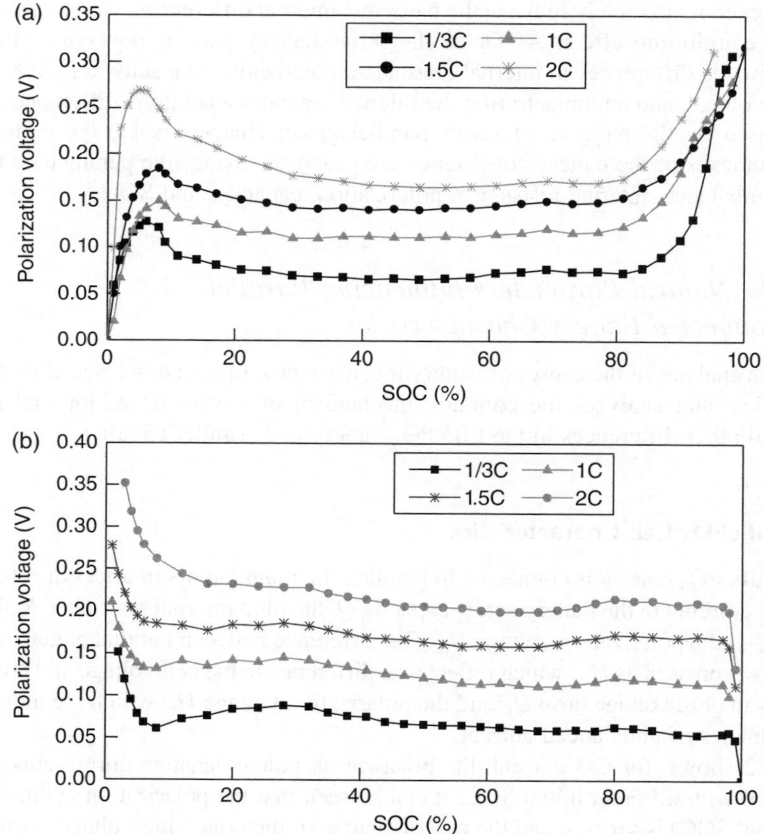

Figure 6.3 Polarization voltage of the $LiFePO_4$ battery under different charge and discharge rates: (a) charging process and (b) discharging process.

the polarizing voltage change of a $LiFePO_4$ cell with different charging and discharging rates. The charging and discharging currents have great influence on the polarization voltage, therefore the parallel unbalanced current causes a change in U_R. At the same time, it causes change in U_p, different polarization rates among parallel batteries affect the unbalanced current in turn, so that the dynamic changes of U_Δ make the parallel battery unbalanced current more complex.

6.2.1.2 $LiFePO_4$ Batteries in Parallel

Figure 6.4 shows the changes in the charging–discharging unbalanced current of four parallel connected $LiFePO_4$ batteries at a charge rate of 1 C, the capacity of the batteries is 99.8 Ah, and in the charging and discharging process it is 95 Ah. For the whole battery pack, the capacity utilization rate of batteries in parallel is higher than that of batteries in series.

The unbalanced current at the end of charging is notable, and the parameters of battery 3, capacity and internal resistance, are only slightly better than others, but there is still a large unbalanced current. The differences between the polarization voltage U_p of the parallel battery

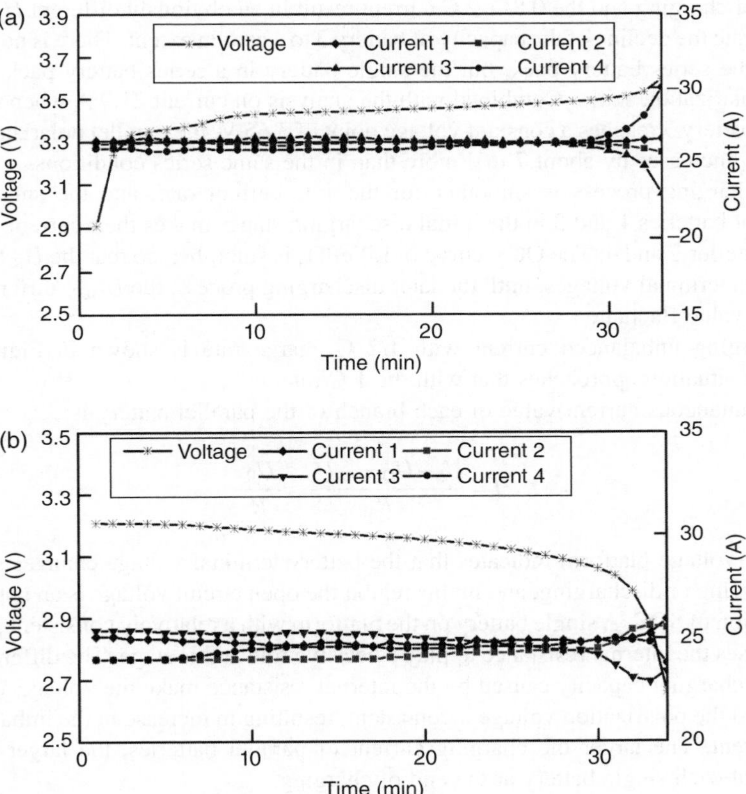

Figure 6.4 Unbalanced current of a $LiFePO_4$ parallel-connected battery with a 1 C charge rate: (a) charging process and (b) discharging process.

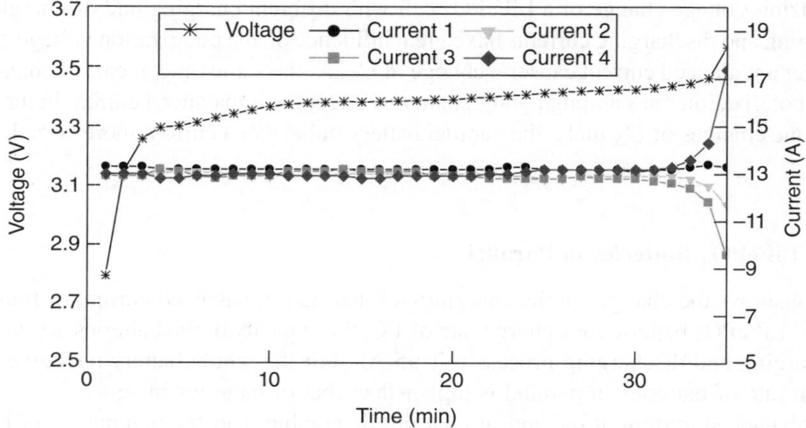

Figure 6.5 The unbalanced charging current of a $LiFePO_4$ parallel-connected battery with a charge rate of C/2.

at the end of charging and the 0.8–1.2 C current result in an obviously different U_Δ value, and also accelerate the decline of the capacity of battery 3 to a certain extent. There is no unbalanced current in the series battery pack, but the single battery in a series battery pack also adopts different polarization states. Combined with the analysis on current 21.9 A when the terminal voltage of battery 3 reaches a constant voltage point of 3.65 V, the parallel polarization voltage of battery 3 increases by about 7 mV more than in the same series conditions.

The discharging process is smoother for the 1 C current rate, and the smaller internal resistance of batteries 1 and 3 in the initial discharging stages makes their current greater than those of batteries 2 and 4. The OCV curve of $LiFePO_4$ is smoother, so that the U_R balances the difference in terminal voltages, until the later discharging process, the U_{OCV} difference makes the current value fluctuate.

The charging unbalanced current with 1/2 C charge rate is shown in Figure 6.5, the discharging situation approaches that with the 1 C rate.

The instantaneous current value of each branch of the parallel battery is:

$$I = \frac{U_o - U_{OCV} - U_p}{R} = \frac{U_R}{R} \tag{6.1}$$

The battery voltage platform indicates that the battery terminal voltage changes slowly with battery charging or discharging, and in this region the open circuit voltage is an approximately linear function of SOC. A single battery on the platform with a relatively consistent polarization voltage causes the internal resistance to play a decisive role at this stage. The differences in the cumulative charging capacity caused by the internal resistance make the voltage at the end of charging and the polarization voltage inconsistent, resulting in increase in the imbalance of the branch current. The larger the charging current of parallel batteries, the larger the current imbalance of each single battery at the end of charging.

The charging current imbalance at a charge rate of 1.5 C is shown in Figure 6.6. The terminal current differences of batteries 3 and 4 do not increase significantly due to the function of

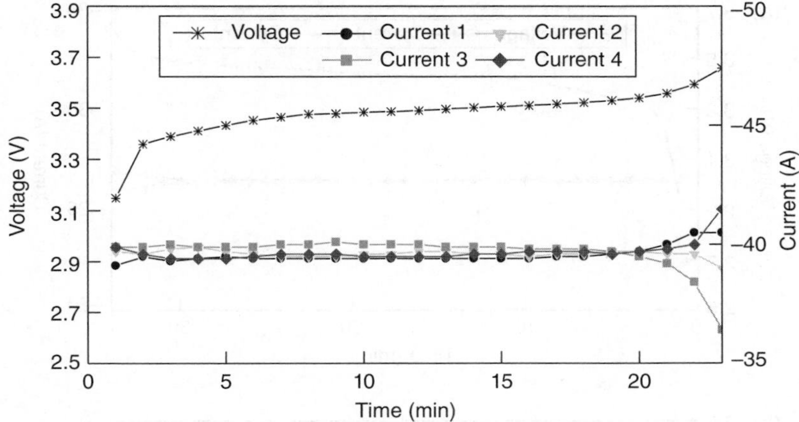

Figure 6.6 The unbalanced current of a $LiFePO_4$ parallel-connected battery with a charge rate of 1.5 C.

batteries 1 and 2, and the spontaneous balanced characteristic of parallel batteries contributes to the consistency of the parallel branch current.

Now, we reduce the number of parallel batteries and connect the batteries (a) 1-2, (b) 1-4, and (c)2-3 in parallel, the unbalanced charging current with a charging rate of 1 C is shown in Figure 6.7. The imbalance of (a) and (c) is smaller than that of (b); because the parallel batteries have different internal resistances and different capacities at the same time, the capacity difference causes the resistance difference in the charging process. Since the current difference for (b) at the end of charging is bigger, and the imbalance is greater than seen in Figure 6.4a, it can be concluded that the increased number of parallel batteries contributes to a decrease in the unbalanced current.

For batteries 1 and 4, shown in Figure 6.7b, the capacity and internal resistance differences lead to the result that the terminal current of battery 1 decreases to 20 A and that of battery 4 increases to 32.7 A. The current increase of 4 shows that its SOC is low at the intersection of the curves, U_{Δ} is large, and it is necessary to balance the SOC difference with battery 1 through current difference. In the battery management system, the actual SOC values of batteries1 and 4 at the intersection are 89 and 87%, respectively. As the capacities of 1 and 4 are basically similar, by the ampere-hour counting principle and combining Figure 6.7b, we can conclude that the area between the current curve 1 and the time axis should be consistent with that of current curve 4, that is, the current differences after the intersection arise from the capacity difference accumulated before the intersection. The actual data indicate that for two parallel $LiFePO_4$ batteries, even a little difference in internal resistance can lead to a significant current imbalance at the end of the whole charging process.

The discharging process of the batteries 1-4 is shown in Figure 6.8. By comparison and analysis, it can be found that the main cause of the increase in unbalanced current in the discharging process is the difference in both internal resistance and internal voltage U_{OCV} of 1 and 4 at the initial discharging time. We can see from Table 6.1 that the internal resistance of battery 1 is smaller. According to the battery management system, the actual SOC values of 1 and 4 are 89 and 87%, respectively, the current of 1 in the early stage of discharging is 28.6 A, which is significantly greater than the 23.2 A of 4. Current alternation will occur in the later stages.

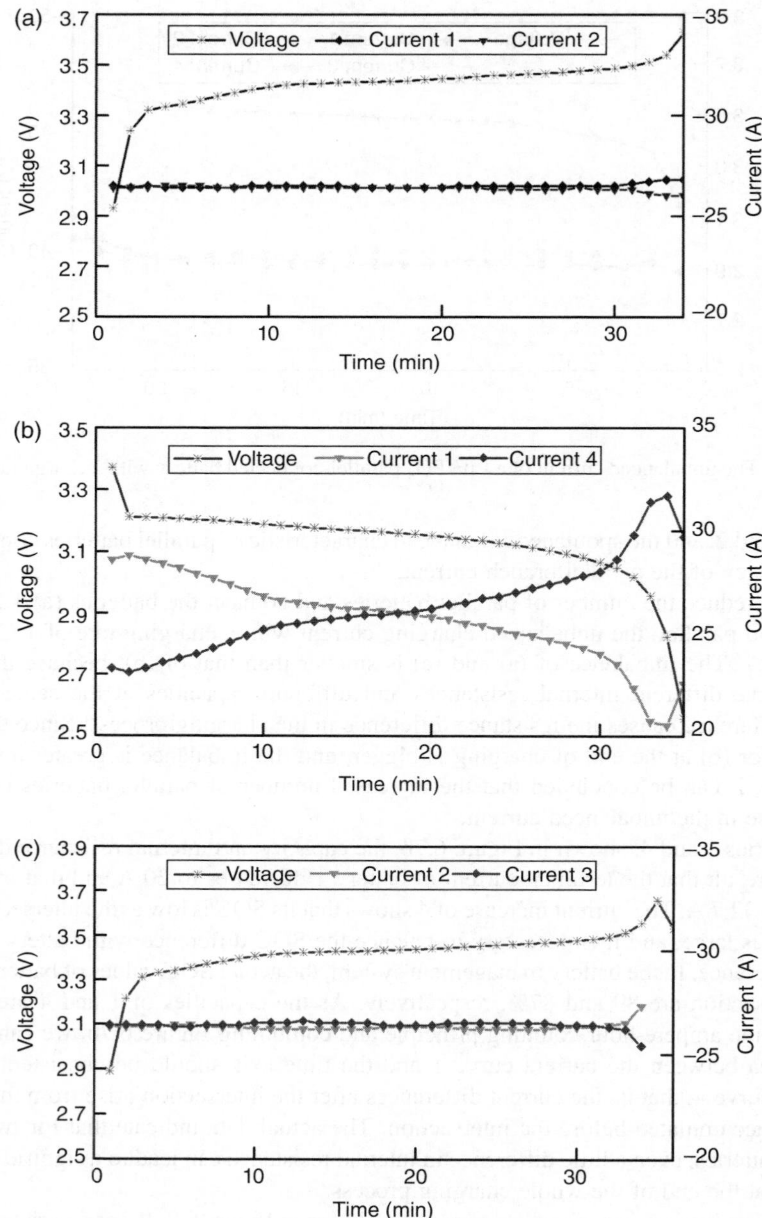

Figure 6.7 The unbalanced charging current of a $LiFePO_4$ parallel-connected battery with a charge rate of 1 C. Batteries connected in parallel: (a) 1 and 2, (b) 1and 4, (c) 2 and 3.

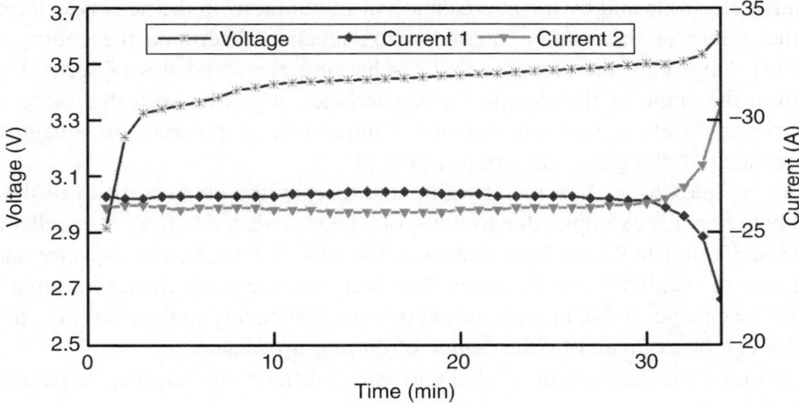

Figure 6.8 The unbalanced charging current of batteries 1 and 4 connected in parallel with a charge rate of 1 C.

Table 6.1 DC resistance of the batteries in the pack.

	Battery 1	Battery 2	Battery 3	Battery 4	Battery 5	Battery 6	Battery 7	Battery 8
$I = 0$ A, voltage (V)	3.536	3.607	3.776	3.765	3.727	3.681	3.671	3.774
$I = 60$A, voltage (V)	3.505	3.572	3.741	3.735	3.694	3.657	3.647	3.744
DC resistance (mΩ)	0.517	0.583	0.583	0.5	0.55	0.4	0.4	0.5

The obvious differences between the ends of the $LiFePO_4$ battery's SOC–OCV curve and the platform increase the unbalanced current after the parallel connection. The actual data indicate that for two parallel $LiFePO_4$ batteries in the charging process, even a little SOC variance can lead to a significant current imbalance at the end of charging.

The test results show that the charging and discharging characteristics of $LiFePO_4$ batteries in parallel include: the battery parameter differences at the initial moment of the charging process will be enlarged at the end of the charging process, and similarly for the discharging process. From comparison of Figures 6.4–6.6, it can be seen that improving the number of batteries in parallel can effectively reduce the initial current differences in the charging and discharging process, and reduce the current differences in the process of charging and discharging.

6.2.1.3 Analyses of Degree of Imbalance

By analysis of the internal electrochemical reactions, the capacity of charging and discharging corresponds to the intercalation of lithium ions in the negative electrode. The $LiFePO_4$ battery voltage platform is formed by the common action of the $FePO_4$–$LiFePO_4$ phase transformation in the positive electrode and by the intercalation of lithium ions in the negative electrode, and the rate and extent of the battery internal redox reaction determine the battery electrical properties [3]. Observing the charging and discharging characteristics of series lithium-ion batteries from the angle of the electrical characteristics, it can be seen that the consistency of the batteries is closely related to the battery's ohm resistance, polarization voltage, capacity, temperature, state of charging, and aging degree [4].

It is more complex to analyze the charging and discharging characteristics of the parallel-connected batteries; for example, due to the spontaneous balanced effect of parallel batteries, the change in U_p and SOC are both related to the branch current, the capacity and energy utilization rate of parallel batteries comprehensively change with change in environmental conditions. The changes in the branch current of a parallel battery in the charging and discharging process can be evaluated by the degree of current imbalance.

Battery parameter differences in parallel will lead to different imbalance, the parallel battery SOC_{par} is expressed as:

$$SOC_{par} = \sum^{n} Q_i \cdot SOC(i) \Big/ \sum^{n} Q_i \tag{6.2}$$

where n is the number of parallel batteries, the SOC(n) difference is formed by the initial SOC differences and the extra formed in the charging and discharging process. The test results show that when the parallel $LiFePO_4$ battery starts to discharge at SOC_{par} =100% or to charge at SOC_{par} =0%, serious polarization voltage differences make the differences in initial SOC(n) relatively larger than those of $LiMn_2O_4$ batteries, and the extra SOC(n) differences caused by the battery internal resistance difference are more obvious. SOC(n) differences also reflect the differences in U_Δ, and more differences will extend the constant voltage stage time at the end of charging and shorten the capacity utilization rate at the end of high rate discharging. Therefore, it is necessary to judge how to control the unbalanced current of each branch of $LiFePO_4$ batteries connected in a parallel circuit, based on the difference D(SOC) of SOC(n), which is expressed as:

$$D(SOC) = \sqrt{\sum_{i=1}^{n} \left(SOC(i) - SOC_{par}\right)^2 \Big/ (n-1)} \tag{6.3}$$

The imbalance reflects the charging and discharging characteristics of parallel batteries, which contribute to the evaluation of the influence of parallel unbalanced current on the parallel battery performance, and help to select the most suitable battery pack types and optimize the control strategy of battery charging and discharging, in order to control the imbalance of parallel batteries, which will effectively guarantee consistency in the parallel battery characteristics and cell characteristics.

Based on the experimental studies of cell charging and discharging characteristics of mass-produced lithium iron phosphate batteries (nominal capacity 25 Ah, working voltage 2.50–3.65 V), and the parallel battery characteristics in different parallel quantity and at different charge and discharge rate, this section analyzes the influence of the parallel batteries on the internal resistance, SOC, polarization, as well as on other battery parameters and the branch current. The results show that: (i) due to the spontaneous balanced effect of parallel batteries, more batteries connected in parallel can effectively reduce the initial current differences and the current difference in the charging and discharging process; (ii) less internal resistance difference and initial SOC differences of a $LiFePO_4$ battery can lead to significant unbalanced current in the later stages. From analysis of the experimental results, the degree of imbalance of the SOC of the parallel battery can be used to evaluate the battery pack mode, and provides a basis for reasonable capacity design of the battery system and improvement of the battery group performance.

6.2.2 *Parameters Influencing the Battery External Voltage*

From the above discussion, we can see that there is a close relationship between the battery external voltage and the voltage drop caused by ohm resistance, the polarization voltage, the maximum available capacity, and the SOC. This section will analyze its effect on U_O.

6.2.2.1 The Influence of Voltage Drop Caused by Ohm Resistance on the Battery External Voltage

From the formula $U_R = IR$, we can see that U_R is directly related to the battery working current and the internal DC resistance.

From the above related test and discussion, we know that the DC resistance is closely related to battery aging and environmental temperature, but has no relation to SOC. In a charging and discharging process under the same conditions, the influence of battery aging on the internal DC resistance can be ignored. In the process of charging, except for a very large variance in temperature, especially low temperature, the influence of temperature on battery DC internal resistance can also be ignored. It should be noted that there may be differences in the DC resistance of different batteries, so although each cell current within the serial battery pack is exactly the same, as a result of DC resistance difference, there will be a voltage difference in the battery charging and discharging process; and the larger the current the greater the difference. Theoretically speaking, in the process of charging, the difference between two kinds of batteries with different DC internal resistance is shown in Figure 6.9a.

While the PEV real vehicle test data also prove this point, in the charging and discharging process, the voltage change curve of three series batteries can be seen in Figures 6.10 and 6.11, respectively.

Before charging, the voltage of each of the three batteries is 3.91 V. After charging, the terminal voltage differences appear with current appearance, battery 3 voltage starts to be higher than that of the others (d (U_3-U_1) is the voltage difference between batteries 3 and 1, d (U_3-U_2) that between batteries 3 and 2). Throughout the charging process, the voltage difference remains unchanged. When the current disappears at the end of charging, the voltage of each battery is again the same, 4.1 V. Apparently, the voltage difference caused by the battery DC resistance appears and disappears with current appearance and disappearance.

During the operation of a vehicle, although the working current is less stable than that in the charging process, because the battery is connected into a series battery pack the current flowing through each cell is exactly the same. When the current is greater, the terminal voltage differences caused by the DC internal resistance will be more obvious, and the voltage fluctuation of battery 3 is significantly greater than that of battery 2. The voltage of the battery with a larger DC internal resistance (such as battery 3) is higher than that of other

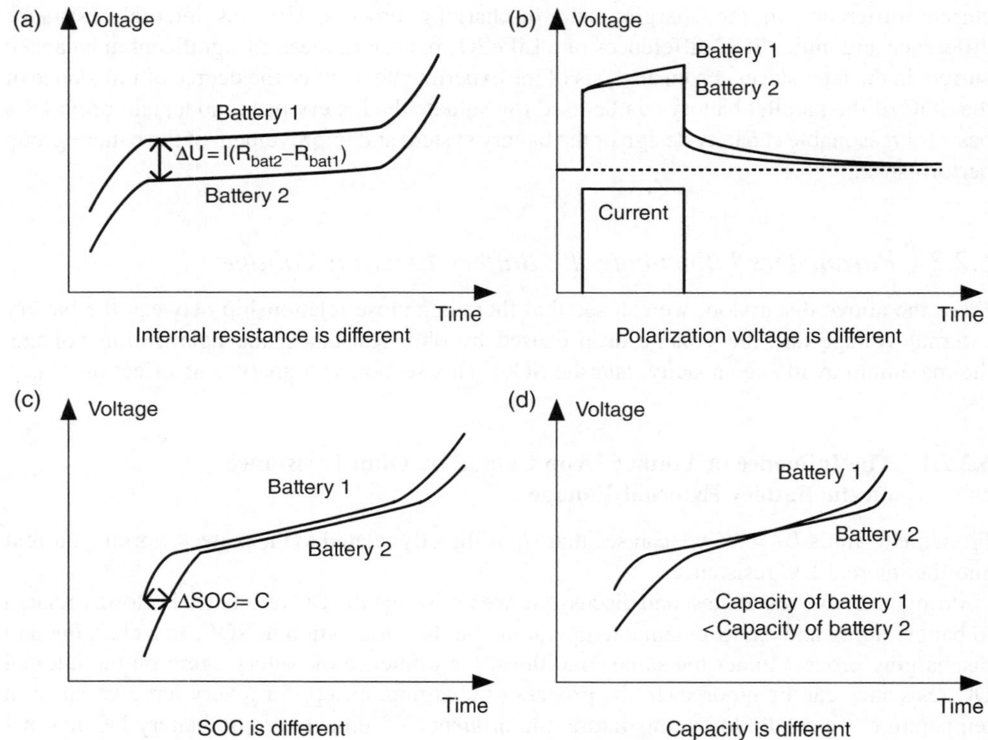

Figure 6.9 The effect of battery parameter differences on external voltage (a) internal resistance, (b) polarization voltage, (c) SOC, and (d) capacity.

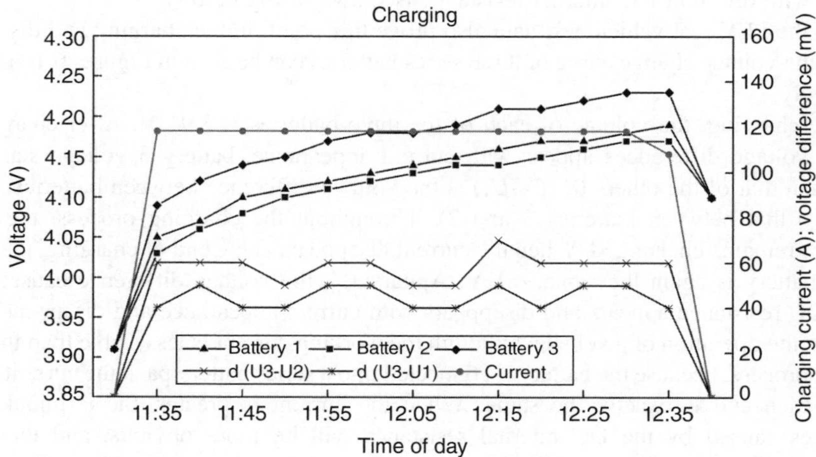

Figure 6.10 The performance of the battery DC internal resistance difference in the charging process.

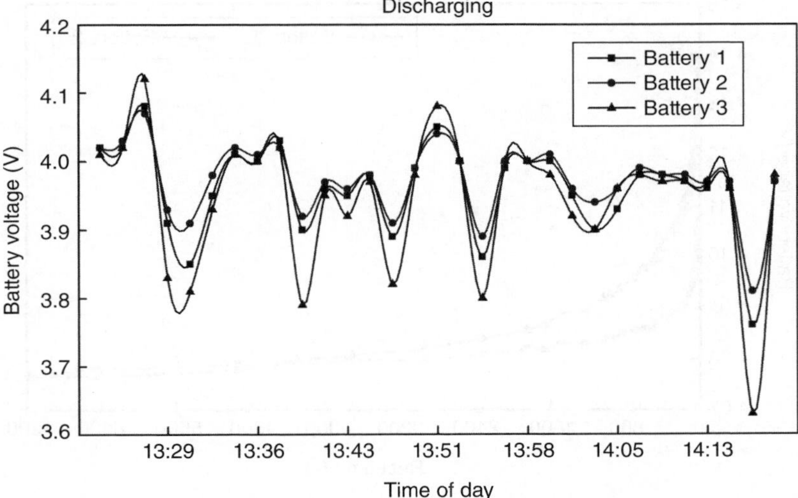

Figure 6.11 The performance of the battery DC internal resistance difference in the discharging process. (Reproduced with permission from Feng Wen, "Study on basic issues of the Li-ion battery pack management technology for Pure Electric Vehicles.", ©2009.)

batteries (such as batteries 1 and 2) in the charging process, and is lower than that of others in the discharging process. It is very easy to see that the external voltage difference caused by the battery DC internal resistance changes instantaneously with the changing of the current.

So, when judging the battery consistency, if the influence of ohm voltage drop on the battery terminal voltage is not taken into consideration, there will be a misleading result: the battery consistency will be worse when the current is larger, while the battery consistency will be better when the current is smaller. This will increase the inaccuracy of battery consistency judgment with changing current.

The batteries in a battery pack can be series or parallel connected and the consistency of the DC internal resistance has a negative impact on both modes. When batteries are connected in series, the current flowing through each battery is the same, so the battery with higher internal resistance will lose more energy, and the heat produced is also greater. This will cause environmental temperature inconsistencies, if the heat does not in dissipate in time, it may cause a serious problem by exceeding the battery limit temperature. When batteries are connected in parallel, while the total current stays within the parallel battery current limit, the current flowing through batteries with small internal resistance will be a greater part of the total current and may exceed the battery current capacity, leading to excessive use of the battery and further increasing the consistency problem.

6.2.2.2 The Influence of the Polarization Voltage on the Battery Terminal Voltage

The polarization voltage is overpotential, formed when the battery internal thermodynamic equilibrium system is broken in the battery charging and discharging process. The overpotential is formed when charging is positive, so that the battery terminal voltage is higher than

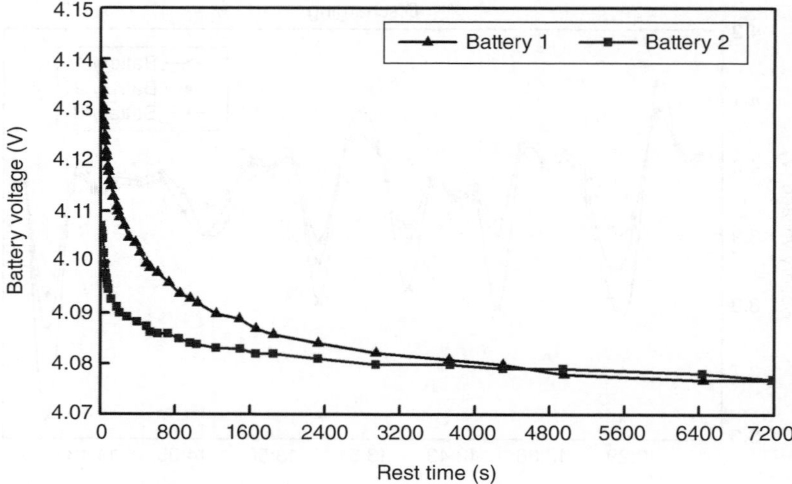

Figure 6.12 The terminal voltage difference caused by the polarization voltage. (Reproduced with permission from Feng Wen, "Study on basic issues of the Li-ion battery pack management technology for Pure Electric Vehicles.", ©2009.)

that in the resting period. A negative voltage will form during discharging. The polarization voltage is closely related to the battery materials and their structure, and is affected by the battery working conditions, such as working current, temperature, SOC, and other factors.

The difference in the polarization voltage will ultimately be reflected in a difference in the process of charging and discharging; when the current recedes, the difference will gradually decrease, and finally subside completely. At this time, the differences will be gradually eliminated. Taking battery charging for example, the recovery curve of two charged batteries with theoretical difference in the rest period is shown in Figure 6.9b. The voltage change curve of two batteries in the rest period after the PEV real vehicle test charging is shown in Figure 6.12.

It is easy to see that in the earlier rest period, the voltage of battery 1 is significantly higher than that of battery 2, and when the rest period ends and the polarization voltage recedes, the voltage of the two batteries will achieve the same level.

So, when judging the battery consistency, if the influence of the polarization voltage on the battery terminal voltage is not effectively solved, there will be an unscientific conclusion that the battery consistency will be worse in the working mode; while it will be better after the rest period finishes, this will increase the inaccuracy of the battery consistency judgment under the conditions of the working and rest status.

The polarization voltage consistency problem has a significant impact on the practical use of the battery pack. Because the polarization voltage increases rapidly in the later phase of the charging or discharging process, it plays an important role at this stage, and the individual battery polarization voltage will affect the charging and discharging of the whole battery pack. As shown in Figure 6.12, the polarization voltage of battery 1 increases rapidly at the end of charging, which forces the battery charging to be completed prematurely in the single battery voltage control mode and causes the available charging capacity of the battery pack to decline.

Also in the discharging process, the polarization voltage will cause the battery discharging to end prematurely, and lead to a decline in the available discharging capacity.

6.2.2.3 The SOC Influence on the Battery Terminal Voltage

According to the statistics data, the initial SOC difference of the same batch of batteries is less than 1%; but because the battery self-discharge rate, charge–discharge efficiency and the actual capacity of the battery are different, the SOC differences of different batteries gradually become larger in use. When using the open-circuit voltage method to estimate the SOC of the batteries, the maximum SOC difference among cells reaches 11%, which directly causes the battery available capacity to decline significantly, so the battery SOC consistency problem is one of the main factors affecting battery performance.

The corresponding relationship between the open-circuit voltage of the battery U_{OCV} and the SOC is shown in Figure 6.13, and the rate of change $dU_{OCV}/dSOC$ at different SOC intervals is shown in Figure 6.14. It can be seen that there is an obvious relationship between the two. The performance of two batteries with different SOC during the charging process is shown in Figure 6.9c.

The rate of change of the OCV for high and low SOC is significantly higher than that at the intermediate stage, its maximum (in a 0–5% interval, the OCV changes by 54 mV for every 1% change in SOC) is 18 times higher than the minimum (over a 75–85% interval, the OCV changes by about 3 mV for every 1% change in SOC). That is to say, under the same conditions (e.g. same capacity, ohm voltage drop, and polarizing voltage), when the SOC of two batteries differs by 1%, the battery terminal voltage difference may reach 54 mV for low SOC ($SOC \in [0\,\%, 5\,\%]$), while the battery terminal voltage's difference may be only 3 mV for mid-range SOC ($SOC \in [75\,\%, 85\,\%]$). So, by just taking the terminal voltage difference of each cell we can conclude that the consistencies of the battery pack cannot effectively describe the battery essential (SOC) difference. This may lead to a confusing conclusion, when the battery SOC is

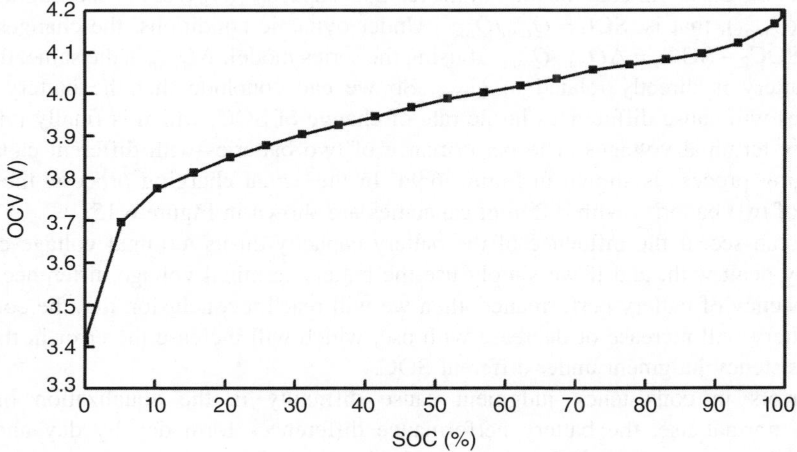

Figure 6.13 The relationship between OCV and SOC.

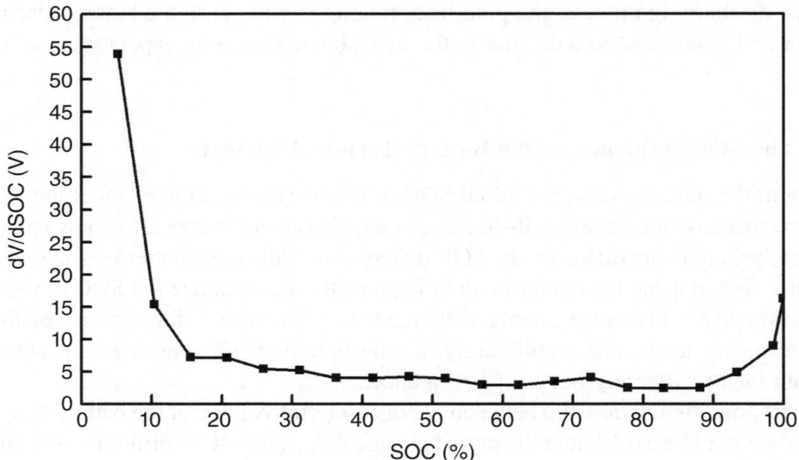

Figure 6.14 The voltage change rate for different SOC intervals.

high or low, the voltage difference is bigger, and the consistency is poor. When the battery SOC is in the mid-range, the voltage difference is smaller, and the consistency is good.

We know that the battery performance difference is accumulated day by day in normal use, and changes gradually. In one cycle of charging and discharging, the difference in battery performance and state stays virtually unchanged, and does not change with change in the SOC, but the battery voltage difference cannot effectively show that. The SOC differences can describe the battery differences well.

6.2.2.4 Influence of the Maximum Available Capacity on the Battery Terminal Voltage

The battery SOC is defined as the ratio of the residual capacity (Q_{rem}) to the maximum available capacity (Q_{max}), that is, SOC = Q_{rem}/Q_{max}. Under dynamic conditions, the changes in SOC $\Delta SOC = SOC_2 - SOC_1 = \Delta Q_{rem}/Q_{max}$. If using the series model, ΔQ_{rem} is the same, the ΔSOC of the battery is directly related to Q_{max}. So we can conclude that the battery capacity differences will cause differences in the rate of change of SOC, which is finally reflected in the battery terminal voltages. The performance of two batteries with different capacities in the charging process is shown in Figure 6.9d. In the actual charging process, the terminal voltages of two batteries with different capacities are shown in Figure 6.15.

So we can see, if the influence of the battery capacity on its terminal voltage cannot be effectively dealt with, and if we simply use the battery terminal voltage difference to judge the consistency of battery performance, then we will reach a conclusion that the consistency of the battery will increase or decrease with use, which will increase the error in the battery pack consistency judgment under different SOC.

The errors in consistency judgment cause difficulty in the equalization of battery packs. In normal use, the battery performance differences form day by day and change gradually. In one process of charging and discharging, the battery performance and state

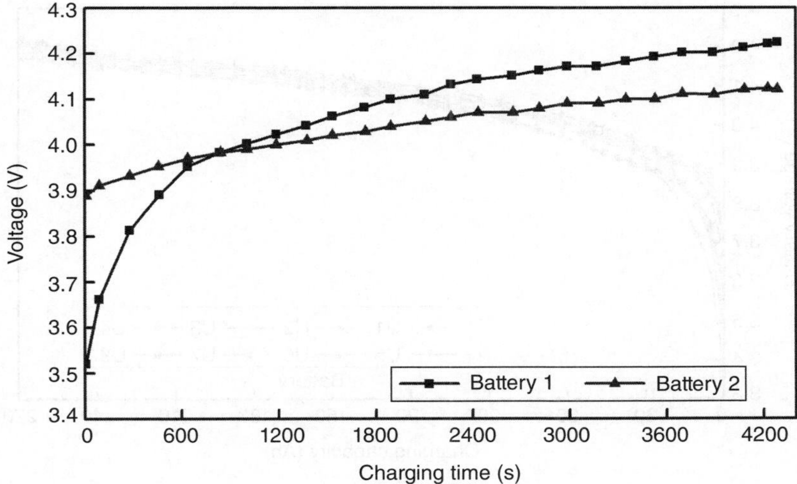

Figure 6.15 The charging curve of two batteries with different capacity.

will not change, no matter whether the battery works or stops, or whether the SOC interval changes or not.

To summarize, the battery terminal voltage is complex in its construction and is related to such factors as DC internal resistance, working current, polarization voltage, SOC, and capacity. So just taking the terminal voltage differences of batteries to judge the consistency of a battery pack cannot effectively describe the performance differences. At the same time, although each single cell's DC internal resistance, polarization voltage, and capacity will all influence the battery external voltage, it is not easy to balance these factors.

6.2.3 Method for Analysis of Battery Consistency

The above analysis and conclusion are based on the argument that the battery's other parameters and state are the same, so the analysis is relatively simple. However, in actual use, the battery terminal voltage differences may result from many causes, which makes the battery consistency analysis more difficult. In this section we will consider a battery pack after a period of operation, and analyze and confirm its consistency by using the four indicators: DC resistance, polarization voltage, SOC, and capacity.

The charging curve of the battery under a C/3 charging rate is shown in Figure 6.16 (where, U_i stands for the external voltage of the ith battery). It is clear that the charging voltage of each battery displays obvious differences, and the difference is a little larger at the start of charging, decreases in the middle period, and then gradually increases at the end of charging. Therefore, when the consistency of the battery pack is described by the difference in the external voltage of each battery, the conclusion is inconsistent.

Here we thoroughly analyze the reason for the difference in the battery external voltage, from the aspect of the difference in the four indicators listed above.

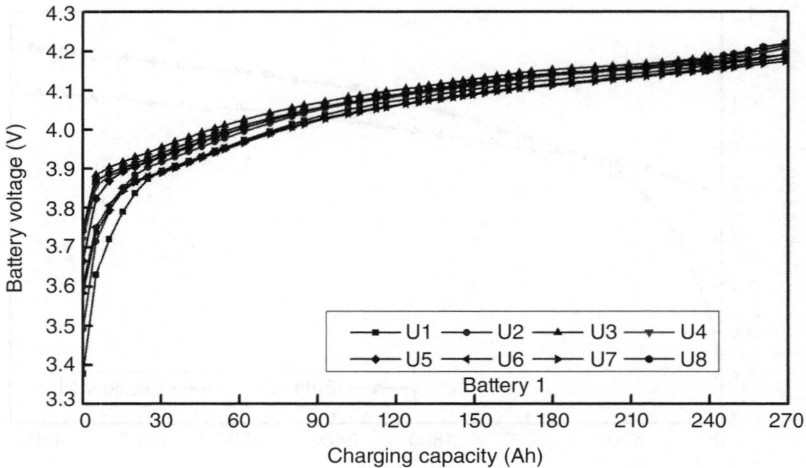

Figure 6.16 Charging curve of the battery pack.

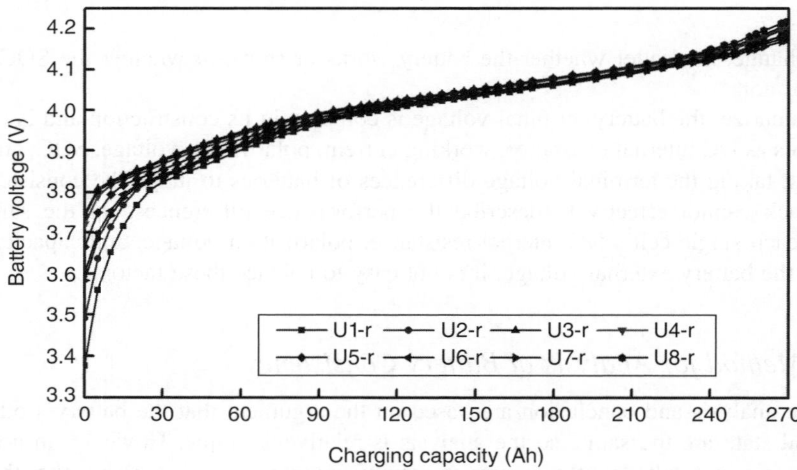

Figure 6.17 The charging curve of the battery after removing the DC resistance effect.

6.2.3.1 Influence of DC Resistance

The DC resistance of each battery can be calculated by using the external voltages of the battery under different currents and formulas (see Table 6.1). The charging curve when removing the ohm voltage drop can be seen in Figure 6.17 (wherein $U_i - r$ stands for the voltage of the *i*th battery after removal of the ohm voltage drop).

It is clear that, after removing the ohm voltage drop, the external voltages of the batteries have better consistency than those from the original data, but there is no great improvement in consistency. This indicates that the ohm voltage drop difference of each battery in the pack has only a trivial influence on the external voltage.

6.2.3.2 Influence of the Polarization Voltage

After fully resting, the polarization voltage of the battery can be effectively removed, thus, the following charging methods are designed:

1. At C/3 rate, discharge the battery, until the lowest voltage of the batteries reaches 3.3 V.
2. Stand for 2 h.
3. Stand for 2 h after the battery is charged for a specified Coulomb.
4. When the voltage of the battery reaches 4.23 V it becomes constant, until the charging current drops to 5 A.

The ohm voltage drop of the battery is removed when the current is zero. The polarization voltage can also be eliminated after standing. The charging voltage curve after standing is seen in Figure 6.18. The set of polarization voltages obtained after 2 h standing is shown in Table 6.2.

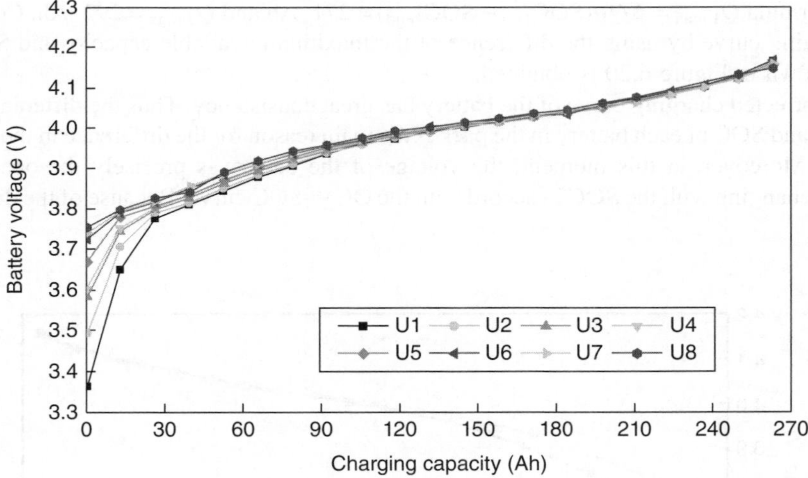

Figure 6.18 The charging curve of the battery after removing the polarization effect.

Table 6.2 The polarization voltage difference (V) of the batteries in the pack.

	Battery 1	Battery 2	Battery 3	Battery 4	Battery 5	Battery 6	Battery 7	Battery 8
Standing for 0 h	3.985	3.99	4.003	3.999	3.995	3.989	3.987	4.004
Standing for 2 h	3.958	3.96	3.974	3.972	3.968	3.966	3.964	3.974
Polarization voltage	0.027	0.03	0.029	0.027	0.027	0.023	0.023	0.03

It is thus clear that, after removing the polarization voltage, especially at the mid to late stages of charging, the external voltages of the batteries have great consistency. However, at the start of charging, the difference in the battery external voltage shows no practical improvement.

6.2.3.3 Influences of the Maximum Available Capacity and SOC

The charging curves of the batteries 1 and 3 with the greatest difference are shown in Figure 6.19. After every 13 Ah is added and the batteries have rested fully, the external voltage of the battery will be the open-circuit voltage and, at this moment, it has a corresponding relationship with the SOC.

At the start of charging, the voltages of batteries 1 and 3 are $U_{1\text{ini}} = 3.436$ V and $U_{3\text{ini}} = 3.76$ V. Based on the relationship of the OCV and the SOC, the initial SOCs of the battery are $\text{SOC}_{1\text{ini}} = 0\ \%$ and $\text{SOC}_{3\text{ini}} = 7.8\ \%$. Adding $\Delta Q = 263$ Ah, the open-circuit voltages of the batteries are $U_{1\text{end}} = 4.175$ V and $U_{3\text{end}} = 4.188$ V, and the SOCs are $\text{SOC}_{1\text{end}} = 97\ \%$ and $\text{SOC}_{3\text{end}} = 98\ \%$. The maximum available capacities of the two batteries can be calculated by the formula $Q_{1\,\text{max}} = \Delta Q_1/(\text{SOC}_{1\text{ini}} - \text{SOC}_{1\text{end}}) = 271$ Ah and $Q_{3\,\text{max}} = 292$ Ah. Correcting the charging curve by using the difference of the maximum available capacity and SOC, the curve shown in Figure 6.20 is obtained.

The corrected charging curve of the battery has great consistency. Thus the difference in the capacity and SOC of each battery in the pack is the main reason for the difference in the external voltage. Moreover, at this moment, the voltage of the battery is precisely the open-circuit voltage, changing with the SOC in accord with the OCV–SOC curve. Because of the difference

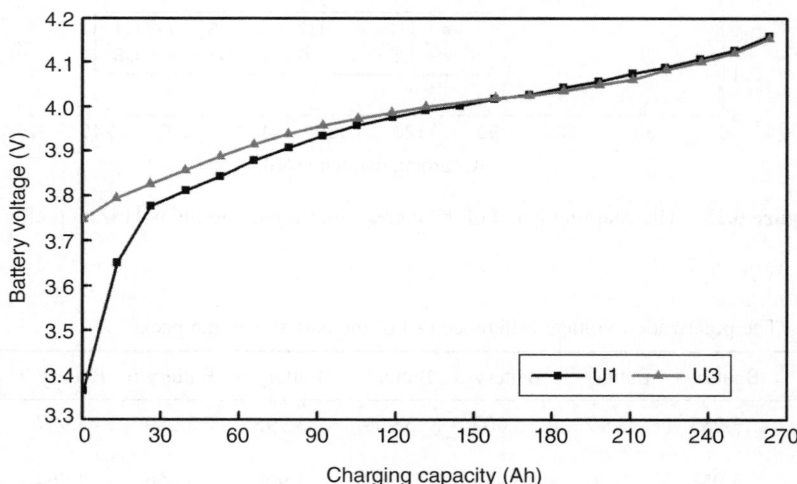

Figure 6.19 The charging curve of batteries 1 and 3 after removing the polarization voltage through charging.

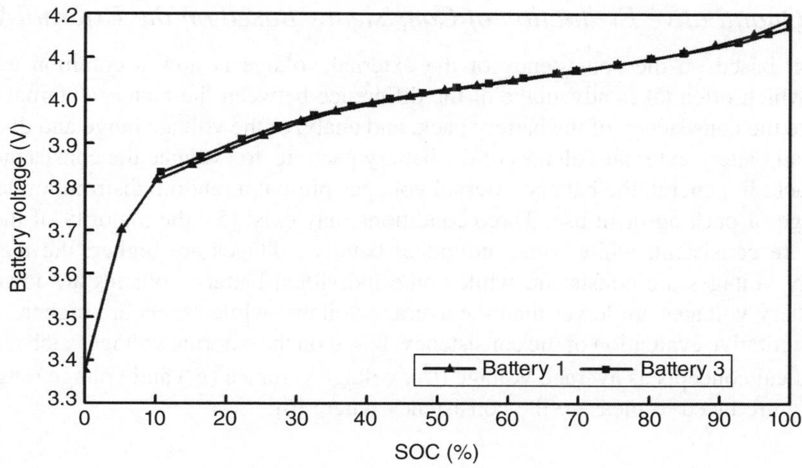

Figure 6.20 The charging curve after correction of the capacity and SOC.

between the SOC–OCV curves of batteries of the same type from the same manufacturer, the charging curve of the battery is bound to coincide.

Different from the method of merely using the external voltage of the battery as the judgment basis for the consistency, in the whole process of charging, when judging the consistency of the batteries from the four aspects of DC resistance, polarization voltage, capacity, and SOC, the performances and states of the batteries are fixed, and do not vary with variation in the charging current and SOC, which provides data support for the evaluation, maintenance, and equalization of the battery.

6.3 Quantitative Evaluation of Battery Consistency

In practice, the provision of data support for the equalization and maintenance of the battery requires quantitative evaluation of the consistency of the battery pack. As previously mentioned, the difference in the batteries is embodied in the DC internal resistance, the polarization voltage, the SOC, and the maximum available capacity. The first three parameters cannot be improved through equalization so equalization of the battery pack is done only by adjusting the SOCs of the batteries and achieving optimization of its configuration. Under the precondition that all batteries will not be overcharged or discharged, the maximization of the battery capacity and energy utilization can be achieved. This section describes a quantitative evaluation of the consistency of the batteries from the aspect of the battery external voltage, capacity utilization rate, and energy utilization rate. Under the conditions that the working mode, required rate, operation ambient, and working condition cannot be determined, the DC internal resistance and the polarization voltage cannot be predicted. Therefore, a quantitative evaluation of the consistency of the battery is made from only the maximum available capacity and the SOC.

6.3.1 Quantitative Evaluation of Consistency Based on the External Voltage

Evaluation based on the consistency of the external voltage is now a common evaluation method, which often takes advantage of the difference between the battery external voltages to measure the consistency of the battery pack, and analyzes the voltage range and distribution in which all battery external voltages of the battery pack lie, to evaluate the consistency of the battery pack. In general, the battery external voltages present a random distribution law in the initial stage of packing or in use. Three conditions may exist [5]: the majority of the battery voltages are consistent, while some individual battery voltages are higher; the majority of the battery voltages are consistent, while some individual battery voltages are lower; some of the battery voltages are lower than the average voltage, while others are higher.

For quantitative evaluation of the consistency, based on the external voltage, such mathematical statistical concepts as average voltage ($\overline{U}$), voltage variance (δ^2) and voltage range (r) are generally introduced to measure the consistency where [6]:

$$\begin{cases} \overline{U} = \dfrac{1}{n}\displaystyle\sum_{i=1}^{n} U_i \\ \delta^2 = \dfrac{1}{n}\displaystyle\sum_{i=1}^{n} \left(U_i - \overline{U}\right)^2 \\ r = \max\{U_i\} - \min\{U_j\} \end{cases} \quad i,j = 1 \ldots n \tag{6.4}$$

The average voltage reflects the general state of energy of the battery pack, and is also the basis for calculating other parameters. The voltage variance embodies the degree of deviation between all the battery voltages and the average voltage of the pack, representing the disparity and evenness of voltage distribution; smaller voltage variance means that the voltage distribution is more concentrated; thus, the consistency of the battery external voltage is better. Using the variance to measure the consistency of the battery external voltage theoretically can achieve a good effect, but it does not facilitate the embedded transplantation. The battery pack charging and discharging control is based on the single cell voltage, and the highest and lowest voltages of the cells determine the use of the battery pack. So, in evaluation of the consistency of the battery pack, the distribution of all the battery voltages does not need to be considered, and controlling the battery variance can improve the consistency of the battery pack. The coefficient of variation and relative range can also be used to evaluate the consistency of the battery pack [6, 7], where the coefficient of variation is the ratio of the square root of the voltage variance to the average voltage, and the relative range is the ratio of the range to the average voltage. In addition, researchers have established the joint model of the overall dispersion and the single model to describe the consistency of the lithium battery pack.

But evaluating the consistency based on the difference in the external voltage cannot reflect the differences of parameters inside the battery. According to previous analysis, the consistency of the battery external voltage is usually affected by the DC resistance, polarization voltage, capacity, and SOC. If the effects of the above factors on the evaluation of the consistency are not comprehensively considered, the equalization evaluation based on the external voltage will be inaccurate.

6.3.2 Quantitative Evaluation of Consistency Based on the Capacity Utilization Rate of the Battery Pack

If consistency problems appear, the capacity of the battery pack cannot be fully utilized. By equalization, the consistency of the battery pack can be improved. However, how many of the maximum available capacities of the battery pack are there before and after equalization? How much is the present capacity utilization of the battery pack? How much of the balanced battery capacity can be improved? These are very important indicators of the efficiency in use of the battery pack. In this section we will deduce the maximum available capacity of the battery pack in the current state, and the maximum available capacity after equalization. We will define this ratio as the capacity utilization of the battery pack, for which the difference is the increase in the capacity after equalization, to provide a theoretical basis and data support for the equalization of the battery pack and to give an expected effect.

6.3.2.1 Equalization Criteria Based on the Capacity of the Battery Pack

Consider a series battery pack, the maximum available capacities of n batteries being $Q_{\max}[1]$, …, $Q_{\max}[n]$ and the present state of charging SOC[1], …, SOC[n]. From the previous relevant discussion it is known that, in the current state, the basic parameters of the battery pack are:

1. The maximum discharging capacity $Q^{\mathrm{B}}_{\mathrm{dch\_max}} = \min\{Q_{\max}[i]\mathrm{SOC}[i]\}$ where $i = 1, \ldots, n$ (6.5)
2. The maximum charging capacity $Q^{\mathrm{B}}_{\mathrm{ch\_max}} = \min\{Q_{\max}[j](1-\mathrm{SOC}[j])\}$, where $j = 1, \ldots, n$ (6.6)
3. The maximum available capacity $Q^{\mathrm{B}}_{\max} = Q^{\mathrm{B}}_{\mathrm{dch\,max}} + Q^{\mathrm{B}}_{\mathrm{ch\,max}}$. (6.7)

It is clear that, when $i = j$, $Q^{\mathrm{B}}_{\max}$ is equal to the maximum available capacity of the battery with the minimum capacity in the battery pack, as shown in Figure 6.21a. The battery pack does not need equalization because this cannot increase the maximum available capacity of the battery pack (equalization cannot increase the maximum available capacity of the cells as well as the DC resistance and the polarization voltage, but can adjust the SOCs of all batteries to be in a more reasonable interval, so as to increase the maximum available capacity of the battery pack under the precondition that there will not be overcharging and discharging of all batteries). When $i \neq j$, $Q^{\mathrm{B}}_{\max}$ is smaller than the maximum available capacity of the battery with the minimum capacity in the battery pack, as shown in Figure 6.21b–d. At this moment, the capacity of the battery pack can be increased and the equalization of the battery pack has much significance.

From the above discussion it can be seen that discharging capacity and available capacity is related to the maximum available capacity and the SOC of the single cell in the pack. The capacity of the battery pack is no more than the capacity of the single cell with the minimum capacity. Therefore, in terms of capacity utilization, the judgment criteria for determining whether the battery pack in use needs equalization include answers to such questions as whether the battery with the minimum capacity reaches full charging and complete discharging at the first time, and whether it achieves sufficient utilization of the battery in the process of charging and discharging.

Therefore, after batteries, with differences in DC resistance, polarization voltage and maximum available capacity, are assembled in series into the pack, in use, the external voltages of the batteries are bound to be different. However, as long as the battery with the minimum capacity first reaches full charging and then complete discharging, the battery pack will not

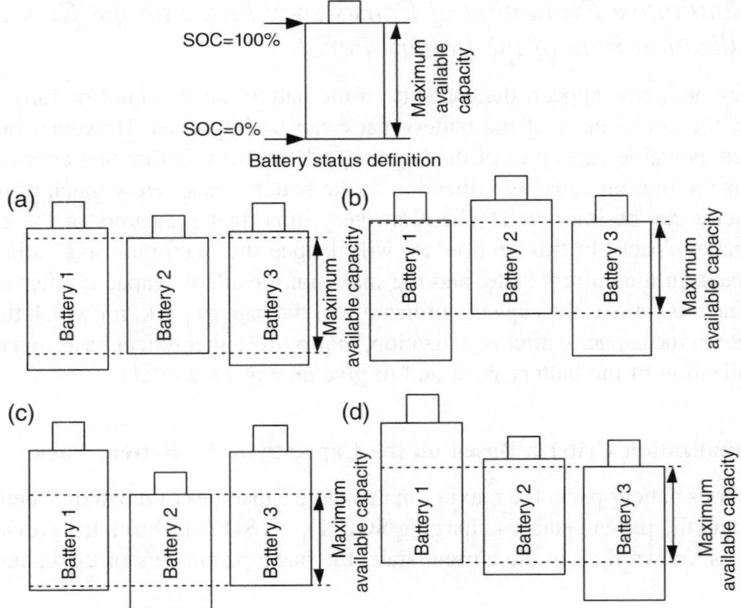

Figure 6.21 Judgment criteria for equalization of the battery.

need equalization, because this cannot increase the maximum available capacity of the battery pack. In other words, in use, the battery pack does not need to ensure the continuous consistency of the external voltages of all batteries in the pack.

The goal of the existing equalizer is to achieve the consistency of each battery's external voltage, but this does not increase the capacity of the battery pack significantly, so both the consistency evaluation method and the equalization control strategy need further improvement. Continuous equalization is not only a waste of energy, but also increases the total charging and discharging capacity of the battery, and cannot effectively increase the maximum available capacity of the battery pack.

6.3.2.2 Capacity Utilization Rate of the Battery Pack

When the single cell with the minimum capacity cannot first reach full charging and complete discharging, the use of the capacity of the battery pack is insufficient, so there is the difficult task of increasing the utilization rate. The capacity utilization rate η_C^B of the battery pack consisting of n batteries in series is the ratio of the maximum available capacity (Q_{max}^B) of the battery pack to the maximum available capacity ($\min\{Q_{max}[k]\}$) of the battery with the minimum capacity in the pack. Because, after equalization, the maximum capacity $Q_{max}^{eq\_B}$ (parameters after equalization are marked with the superscript eq) of the battery pack is $\min\{Q_{max}[k]\}$, that is:

$$\eta_C^B = \frac{Q_{max}^B}{\min\{Q_{max}[k]\}} = \frac{Q_{max}^B}{Q_{max}^{eq\_B}}, \text{ where } k = 1, \ldots, n \tag{6.8}$$

If the battery m is the battery with the minimum capacity in the pack, namely $k = m$, then $Q_{\max}^{\text{eq\_B}} = Q_{\max}[m]$, at this moment:

$$\eta_{\text{C}}^{\text{B}} = \frac{Q_{\max}^{\text{B}}}{Q_{\max}[m]} \tag{6.9}$$

The discharging capacity of the battery m is

$$Q_{\text{dch\_max}}[m] = Q_{\max}[m]\text{SOC}[m] \geq \min\{Q_{\max}[i]\text{SOC}[i]\}, \text{where } i = 1, \ldots, n \tag{6.10}$$

The charging capacity is

$$Q_{\text{ch\_max}}[m] = Q_{\max}[m](1 - \text{SOC}[m]) \geq \min\{Q_{\max}[j](1 - \text{SOC}[j])\}, \text{ where } j = 1, \ldots, n. \tag{6.11}$$

Therefore: $Q_{\max}^{\text{B}} = \min\{Q_{\max}[i]\text{SOC}[i]\} + \min\{Q_{\max}[j](1 - \text{SOC}[j])\} \leq Q_{\max}[m]$ (6.12)

That is $\eta_{\text{C}}^{\text{B}} \leq 1$. (6.13)

And when $i = j = m$, the capacity utilization rate of the battery pack is 1, otherwise it will be less than 1. The capacity utilization rate of the battery pack is the basis for the equalization of the battery pack.

6.3.2.3 Capacity Increase in Equalization

The increase in the maximum available capacity after equalizing the battery pack is:

$$\Delta Q_{\text{ea}}^{\text{B}} = Q_{\max}^{\text{eq\_B}} - Q_{\max}^{\text{B}} = \left(1 - \eta_{\text{Cap}}^{\text{B}}\right) Q_{\max}^{\text{eq\_B}} \tag{6.14}$$

For example, after the two batteries, with parameters $Q_{\max}[1] = 100\text{Ah}$, $\text{SOC}[1] = 75\,\%$, and $Q_{\max}[2] = 90\text{Ah}$, $\text{SOC}[2] = 50\,\%$, are constituted in series into the pack, the maximum discharging capacity $Q_{\text{dch\_max}}^{\text{B}}$ of the battery pack is:

$$Q_{\text{dch\_max}}^{\text{B}} = \min\{Q_{\max}[1]\text{SOC}[1], Q_{\max}[2]\text{SOC}[2]\} = 45 \text{ Ah}$$

The maximum charging capacity $Q_{\text{ch\_max}}^{\text{B}}$ is:

$$Q_{\text{ch\_max}}^{\text{B}} = \min\{Q_{\max}[1](1 - \text{SOC}[1]), Q_{\max}[2](1 - \text{SOC}[2])\} = 25 \text{ Ah}$$

So the maximum available capacity of the battery pack is:

$$Q_{\max}^{\text{B}} = Q_{\text{dch\_max}}^{\text{B}} + Q_{\text{ch\_max}}^{\text{B}} = 70 \text{ Ah}$$

After equalization, the maximum available capacity of this battery pack is increased to $Q_{\max}^{\text{eq\_B}} = \min\{100 \text{ Ah}, 90 \text{ Ah}\} = 90 \text{ Ah}$, so the capacity utilization rate of the battery pack is:

$$\eta_{\text{C}}^{\text{B}} = \frac{Q_{\max}^{\text{B}}}{Q_{\max}^{\text{eq\_B}}} = \frac{70 \text{ Ah}}{90 \text{ Ah}} = 78\%.$$

After equalization, the maximum available capacity of this battery pack is increased by

$$\Delta Q_{\text{ea}}^{\text{B}} = Q_{\max}^{\text{eq\_B}} - Q_{\max}^{\text{B}} = \left(1 - \eta_{\text{C}}^{\text{B}}\right) Q_{\max}^{\text{eq\_B}} = 20 \text{ Ah}.$$

6.3.3 Quantitative Evaluation of Consistency Based on the Energy Utilization Rate of the Battery Pack

From the perspective of energy utilization, when a single cell is used independently, its SOC can vary in the range 0–100%, so that the maximum available energy of the battery can be fully utilized. But when the pack contains many batteries in series, due to the inconsistency of the battery capacity as well as the state of charging, when being charged, some of the batteries will be fully charged first; reaching the charging cut-off voltage. So the battery pack cannot be continuously charged and other batteries cannot be fully charged. Similarly, when discharging, some of the batteries are fully discharged first, so the remaining batteries cannot fully discharge. Therefore, for a single battery in series in the pack the variation of its SOC is not from 0 to 100%, and not all batteries in the pack can be fully charged or discharged, that is, cannot achieve maximization of energy storage and release. Thus, the energy of the battery pack cannot be fully utilized.

6.3.3.1 Equalization Criterion Based on the Energy of the Battery Pack

From the previous relevant discussion it is known that, when the maximum variation of the battery m is $\text{SOC}_1[m] \sim \text{SOC}_2[m]$, the maximum available capacity of the battery in the process of charging and discharging is the maximum available capacity of the battery pack, that is, $Q_{\max}[m](\text{SOC}_2[m] - \text{SOC}_1[m]) = Q_{\max}^{\text{B}}$. So the maximum available energy of the battery when using the battery pack is:

$$E_{\max}[m] = Q_{\max}[m](\text{SOC}_2[m] - \text{SOC}_1[m]) U_{\text{OCV}|\text{SOC}=[\text{SOC}_1[m],\text{SOC}_2[m]]} \tag{6.15}$$

From the previous relevant discussion it is known that, after n batteries with the maximum available capacity $Q_{\max}[1], \ldots, Q_{\max}[n]$, and SOCs $\text{SOC}_0[1], \ldots, \text{SOC}_0[n]$, are put in series in the pack then, in the current state, the main parameters of the battery pack are:

1. The maximum discharging capacity $Q_{\text{dch\_max}}^{\text{B}} = Q_{\text{rem}}^{\text{B}} = \min\{Q_{\max}[i]\text{SOC}[i]\}$, where $i = 1, \ldots, n$; (6.16)
2. the maximum charging capacity $Q_{\text{ch\_max}}^{\text{B}} = \min\{Q_{\max}[j](1 - \text{SOC}[j])\}$, where $j = 1, \ldots, n$; (6.17)
3. the maximum available capacity $Q_{\max}^{\text{B}} = Q_{\text{dch\_max}}^{\text{B}} + Q_{\text{ch\_max}}^{\text{B}}$; (6.18)
4. the maximum available energy $E_{\max}^{\text{B}} = \sum_{m=1}^{n} E[m] = Q_{\max}^{\text{B}} \sum_{m=1}^{n} U_{\text{OCV}|\text{SOC}=\left[\text{SOC}_{\text{dch\_end}}[m],\text{SOC}_{\text{ch\_end}}[m]\right]}$, where $m = 1, \ldots, n$. (6.19)

$$\text{SOC}_{\text{dch\_end}}[m] = \text{SOC}_0[m] - \frac{Q_{\text{rem}}^{\text{B}}}{Q_{\max}[m]}, \text{SOC}_{\text{ch\_end}}[m] = \text{SOC}_0[m] + \frac{Q_{\text{ch\_max}}^{\text{B}}}{Q_{\max}[m]} \tag{6.20}$$

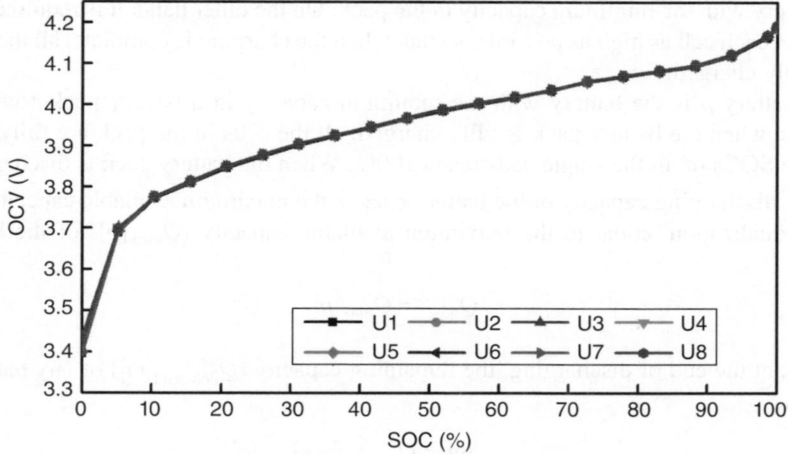

Figure 6.22 The OCV–SOC curve of different batteries from the same batch.

For the same batch of batteries, their SOC–OCV curves can be considered consistent (The SOC–OCV testing curves of 8 batteries of the same batch are shown in Figure 6.22). Therefore

$$U_{\mathrm{OCV}}\Big|\mathrm{SOC}=\left[\mathrm{SOC}_0[m]-\frac{Q^{\mathrm{B}}_{\mathrm{rem}}}{Q_{\max}[m]},\ \mathrm{SOC}_0[m]+\frac{Q^{\mathrm{B}}_{\mathrm{ch\_max}}}{Q_{\max}[m]}\right] \tag{6.21}$$

is mainly directly related to the SOC of the battery. The maximum available energy of the battery pack has a close relation with the maximum available capacity and the SOCs of each single cell.

It is clear that, after the cells are put in series in the pack, the maximum available energy of the battery pack is the sum of the maximum available energy of each single cell, and the maximum available energy of the single cell is related to the maximum available capacity of the battery pack and the SOCs of each single battery. Therefore, the full utilization of the capacity of the battery pack is just one aspect of the full utilization of energy, and each cell with the higher SOC also has great significance in the utilization of the battery energy. That is, in order to achieve maximization of the energy storage, the equalization of the battery pack is not only required to achieve the full charging and complete discharging of the battery with the minimum capacity (i.e. the capacity of the battery is fully utilized), but also to make the SOCs of all the cells in the battery pack reach a level as high as possible, which may help to ensure that all batteries are fully charged at the end of charging. This is the judgment criteria of the equalization of the battery pack from the perspective of the energy.

6.3.3.2 Energy Utilization Rate of the Battery Pack

As described above, in order to achieve maximization of the battery pack energy utilization, on the one hand, it is required to achieve maximization of the battery pack capacity utilization, that is, the maximum available capacity of the battery pack reaches the capacity of the battery with the minimum capacity in the pack. On the other hand, it is required to make the SOC of each cell as high as possible, so that when the charging is complete, all the batteries can be fully charged.

If the battery p is the battery with the minimum capacity in a battery pack consisting of n batteries, when the battery pack is fully charged, all the cells in the pack are fully charged, that is, the SOCs of all the single cells reach 100%. When the battery pack is discharging, the maximum discharging capacity of the battery pack is the maximum available capacity ($Q_{\max}^{\mathrm{eq\_B}}$) after its equalization, equal to the maximum available capacity ($Q_{\max}[p]$) of the battery p, that is:

$$Q_{\max}^{\mathrm{eq\_B}} = Q_{\max}[p] \tag{6.22}$$

Therefore, at the end of discharging, the remaining capacity ($Q_{\mathrm{dch\_end}}^{\mathrm{eq}}[m]$) of any battery m in the pack is:

$$Q_{\mathrm{dch\_end}}^{\mathrm{eq}}[m] = Q_{\max}[m] - Q_{\max}[p] \tag{6.23}$$

At this moment, the SOC ($\mathrm{SOC}_{\mathrm{dch\_end}}^{\mathrm{eq}}[m]$) of this battery is:

$$\mathrm{SOC}_{\mathrm{dch\_end}}^{\mathrm{eq}}[m] = \frac{Q_{\max}[m] - Q_{\max}[p]}{Q_{\max}[m]} \times 100\% = \frac{Q_{\max}[m] - Q_{\mathrm{man}}^{\mathrm{eq\_B}}}{Q_{\max}[m]} \times 100\% \tag{6.24}$$

Therefore, in the process of charging and discharging, the maximum available energy ($E_{\max}^{\mathrm{eq}}[m]$) of the battery m is:

$$E_{\max}^{\mathrm{eq}}[m] = Q_{\max}^{\mathrm{eq\_B}} U_{\mathrm{OCV}} \Big|_{\mathrm{SOC} = \left[\mathrm{SOC}_{\mathrm{dch\_end}}^{\mathrm{eq}}[m],\, 100\%\right]} \tag{6.25}$$

The maximum available energy ($E_{\max}^{\mathrm{eq\_B}}$) of the battery pack is:

$$E_{\max}^{\mathrm{eq\_B}} = Q_{\max}^{\mathrm{eq\_B}} \sum_{k=1}^{n} U_{\mathrm{OCV}} \Big|_{\mathrm{SOC} = \left[\mathrm{SOC}_{\mathrm{dch\_end}}^{\mathrm{eq}}[k],\, 100\%\right]}, \quad k = 1, \ldots, n \tag{6.26}$$

Therefore, the energy utilization rate $\eta_{\mathrm{E}}^{\mathrm{B}}$ of the battery is:

$$\eta_{\mathrm{E}}^{\mathrm{B}} = \frac{E_{\max}^{\mathrm{B}}}{E_{\max}^{\mathrm{eq\_B}}} = \frac{Q_{\max}^{\mathrm{B}} \sum_{m=1}^{n} U_{\mathrm{OCV}} \Big|_{\mathrm{SOC} = \left[\mathrm{SOC}_{\mathrm{dch\_end}}[m],\, \mathrm{SOC}_{\mathrm{ch\_end}}[m]\right]}}{Q_{\max}^{\mathrm{eq\_B}} \sum_{k=1}^{n} U_{\mathrm{OCV}} \Big|_{\mathrm{SOC} = \left[\mathrm{SOC}_{\mathrm{dch\_end}}^{\mathrm{eq}}[k],\, 100\%\right]}}$$

$$= \eta_{\mathrm{C}}^{\mathrm{B}} \frac{\sum_{m=1}^{n} U_{\mathrm{OCV}} \Big|_{\mathrm{SOC} = \left[\mathrm{SOC}_{\mathrm{dch\_end}}[m],\, \mathrm{SOC}_{\mathrm{ch\_end}}[m]\right]}}{\sum_{m=1}^{n} U_{\mathrm{OCV}} \Big|_{\mathrm{SOC} = \left[\mathrm{SOC}_{\mathrm{dch\_end}}^{\mathrm{eq}}[k],\, 100\%\right]}} \tag{6.27}$$

Because, after equalization, for any single cell:

$$Q_{\max}^{\text{eq\_B}} U_{\text{OCV}} \Big|_{\text{SOC}=\left[\text{SOC}_{\text{dch\_end}}^{\text{eq}}[k],\, 100\%\right]} \geq Q_{\max}^{\text{B}} U_{\text{OCV}} \Big|_{\text{SOC}=\left[\text{SOC}_{\text{dch\_end}}[k],\, \text{SOC}_{\text{ch\_end}}[k]\right]} \tag{6.28}$$

$\eta_{\text{E}}^{\text{B}}$ is always not more than 1, thus, in use, the poorer the consistency of the battery pack, the smaller $\eta_{\text{E}}^{\text{B}}$. This provides data support for the equalization of the battery pack.

The inconsistency issue decreases the maximum available energy of the battery pack from $E_{\max}^{\text{eq\_B}}$ to $E_{\max}^{\text{B}}$, so, after reasonable equalization, the maximum available energy of the battery pack can increase by $\Delta E_{\text{ea}}^{\text{B}}$, and:

$$\Delta E_{\text{ea}}^{\text{B}} = E_{\max}^{\text{eq\_B}} - E_{\max}^{\text{B}} \tag{6.29}$$

6.4 Equalization of the Battery Pack

As previously mentioned, after a period of use, the consistency problem of the battery pack will gradually appear, preventing the efficient utilization of the capacity and energy of the battery pack. By the same line of reasoning, for PEVs, the operational efficiency will decrease, and the continuous driving mileage of one-single charging will be shortened. In order to solve this problem, the most generally used method is to equalize the battery pack. The equalization has great significance for the more efficient utilization of the energy battery pack, so it has received widespread attention when using the battery pack in series, and various equalization control strategies and circuit topologies have been proposed. But the control basis for these methods is the external voltage differences of single cells in the pack, and the target for equalization is to make the external voltage difference of each single cell reach its listed range. The equalized capacity of each single cell is not obtained in advance, and the external voltage difference of each battery will be more obvious at the end of charging; therefore, the action time of the equalizer is a little limited, which causes much difficulty in choosing the equalizer capacity and the equalization process. When these methods are applied to PEVs there will be some tough problems: since the capacity of the battery is a little large, the equalization capacity is not large enough to effectively achieve the equalization effect, or the capacity, volume, and cost of the equalizer cannot meet the practical requirements.

Based on the above analysis on maximum available capacity, the estimation of SOC and the evaluation of the consistency of the battery, this section discusses the basis and methods of the equalization of the battery pack, and the capacity computing method in the process of equalization, and compares the three equalization strategies, which provide the theoretical basis and data support for the equalization of the battery pack from the aspects of the external voltage of the single battery, the maximum available capacity of the battery pack, the maximum available energy of the battery pack and the SOC of single batteries.

6.4.1 *Equalization Based on the External Voltage of a Single Cell*

The traditional equalization method is based on an insufficiently thorough understanding of the consistency problems of the battery pack, at the same time the identification method of the state

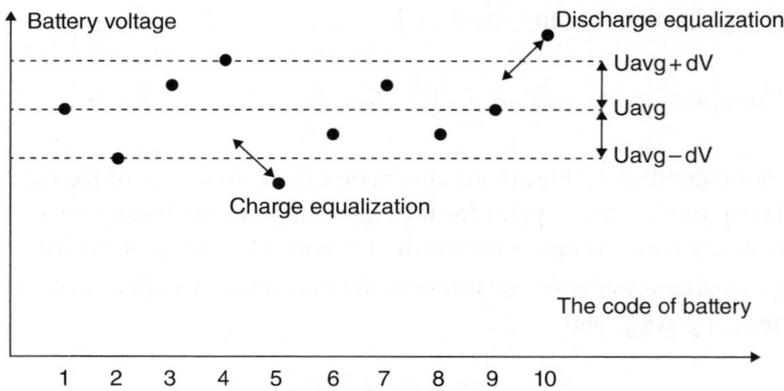

Figure 6.23 Schematic diagram for equalization based on the external voltage.

of the battery cannot do the work by itself. Therefore, in practice, the external voltage of the battery is used directly to evaluate the consistency of the battery pack, and the external voltage is taken as the control object of the equalization to achieve equalization.

Since the equalization strategies are based on the external voltage, in the process of charging and discharging, the external voltage of the battery is always taken as the control object of the equalization, that is, discharging the batteries with a higher voltage and charging the batteries with a lower voltage in the pack. Thus, the voltage of the battery pack is adjusted to be consistent. Because the voltage range can reflect the consistency of the battery pack in the consistency evaluation system of the external voltage, the control zone (dU) is determined in advance in the equalization module, and the batteries within the control zone are not considered to be equalized; at the same time, the control zone can also prevent the repeated fluctuation of the equalization of the same battery, shown in Figure 6.23. When the equalization judgment of the battery pack is being carried out, we should first calculate the average voltage of the battery pack, then determine the equalization control zone, and perform the discharge equalization for the battery above the control zone, namely the charge equalization. The detailed judgment procedure is shown in Figure 6.24.

This equalization strategy is easy to achieve, the equalization judgment can be carried out by measuring only the external voltages of each cell in the pack. However, there are certain drawbacks. For one thing, it can easily be affected by the battery's own parameters and conditions, so that inaccuracy of the equalization judgment occurs. According to the aforementioned analysis of the consistency of the battery pack, the external voltage of the battery is influenced by a plurality of parameters, including the SOC, the DC resistance, the polarization voltage, and so on. When two power batteries with different internal resistances are equalized based on the equalization strategy of external voltage, in the process of charging, because the voltage of battery 2 is lower, it requires charging equalization (or discharging equalization for battery 1). However, in the process of discharging, the voltage of battery 1 is lower than that of battery 2; therefore, based on the equalization of the external voltage, it is necessary to carry out discharging equalization for battery 2 (or charging equalization for battery 1). During the charging and discharging process, the equalization object constantly changes, which does not in fact

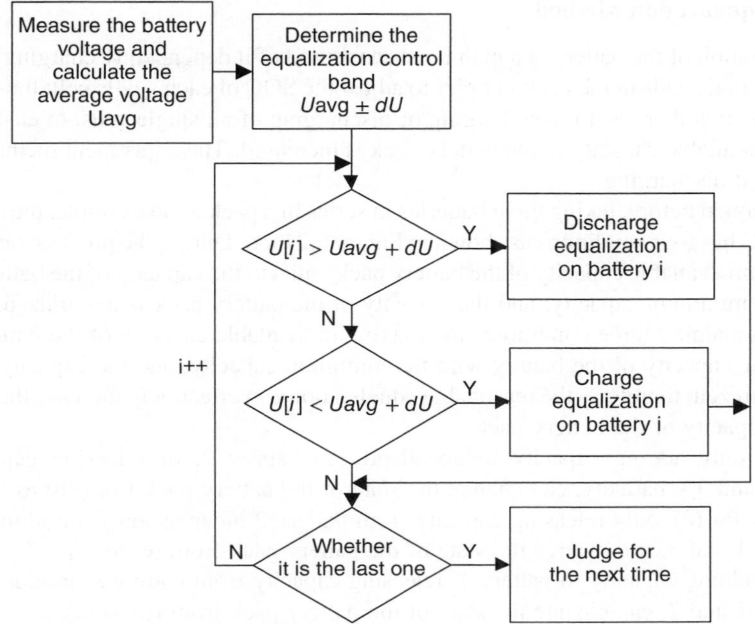

Figure 6.24 Judgment procedures for equalization based on the external voltage.

change the state of each cell. This kind of equalization is not only ineffective, but also increases the burden of the equalizer.

Therefore the equalization strategy based on the battery external voltage cannot effectively deal with the essential factors which the consistency problems of the battery pack cause. In the process of the equalization judgment, due to the differences in DC resistance, polarization voltage and capacity of the battery, the results change with variation of the working condition of the battery pack, producing instability. Meanwhile, the equalization strategy aims at the consistency problems of the battery external voltage, and does not effectively increase the available capacity of the battery pack, so it cannot reduce the adverse impact of the consistency problem in the use of the battery packs.

6.4.2 *Equalization of the Battery Pack Based on the Maximum Available Capacity*

The maximum available capacity of the battery pack is not only related to the maximum available capacity of each single cell in the pack, but is also closely connected with the SOC. After the batteries are put in series in a pack, the maximum available capacity of the pack is no more than that of the battery in the pack with the minimum capacity. Therefore, the equalization target is to make the maximum available capacity of the battery pack equal to the maximum available capacity of the battery with the minimum capacity.

6.4.2.1 Equalization Method

The equalization of the battery is a maintenance method of independently charging or discharging some single cells in the pack in order to adjust the SOC of each single cell; thus, under the precondition that there is no overcharging or discharging of all single cells, to ensure that the maximum available capacity of the battery pack is increased. The adjustment methods include charging and discharging.

As mentioned before, taking three batteries in series in a pack as an example, the distribution of their SOC has four conditions as shown in Figure 6.21a–d. During the process, only for (a) is the maximum available capacity of the battery pack equal to the capacity of the battery (battery 2) with the minimum capacity, and the capacity of the battery pack is maximized. However, under the remaining three conditions, the maximum available capacity of the battery pack is less than the capacity of the battery with the minimum capacity, and the capacity utilization is not maximized; therefore, the reasonable equalization can effectively increase the maximum available capacity of the battery pack.

For (b), only adding capacity independently to battery 2, or releasing capacity from batteries 1 and 3 separately, can change the state of the battery pack from (b) to (a).

Similarly for (c), only releasing capacity from battery 2 independently, or adding capacity to batteries 1 and 3, can change the state of the battery pack from (c) to (a).

For (d), adding capacity to battery 1, releasing capacity from battery 3 or adding capacity to batteries 1 and 2, can change the state of the battery pack from (d) to (a).

It is clear that the equalization is not achieved by working on one cell in the battery pack; either discharging the battery with a higher SOC, or charging the battery with a lower SOC, can just get a different equalization object. Through equalization, the maximization of the capacity of the battery pack can be achieved. In order to improve the utilization rate, in the following context we take the charging equalization as an example, in which the equalization capacity of the battery is calculated as what needs to be done in the discharging equalization.

6.4.2.2 Calculation of the Equalization Capacity

Taking n batteries in series in a pack as an example, if the n batteries have maximum available capacity $Q_{\max}[1], \ldots, Q_{\max}[n]$, SOCs SOC[1], …, SOC[$n$], and if battery m is the battery with the minimum capacity in the pack, then:

1. $Q_{\max}[m] = \min\{Q_{\max}[1], \ldots, Q_{\max}[n]\}$; (6.30)
2. the maximum charging capacity of the battery m is:

$$Q_{\text{ch\_max}}[m] = Q_{\max}[m](1-\text{SOC}[m]) \tag{6.31}$$

3. the maximum discharging capacity of the battery m is:

$$Q_{\text{dch\_max}}[m] = Q_{\max}[m]\text{SOC}[m] \tag{6.32}$$

4. the maximum charging capacity of the battery pack is:

$$Q^{\text{B}}_{\text{ch\_max}} = \min\{Q_{\text{ch\_max}}[i]\}, \quad i = 1, \ldots, n \tag{6.33}$$

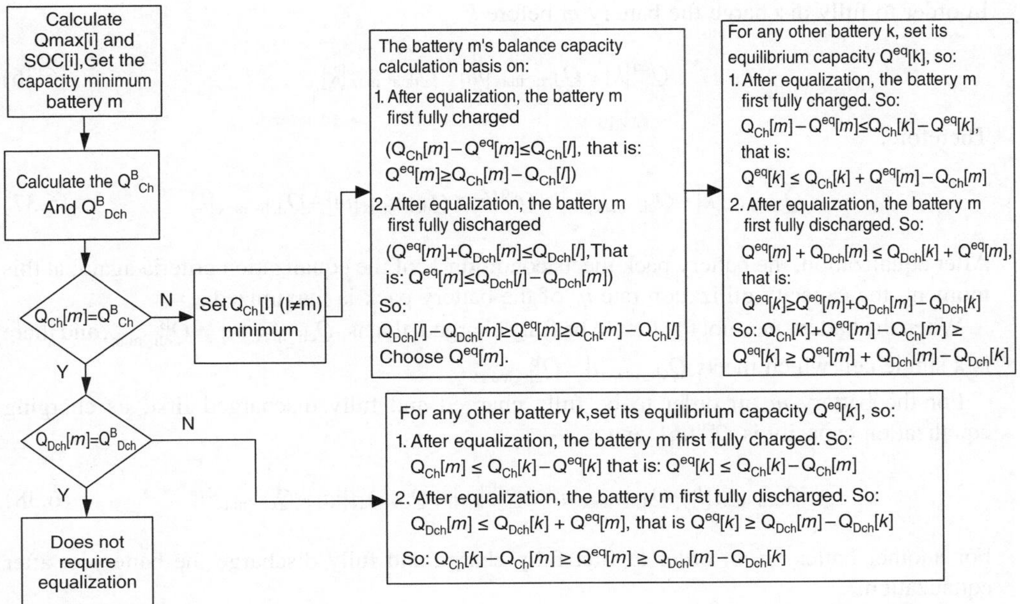

Figure 6.25 The computing method for calculation of the equalization capacity and procedures for charging and discharging.

5. the maximum discharging capacity of the battery pack is:

$$Q^{\mathrm{B}}_{\mathrm{dch\_max}} = \min\{Q_{\mathrm{dch\_max}}[j]\},\quad j = 1,\ldots,n. \tag{6.34}$$

Based on the above equalization criteria, in order to achieve the utilization maximization of the capacity of the battery pack, it is required that the battery with minimum capacity is fully charged or discharged in the process of charging and discharging. The method of calculation of the charging and discharging equalization of the battery is shown in Figure 6.25.

In accordance with the aforementioned definition of the capacity and the SOC of the battery pack with n batteries in series based on the evaluation of the consistency evaluation of the capacity utilization rate, in use, the distribution of the state of the battery has the following several conditions:

When the battery m is fully charged and fully discharged, that is, $Q_{\mathrm{ch\_max}}[m] = Q^{\mathrm{B}}_{\mathrm{ch\_max}}$ and $Q_{\mathrm{dch\_max}}[m] = Q^{\mathrm{B}}_{\mathrm{dch\_max}}$, the capacity utilization rate of the battery pack $\eta_{\mathrm{c}} = \frac{Q^{\mathrm{B}}_{\mathrm{max}}}{Q_{\mathrm{max}}[m]} = 1$; so, there is no need for equalization.

When the battery m is the first fully charged, but not the first fully discharged, that is, $Q_{\mathrm{ch\_max}}[m] = Q^{\mathrm{B}}_{\mathrm{ch\_max}}$ and $Q_{\mathrm{dch\_max}}[m] > Q^{\mathrm{B}}_{\mathrm{dch\_max}}$, obviously $\eta_{\mathrm{c}} \le 1$: so, for any single cell, if its charging equalization capacity is $Q^{\mathrm{eq}}[k]$, then:

In order to fully charge the battery m before k:

$$Q^{\mathrm{eq}}[k] \le Q_{\mathrm{ch\_max}}[k] - Q_{\mathrm{ch\_max}}[m] \tag{6.35}$$

In order to fully discharge the battery m before k:

$$Q^{\mathrm{eq}}[k] \geq Q_{\mathrm{dch\_max}}[m] - Q_{\mathrm{dch\_max}}[k] \tag{6.36}$$

Therefore:

$$Q_{\mathrm{ch\_max}}[k] - Q_{\mathrm{ch\_max}}[m] \geq Q^{\mathrm{eq}}[k] \geq Q_{\mathrm{dch\_max}}[m] - Q_{\mathrm{dch\_max}}[k] \tag{6.37}$$

After equalization, the battery pack meets condition 1. of the equalization criteria again; at this moment, the capacity utilization rate η_c of the battery pack is maximized.

When the battery m is not the first to be fully charged, that is, $Q_{\mathrm{ch\_max}}[m] \neq Q^{\mathrm{B}}_{\mathrm{ch\_max}}$, and there is a single cell which meets $Q_{\mathrm{ch\_max}}[l] = Q^{\mathrm{B}}_{\mathrm{ch\_max}}$.

For the battery m, in order to be fully charged and fully discharged first, its charging equalization capacity is $Q^{\mathrm{eq}}[m]$, and:

$$Q_{\mathrm{dch\_max}}[l] - Q_{\mathrm{dch\_max}}[m] \geq Q^{\mathrm{eq}}[m] \geq Q_{\mathrm{ch\_max}}[m] - Q_{\mathrm{ch\_max}}[l] \tag{6.38}$$

For another battery p, in order to first fully charge and fully discharge the battery m after equalization:

$$Q_{\mathrm{ch\_max}}[p] + Q^{\mathrm{eq}}[m] - Q_{\mathrm{ch\_max}}[m] \geq Q^{\mathrm{eq}}[p] \geq Q^{\mathrm{eq}}[m] + Q_{\mathrm{dch\_max}}[m] - Q_{\mathrm{dch\_max}}[p] \tag{6.39}$$

In the above expression, $Q_{\mathrm{ch\_max}}[p] + Q^{\mathrm{eq}}[m] - Q_{\mathrm{ch\_max}}[m]$ is the capacity which can be charged into the battery p after fully charging the battery pack.

$$Q_{\mathrm{dch\_max}}[p] - Q^{\mathrm{eq}}[m] - Q_{\mathrm{dch\_max}}[m] \tag{6.40}$$

is the capacity which can be discharged from battery p after fully discharging the battery pack.

It is clear that after equalization, the battery m with the minimum capacity in the pack, is first to be fully charged and fully discharged in the process of charging and discharging, achieving the maximization of the capacity utilization rate of the battery pack.

6.4.2.3 Confirmations

For the condition (b) in Figure 6.21, the capacity of battery 2 is the smallest. When the battery pack is fully discharged, the battery 2 is fully discharged, but the other two batteries only keep a part of the capacity. Now assume that the parameters of these three batteries are:

Battery 1: $Q_{\max}[1] = 100$ Ah, SOC[1] = 15 %
Battery 2: $Q_{\max}[2] = 90$ Ah, SOC[2] = 0 %
Battery 3: $Q_{\max}[3] = 100$ Ah, SOC[3] = 5 %

The analysis shows that the maximum available capacity of the battery pack is 85 Ah, less than that of battery 2. Further calculation shows that, among these three batteries, the discharging capacity of battery 2 is the smallest, so it is first to be fully discharged. However, because the charging capacity of battery 1 is less than that of battery 2, when charging this battery pack, battery 1 will be the first to be fully charged. Therefore, the discharging equalization should be done to battery 1, and the discharging capacity can be obtained from $Q_{ch\_max}[p] + Q^{eq}[m] - Q_{ch\_max}[m] \geq Q^{eq}[p] \geq Q^{eq}[m] + Q_{dch\_max}[m] - Q_{dch\_max}[p]$ which should meet: 15 Ah $\geq \Delta Q^{eq}_{dch}[1] \geq 5$ Ah.

For battery 3, because $Q_{dch\_max}[3] > Q_{dch\_max}[2]$, there is no need for equalization.

After equalization, the maximum available capacity of the battery pack can be increased to 90 Ah, equal to the capacity of the battery with the minimum capacity in the pack. A similar analysis method can be applied to (c) and (d).

Therefore, though the equalization of the battery pack cannot increase the maximum available capacity of any single cell, by adjusting the SOC of the batteries, the maximum available capacity of the battery pack can be increased. This is the significance and role of equalization.

The equalization strategy based on capacity takes the single cell with the minimum capacity in the pack as the reference standard, and takes the capacity of full charge and full discharge of the battery as the basis to calculate the limits of the equalization capacity needed by each cell, which provides effective data support for achieving equalization. After being equalized based on capacity, the capacity utilization rate of the whole pack is maximized, that is, the battery pack reaches a good consistent state. Nevertheless, this strategy requires the recognition of SOC and the maximum available capacity of each cell in the pack, to determine the position of the battery with the minimum capacity and the equalization capacity of each cell. However, the maximum available capacity of the lithium battery can only be measured offline. But when affected by the consistency of the battery capacity and the reduction in the battery capacity, under online conditions, it is difficult to measure the maximum available capacity. Therefore, this strategy is not suitable for an online equalization scheme.

6.4.3 *Equalization of the Battery Pack Based on the Maximum Available Energy*

In order to maximize the stored energy of the battery, the equalization of the battery not only needs full charging and discharging of the battery with the minimum capacity (i.e., the capacity of the battery is fully utilized), but also need to fully charge all cells in the pack, which provides a theoretical basis for the equalization of the series battery pack. Then, for each cell, only ensuring that all cells can be fully charged, the maximization of the stored energy of the battery pack can be achieved.

6.4.3.1 Calculation of the Equalization Capacity

Similarly assume that, n batteries with maximum available capacity $Q_{max}[1], \ldots, Q_{max}[n]$, and SOC SOC[1], …, SOC[n], are connected in series in a pack. Then, before equalization, the maximum charging capacity ($Q^{B}_{ch\_max}$) of the battery pack is:

$$Q^{B}_{ch\_max} = \min\{Q_{max}[1](1-\text{SOC}[1]), \ldots, Q_{max}[n](1-\text{SOC}[n])\} \tag{6.41}$$

If the charging equalization strategy is used for the batteries with the maximum charging capacity, then for any battery *i* in the pack:

$$Q_{\max}[i](1-\mathrm{SOC}[i]) \geq Q^{\mathrm{B}}_{\mathrm{ch\_max}} \tag{6.42}$$

That is, when the charging capacity of the battery pack reaches $Q^{\mathrm{B}}_{\mathrm{ch\_max}}$ without equalization, there is more than one cell fully charged; the necessary equalization charging capacity of the whole pack ($\Delta Q^{\mathrm{eq}}_{\mathrm{ch}}[i]$) is

$$\Delta Q^{\mathrm{eq}}_{\mathrm{ch}}[i] = Q_{\max}[i](1-\mathrm{SOC}[i]) - Q^{\mathrm{B}}_{\mathrm{ch\_max}} \tag{6.43}$$

If the discharging equalization strategy is used for the batteries with the least maximum charging capacity, then for any battery *i* in the pack:

$$Q^{\mathrm{B}}_{\mathrm{ch\_max}} = \min\{Q_{\max}[1](1-\mathrm{SOC}[1]),\ldots,Q_{\max}[n](1-\mathrm{SOC}[n])\} \tag{6.44}$$

That is, if charging directly without equalization, then when the charging capacity of the battery pack reaches $\max\{Q_{\max}[1](1-\mathrm{SOC}[1]), \ldots, Q_{\max}[n]\times(1-\mathrm{SOC}[n])\}$, for this battery, the capacity has been overcharged, that is, the necessary discharging equalization capacity ($\Delta Q^{\mathrm{eq}}_{\mathrm{dch}}[i]$) is:

$$\Delta Q^{\mathrm{eq}}_{\mathrm{dch}}[i] = \max\{Q_{\max}[1](1-\mathrm{SOC}[1],\ldots,Q_{\max}[n](1-\mathrm{SOC}[n]))\} - Q_{\max}[i](1-SOC[i]) \tag{6.45}$$

This is the discharging equalization capacity needed by this battery.

6.4.3.2 Confirmations

Similarly, taking the condition (b) in Figure 6.21 as an example, the parameters of each cell are as given previously. The maximum charging capacity of the three batteries is:

$$Q_{\mathrm{ch\_max}}[1] = 85\ \mathrm{Ah};$$
$$Q_{\mathrm{ch\_max}}[2] = 90\ \mathrm{Ah};$$
$$Q_{\mathrm{ch\_max}}[3] = 95\ \mathrm{Ah}.$$

Therefore,

$$Q^{\mathrm{B}}_{\mathrm{ch\_max}} = 85\ \mathrm{Ah}.$$

Only battery 1 does not need extra charging, while both batteries 2 and 3 need equalization charging, and the equalization charging capacity is $\Delta Q_{\mathrm{ch}}[2] = Q_{\max}[2](1-\mathrm{SOC}[2]) - Q^{\mathrm{B}}_{\mathrm{ch\_max}} = 5$ Ah, $\Delta Q_{\mathrm{ch}}[3] = 10$ Ah. After equalization, when the battery pack is fully charged, the three batteries are fully charged.

From the above analysis it can be concluded that when using the maximum available energy of the battery pack as the basis for equalization, after equalization, the maximum available capacity of the battery pack is also maximized, reaching 90 Ah; at the same time all batteries are fully charged and the stored energy is also maximized.

6.4.4 *Equalization Based on the SOC of the Single Cells*

According to the above analysis of the consistency problem of the battery pack, the problems of the battery pack are mainly embodied in the SOC, DC resistance, polarization voltage, and the maximum available capacity. Because the SOC, DC resistance, polarization voltage, and maximum available capacity do not change dramatically in one charging and discharging process, the equalization of the battery pack is achieved by adjusting the SOCs of all the single cells. In research, when taking SOC as the reference object of equalization, the equalization circuit has higher efficiency.

The equalization strategy based on the SOC achieves equalization of the battery pack by narrowing the SOC differences between all the single batteries. According to the judgment based on the capacity equalization strategy, it can be determined that the equalization method based on SOC can also achieve an increase in the capacity utilization rate of the battery pack. The confirmation is as follows:

According to the different states of the cells in the estimation of the required equalization capacity:

For the first condition, when the SOC of each cell in the pack is approximately consistent, that is:

$$\mathrm{SOC}[1]=\mathrm{SOC}[2]=\cdots=\mathrm{SOC}[n]=\mathrm{SOC}' \tag{6.46}$$

Based on the definition of the maximum available capacity of the battery pack, obviously:

$$Q^{\mathrm{B}}_{\mathrm{ch\_max}}=\min\{Q_{\max}[i]\mathrm{SOC}'\}=Q_{\mathrm{ch\_max}}[m];\ \ i=1,\ldots,n \tag{6.47}$$

$$Q^{\mathrm{B}}_{\mathrm{dch\_max}}=\min\{Q_{\max}[i](1-\mathrm{SOC}')\}=Q_{\mathrm{dch\_max}}[m];\ \ j=1,\ldots,n \tag{6.48}$$

The capacity utilization rate of the whole pack $\eta_{\mathrm{c}}=\dfrac{Q^{\mathrm{B}}_{\max}}{Q_{\max}[m]}=1$ has been maximized, so there is no need to conduct the equalization of the battery pack.

For the second condition, any battery *k*, if it is equalized to the same SOC level with *m*, the equalization capacity is:

$$Q^{\mathrm{eq}}_{\mathrm{s}}[k]=Q_{\max}[k](\mathrm{SOC}[m]-\mathrm{SOC}[k]) \tag{6.49}$$

After substituting the formula of the equalization capacity judgment:

$$\begin{cases} Q_{\max}[k](\mathrm{SOC}[m]-\mathrm{SOC}[k])\le Q_{\max}[k](1-\mathrm{SOC}[k])-Q_{\max}[m](1-\mathrm{SOC}[m]) \\ Q_{\max}[k](\mathrm{SOC}(m)-\mathrm{SOC}[k])\ge Q_{\max}[m]\mathrm{SOC}(m)-Q_{\max}[k]\mathrm{SOC}[k] \end{cases} \tag{6.50}$$

Because $Q_{\max}[k] \ge Q_{\max}[m]$, the above set of inequalities is permanent, that is, $Q_s^{eq}[k]$ fits the relation:

$$Q_{\text{ch\_max}}[k] - Q_{\text{ch\_max}}[m] \ge Q_s^{eq}[k] \ge Q_{\text{dch\_max}}[m] - Q_{\text{dch\_max}}[k] \tag{6.51}$$

Similarly, it can be confirmed under the third condition, $Q_s^{eq}[k]$ also fits the relation:

$$Q_{\text{ch\_max}}[k] + Q^{eq}[m] - Q_{\text{ch\_max}}[m] \ge Q^{eq}[k] \ge Q^{eq}[m] + Q_{\text{dch\_max}}[m] - Q_{\text{dch\_max}}[k] \tag{6.52}$$

In conclusion, the equalization strategy based on SOC enables the SOC of all the single batteries to be consistent by adjusting them, at the same time improving the capacity utilization rate of the battery pack.

The judgment content of the equalization based on SOC contains such procedures as taking the SOC of the battery as the control object, narrowing the SOC difference between cells by charging and discharging the single cell. First, it requires the recognition of the SOC of each single cell in the pack, and then chooses one of the SOCs of the batteries as the equalization target. Generally speaking, in order to improve the efficiency of the equalization and take full advantage of the strengths of the charging and discharging equalization, it is necessary to set the average ($\overline{\text{SOC}}$) of the SOCs of the battery pack as this target. Similarly, set the equalization control zone to curb the fluctuation of the equalization, and for the battery with higher SOC, do the discharging equalization; otherwise do the charging equalization. Then the difference (ΔSOC) between the SOC and the average SOC, and the rated capacity can be used to calculate the equalization capacity needed by each cell; and the equalization is completed by measuring the capacity. The judgment procedures of this equalization strategy are shown in Figure 6.26.

The equalization strategy based on SOC only needs recognition of the SOC of the battery, and has no requirements for recognition of the maximum available capacity or for the determination of the position of the battery with the minimum capacity; therefore, it differs from the aforementioned capacity equalization strategy. Because the maximum available capacities of the cells are different, it is unrealistic to ensure the permanent consistency of the SOC of the battery, and that raises harsh requirements for the equalizer. In the process of equalization based on SOC, it is necessary to determine the SOC of all batteries at once, and then carry out the equalization aiming at the static SOC of this pack, to complete the equalization with the above method of measuring capacity. However, if in different charging and discharging cycles, the selected time of the SOC judgment differs a lot, it may lead to different results of the equalization in different cycles, which may cause the battery pack to convert repeatedly in different equalization states, resulting in the ineffective working of the equalizer. Therefore, it is essential to determine a certain moment to carry out equalization in the process of charging or discharging, for example, the start of charging, or the end of discharging, and so on.

The equalization strategy based on SOC not only improves the capacity utilization rate of the battery pack, but also solves the problem that the consistency affects the recognition of the state of the battery pack. Because, after equalization, the SOC of each cell comes near to being consistent, and the SOC of the battery pack is the SOC of the single cell, then using this method to correct SOC can greatly reduce the complexity of the estimation of the SOC of the battery pack.

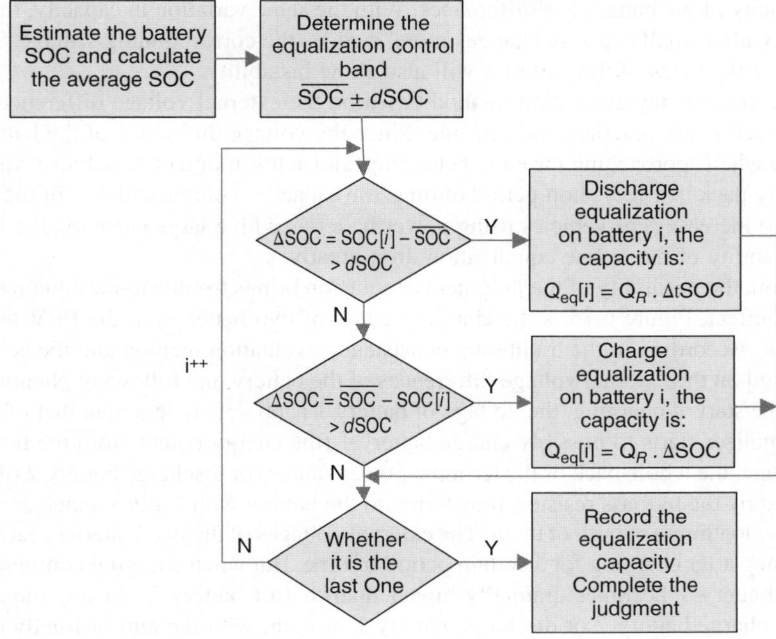

Figure 6.26 Procedures for equalization judgment based on SOC.

6.4.5 Control Strategy for the Equalizer

6.4.5.1 Equalization Control Strategy Based on the External Voltage Difference and the Main Problems

Based on the external voltage difference among the batteries in the series battery pack, the main problems in the equalization control strategy and judgment are listed below.

Because when the SOC of the battery is in the mid-range, every 1% change in SOC only results in a several mV change in the battery voltage; therefore, even if the SOC of the battery has some differences, it cannot be effectively reflected in the terminal voltage. Only when the battery is almost fully charged, because the voltage difference increases for every 1% SOC, and the voltage difference between the batteries can be effectively reflected in the terminal voltage of the battery, can it be determined that there will be problems with the consistency of the battery and the equalizer will be started to achieve equalization. However, because the battery has been almost fully charged, the equalizer has to have a large capacity in order to achieve equalization of a battery pack late in the charging within a comparatively short time. This increases the cost and size of the equalizer. In addition, the heat generated in the equalization process will lead to deterioration of the battery working environment, affecting the battery life.

There are differences between the DC resistances and the polarization voltages of the batteries, which will result in voltage differences of the batteries; and with the variation of the battery operating conditions (such as the current and the work mode), the consistency of the battery will be changing, showing instability.

The capacity of the battery has differences. With the same variation in capacity, the SOC of the battery with a small capacity changes more, as does the corresponding voltage. Therefore the voltage differences of the batteries will also show instability.

Thus, the consistency evaluation method based on the external voltage differences of each cell in the pack is not practical and sensible. Since the voltage difference of the batteries will appear only when approaching the end of charging, and at this moment, to achieve equalization of the battery pack in a very short period of time, the capacity, volume and cost of the equalizer are bound to increase. For vehicles using power batteries with a large capacity, the feasibility and practicability of the online equalization drop greatly.

In addition, the instability of the judgment conclusion brings trouble to the equalization control and its effect. Figure 6.15 is the charging curve of two batteries in the PEV lithium-ion battery pack. According to the traditional consistency evaluation method and the equalization method based on the external voltage differences of the battery, the following phenomena will appear; at the start of charging, the voltage of battery 1 is obviously less than that of battery 2, thus the equalizer starts to possibly charge battery 1 (the energy comes from the battery with higher voltage, the whole pack or the terminal power source) or discharge battery 2 (the energy is consumed by the by-pass resistor, transferred to the battery with lower voltage or the whole pack or other loadings) or both of them. The external voltages of the two batteries reach the state of consistency after charging for a certain period of time. But when charging continuously, the voltage of battery 1 becomes gradually higher than that of battery 2. At this moment, it is required to charge battery 2 or discharge battery 1 or both, with the aim of finally achieving the consistency of the two batteries. But after carefully analyzing the above process, it can be inferred that the aforementioned equalization method has the following problems:

1. In the process of charging, the equalization object changes, resulting in increase in the exchanging energy of the equalizer. At the start of charging, the voltage of battery 1 is lower, so the first-round equalization begins. Later, the voltage of battery 2 is lower, so the second-round equalization begins. But the equalization objects of the two rounds are different, thus increasing the burden of the equalizer. The capacity, volume and cost of the equalizer increase correspondingly.
2. The equalizer produces a large amount of heat. No matter whether the resistor bypass is used or the power electronic circuit is transformed, there will be more or less energy consumption, generating heat. On the one hand, this will lead to waste of energy; on the other hand, more importantly, if this heat is not immediately lost, it will result in increasing the temperature in the working environment of the battery and speed up the aging of the battery.
3. The life of the battery is affected. In order to make sure battery 2 is fully charged, at the end of the charging, the voltage of battery 1 is bound to be at a high level for a long time, which will speed up the loss of performance of the battery.

It is clear that, from the aspect of capacity utilization, this group of batteries do not need equalization. From the charging curve it can be clearly seen that the capacity of battery 1 is less than that of battery 2, and the SOC of battery 1 is low at the end of discharging, high at the end of charging; therefore, battery 1 (the battery with the minimum capacity) is fully charged and fully discharged first. However, from the aspect of energy, in order to fully charge the two batteries at the end of charging, and achieve the maximization of the available energy of the battery pack, the reasonable equalization to use is the continuous equalization charging for battery 2 in the process of charging, thus effectively reducing the capacity, volume and cost of the equalizer.

6.4.5.2 Equalization Strategy Based on the Capacity, Energy and SOC

As previously mentioned, the consistency evaluation system based on the DC resistance, polarization voltage, capacity, and SOC can effectively make a quantitative evaluation of the consistency of the battery from the aspect of available capacity and energy. In the whole process of charging, the consistency evaluation system based on this has good stability, and its evaluation conclusion does not change with the variation in the SOC and working current of the battery. In other words, the conclusion as to whether the battery pack needs equalization at the start of charging and at the end of discharging is consistent, and the charging capacity difference of the batteries is also fixed, thus effectively preventing changing of the equalization object.

The above analysis provides the method for deciding whether the battery pack needs equalization and formulates the method for calculating the capacity which will be provided by the equalizer of the battery from the aspect of capacity and energy utilization. By the SOC and capacity estimation methods of the battery, at the start of charging, the SOC and capacity of all batteries can be effectively obtained, then we can achieve a quantitative evaluation of the consistency of the battery pack, determine whether this battery needs equalization, and, in addition, determine how to equalize and how much capacity needs equalization. Thus, this provides the theoretical basis and data support for the equalization of the battery pack. Its advantages are:

In the whole process of charging, the equalizer can continuously and objectively charge or discharge the battery which needs equalization, thus providing enough time for equalization, improving the utilization rate of the equalizer, reducing the design capacity, and then reaching the goal of reducing the cost and volume.

The difference in the batteries gradually develops over repeated cycles. Therefore, for the battery with large capacity, with every cycle, the difference in the batteries changes a little. Thus, the capacity of the battery equalizer does not need to be very large. This makes it possible for a small equalizer to be applied to the batteries with large capacity, effectively reaching the equalization effect.

From the previous relevant discussion and the formula of the SOC of the battery, it is known that the present estimation of the SOC of the battery pack is converted into the SOC estimation of the single battery with the minimum capacity, which greatly reduces the complexity of estimation of the SOC of the battery pack.

6.4.6 *Effect Confirmation*

By testing the selected four batteries in series in the battery pack after 6 months of actual loading operation, in accordance with the test and calculation method of Chapter 3, the maximum available capacity of the battery and the SOC at the end of charging could be obtained. Based on the above analysis, the maximum charging and discharging capacity of each cell can be obtained by calculation. The test and calculation results are shown in Table 6.3.

Table 6.3 The main parameters of each battery and the SOC at the end of charging.

	Battery 1	Battery 2	Battery 3	Battery 4
SOC (%)	94	96	96	100
Maximum available capacity (Ah)	84.1	84.2	81.2	84.2
Maximum discharging capacity (Ah)	79.2	80.9	78	84.2
Maximum charging capacity (Ah)	5.1	3.4	3.2	0

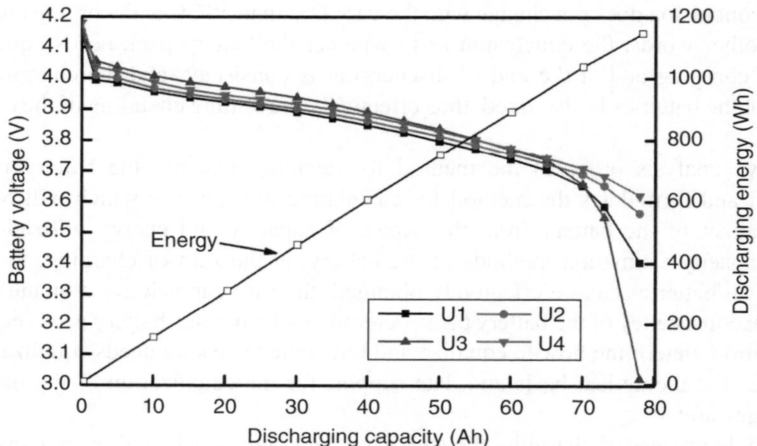

Figure 6.27 The discharging curve of the battery pack before equalization. (Reproduced with permission from Feng Wen, "Study on basic issues of the Li-ion battery pack management technology for Pure Electric Vehicles.", ©2009.)

Before equalization, performing the charging test on the battery pack produces the discharging curve shown in Figure 6.27. The discharging capacity of the battery pack is 78 Ah, and the discharging energy is 1145 Wh.

In order to achieve maximization of the capacity and energy utilization of the battery pack, it is required that at the end of charging all batteries should be fully charged. Therefore, in the following process, the equalizer will add the extra 5.1, 3.4, and 3.2 Ah to batteries 1, 2, and 3, respectively. After equalization, the discharging curve of the battery is shown in Figure 6.28.

It is clear that, at the start of charging, the voltages of the batteries are consistent and reach the state of being fully charged, that is, achieve high-end alignment. The discharging capacity of the battery reaches 81 Ah, the discharging energy being 1182 Wh. Compared with that before equalization, the discharging capacity and energy are increased by 3 Ah and 37 Wh, respectively, with the relative increments of 3.7 and 3.1%.

This experiment also proves the effectiveness of the discussions about the aforementioned content once again, including battery capacity estimation, SOC estimation, consistency evolution and equalization.

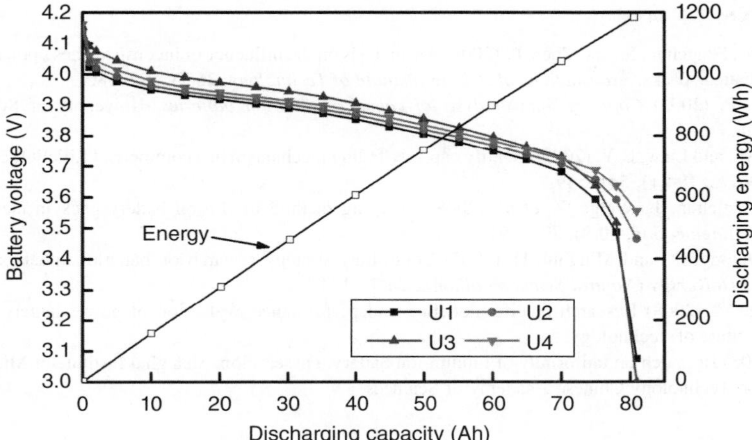

Figure 6.28 The discharging curve of the battery after equalization. (Reproduced with permission from Feng Wen, "Study on basic issues of the Li-ion battery pack management technology for Pure Electric Vehicles.", ©2009.)

6.5 Summary

This chapter aims to curb the equalization effect degradation caused by the traditional consistency evaluation method based on the external voltage differences of the battery, which cannot indicate internal differences of the batteries. It analyzes and discusses the consistency and equalization of the battery pack, mainly from the following aspects.

1. The consistency of the battery pack can be evaluated comprehensively from four aspects: DC resistance, polarization voltage, maximum available capacity and SOC, so the consistency judgment of the battery pack becomes more stable and the corresponding equalization more objective.
2. The different patterns of manifestation of each parameter in the process of charging and discharging are analyzed by using theoretical analysis and practical vehicle testing data, and the feasibility and stability of the consistency evaluation and analysis method proposed has been confirmed by a set of batteries used in practical loading operation.
3. The concept and the calculation method of the capacity (energy) of the battery pack are proposed based on the maximum available capacity (energy) before and after equalization of the battery pack and quantitative evaluation of the consistency of the battery pack has been carried out, producing a criterion for deciding whether a battery pack needs to be equalized, and the energy increase brought by the equalization is deduced and quantified.
4. Based on the maximization utilization of the capacity and energy of the battery pack, a calculation method and a more reasonable equalization control strategy have been proposed to improve the utilization rate of the equalizer. Thus, under the precondition of meeting the equalization requirements, the capacity, the volume and the cost of the equalizer are reduced effectively, so that the online equalization of power batteries with large capacity becomes possible.

References

[1] Zhenpo, W., Fengchun, S., and Chen, L. (2006) An analysis on the influence of inconsistencies upon the service life of power battery packs. *Transactions of Beijing Institute of Technology*, **26**(7), 577–580.

[2] Ramadass, P. (2003) *Capacity fade analysis of commercial Li-ion batteries*, University of South Carolina, Dissertation.

[3] Dubarry, M., and Liaw, B.Y. (2009) Identify capacity fading mechanism in a commercial $LiFePO_4$ cell. *Journal of Power Sources*, **194**(1), 541–549.

[4] Feng, W., Jiuchun, J., Weige, Z. *et al.* (2008) Charging method for Li-ion battery pack in electric vehicles. *Automotive Engineering*, **30**(9), 792–795.

[5] Rui, T., Datong, Q., and Minghui, H. (2005) Controlling strategy research on batteries imbalance. *Journal of Chongqing University (Natural Science Edition).*, **28**(7), 1–4.

[6] Xiangfeng, M. (2008) Research on life modeling and performance evaluation of power battery. Dissertation. Beijing Institute of Technology.

[7] Yi, D. (2008) Research on uniformity of lithium-ion battery. Dissertation. Shanghai Institute of Microsystem and Information Technology Chinese Academy of Sciences.

7

Technologies for the Design and Application of the Battery Management System

7.1 The Functions and Architectures of a Battery Management System

Electric vehicles always work in a complex operating environment with a variety of variables, such as temperature and humidity, load capacity, pressure, atmospheric corrosion, vibratory shock, input and output power, static placement, and so on. Therefore, the LIBs also work in a complex environment and these factors pose great challenges to their security, cycle life and effective use. It is important to manage the battery to reduce its high cost and improve its anti-abuse ability. The abuse of the battery (including over-charging, over-discharging, over-heating, etc.) could cause reduction in the battery cycle life, or even security accidents. These issues are more prominent in the battery pack because of the consistency differences of single cells. Ensuring the safety and cycle life of the battery pack in its use and management has become urgent.

Technologies for scientific management include measures for monitoring battery parameters, which can estimate battery states, ensure safe use and avoid fast reduction in cycle life in practical application.

7.1.1 The Functions of the Battery Management System

In EVs, the battery management system (BMS) has a great impact on safe operation, optimization of the strategy for vehicles, choice of the charging mode and reduction of operating costs. Either in the operation process or in the static charging process, the BMS should provide real-time monitoring of battery states and fault diagnosis, and inform the vehicle control unit (VCU) or charger through communication. Then the VCU or charger can adopt the corresponding control strategy to achieve effective and safe use of the battery. The operation conditions of the battery pack vary according to vehicle type (battery electric vehicle, BEV or hybrid electric vehicle, HEV), and the relevant functions and parameters of the BMS will also be different.

Fundamentals and Applications of Lithium-ion Batteries in Electric Drive Vehicles, First Edition.
Jiuchun Jiang and Caiping Zhang.

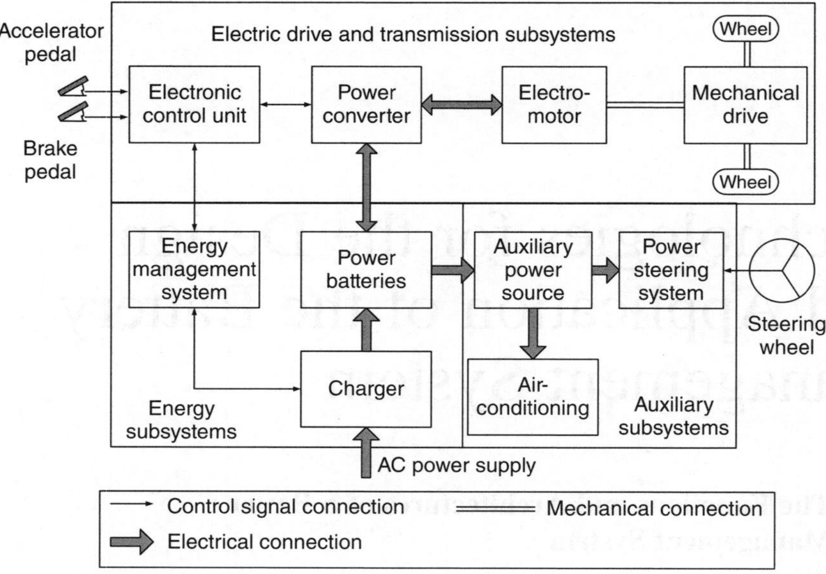

Figure 7.1 The system structure of a BEV.

Figure 7.1 shows the structure of the BEV. All the power in the BEV comes from a traction battery, which is required to have large capacity and high power in the battery pack. During the operating process, the battery pack works in a discharge mode or a feedback mode.

Figure 7.2 shows the structure of the HEV. Besides the battery pack, the power of HEVs also comes from an internal combustion engine (ICE). The working modes of the battery pack are various, including pure electric operation, charging, discharging and feedback.

In order to achieve the efficient operation of vehicles as well as to extend the cycle life of batteries, the BMS should include the following functions:

- Battery cell voltage measurement;
- Battery temperature measurement;
- Battery pack current measurement;
- Battery total voltage measurement;
- Insulation resistance measurement;
- Thermal management;
- Battery pack state of health (SOH) estimation;
- Battery pack SOC estimation;
- Analysis of battery fault and online alarm;
- Communication with on-board equipment;
- Communication with on-board monitoring equipment;
- Communication with battery charger which can realize the safe charge;
- Evaluation of battery discreteness in each pack;
- Recording of discharge and charge times.

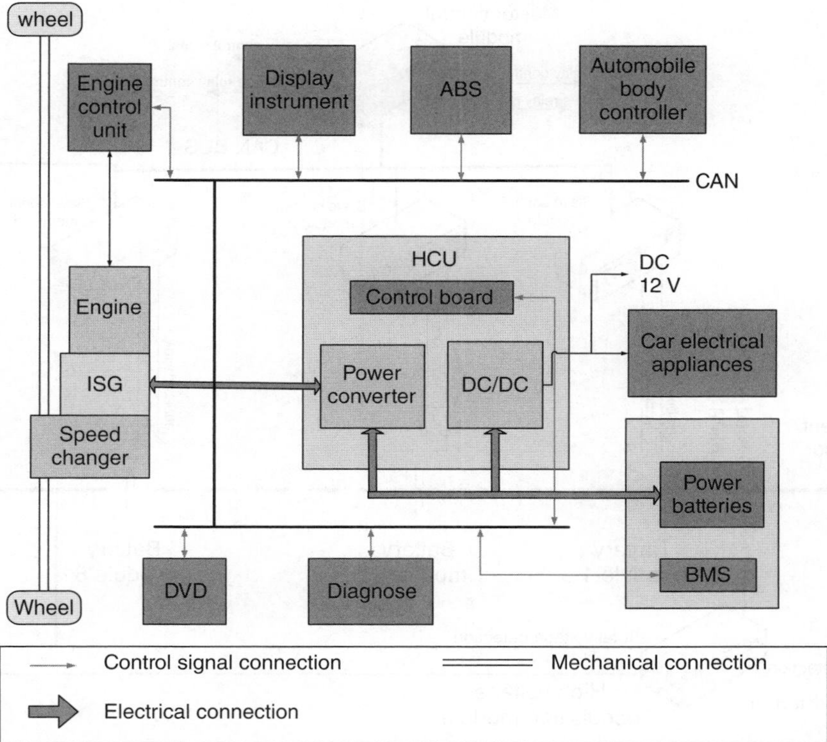

Figure 7.2 The system structure of an HEV.

7.1.2 Architecture of the Battery Management System

The BMS is constructed by using two basic architectures: the centralized architecture and the distributed architecture. In a centralized architecture, the functions of the BMS central processor are the measurement of voltage, temperature, current, insulation resistance, states estimation and communication, and so on. It requires relatively concentrated sample points and has advantages like simple connection, low cost and easy maintenance. However, due to the large number of traction batteries and the limited space in the vehicles, the batteries are always distributed in different areas in the EVs so that the centralized architecture has gradually disappeared in EVs.

The distributed architecture of the battery control unit (BCU) and the battery measurement unit (BMU) is shown in Figure 7.3.

The BCU is mainly used to process the battery parameters of each BMU, and estimate the states of the battery pack, including SOC, SOE (state of energy), SOH, and SOF (state of function), which provide the basic data for vehicle control and charge control, and also sends a control order to the BMUs. The hardware circuit structure is shown in Figure 7.4.

The BMU is mainly used to measure cell voltage, total voltage, current, insulation resistance and temperature, as well as to carry out the control order of the BCU. The BMU can be

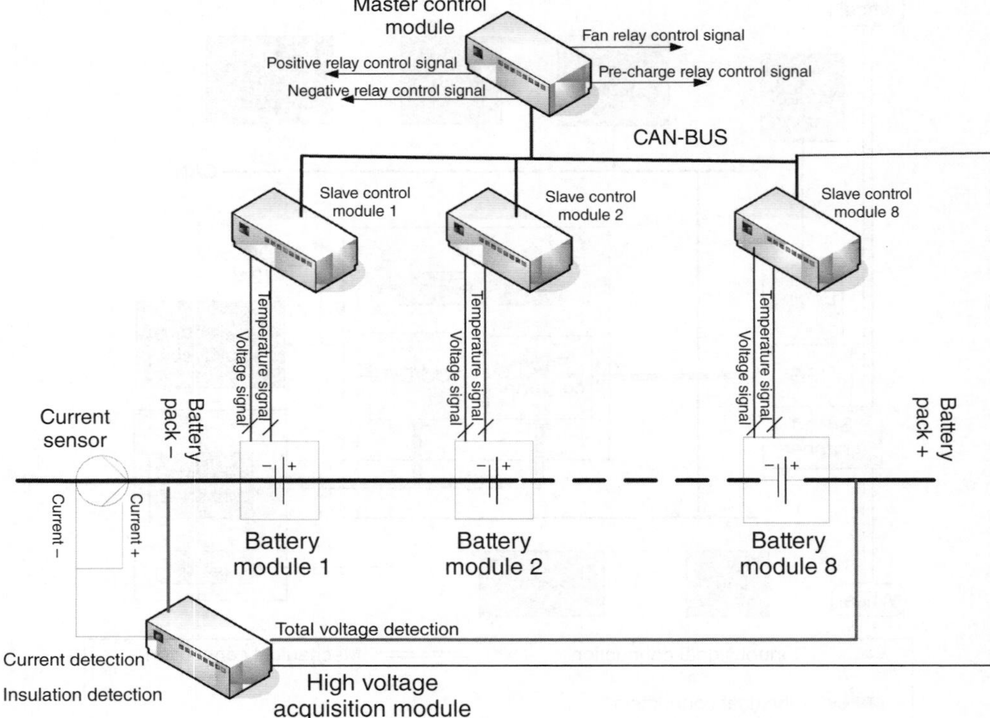

Figure 7.3 The distributed BMS architecture.

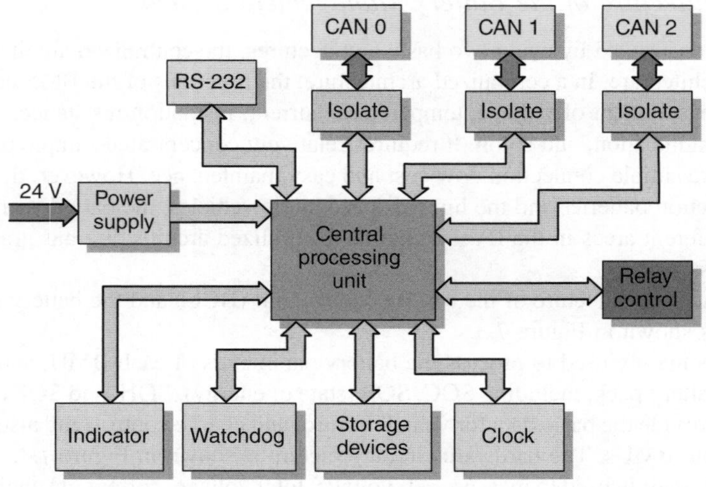

Figure 7.4 The hardware circuit structure of the BCU.

classified into a slave module and a high voltage measurement module according to their functions. The slave module generally is installed in the battery box and performs the following functions: cell voltage measurement, temperature measurement, balance control and thermal management (as shown in Figure 7.5). The high voltage measurement model is always installed in the vehicle central box and performs the following functions: the total voltage, current, energy and insulation resistance measurement and high voltage relay control (as shown in Figure 7.6). This model can prevent the mutual interference of high and low voltage, increase the flexibility of design and arrangement, and improve the reliability and safety of the system.

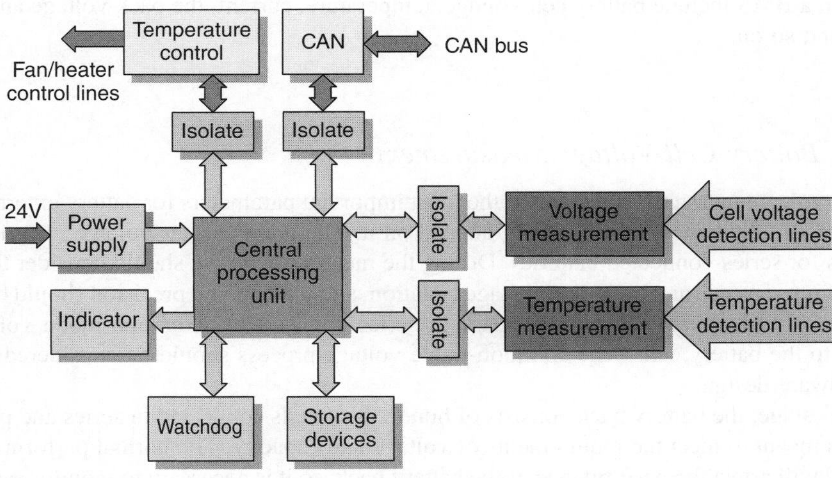

Figure 7.5 The hardware circuit structure of the slave module.

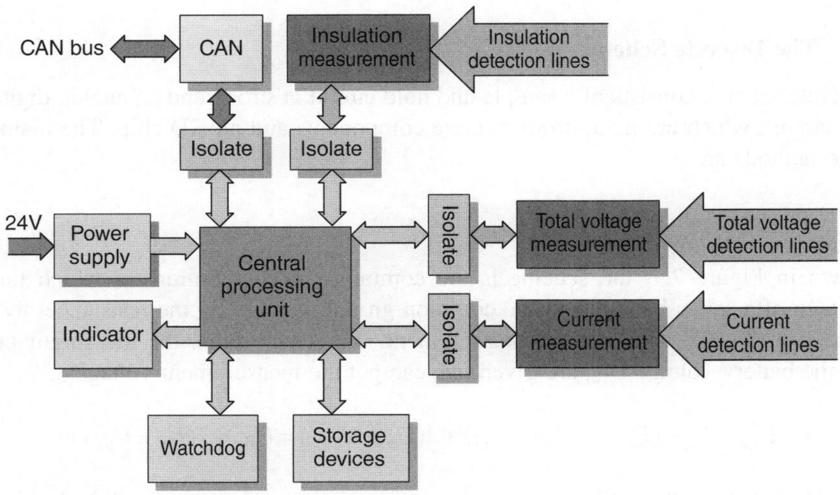

Figure 7.6 The hardware circuit structure of the high voltage measurement module.

The BCU and the BMU both have the function of communication by the controller area network (CAN) bus or RS485 bus which solves communication problems in the now widely applied distributed architecture.

7.2 Design of the Battery Parameters Measurement Module

The battery parameters measurement module is the key module of BMS for obtaining battery states and also is the core of the hardware design. Real-time measurements of EV battery parameters in a BMS include battery cell voltage, temperature, current, the pack voltage and insulation, and so on.

7.2.1 *Battery Cell Voltage Measurement*

The cell voltage measurement is one of the most important parameters for battery external performance and the battery states largely depend on it. There are various voltage measurement methods for series-connected batteries. During the measurement, we should consider the problems of interference and high–low voltage isolation, and the cost and precision should be carefully considered. Along with the increase in series connections, common-mode voltage is applied to the battery cells. The common-mode voltage process should be considered during the hardware design.

At this stage, the battery pack consists of hundreds of cells connected in series and parallel, which is meant to meet the requirements of voltage and capacity. The normal performance of each cell will affect the performance of the battery pack, so it is necessary to monitor each cell. The cell voltage measurement is generally carried out by two schemes: discrete and integrated.

7.2.1.1 The Discrete Scheme

The discrete scheme consists of a sample and hold circuit, a strobe and an analog digital conversion circuit, which are made from discrete components and an AD chip. The main measurement methods are:

The Resistance Divider

As shown in Figure 7.7, the scheme is the common-mode measurement which turns the cell voltage of a series battery into the common ground voltage by the resistance divider. If the relative error λ of each voltage measurement, the actual value, the measurement value U'_n and the battery voltage $U_{\mathrm{B}n}$ are given, we can get the measurement voltage:

$$U'_{\mathrm{B}n} = U'_n - U'_{n-1} = (U_n - U_{n-1}) + \lambda(U_n \pm U_{n-1}) = U_{\mathrm{B}n} + \lambda(U_n \pm U_{n-1}) \tag{7.1}$$

If there are a large number of batteries in series, the absolute value of U_n will be large and the relative error of $\lambda(U_n \pm U_{n-1})$ will also be large, causing decline in measurement accuracy.

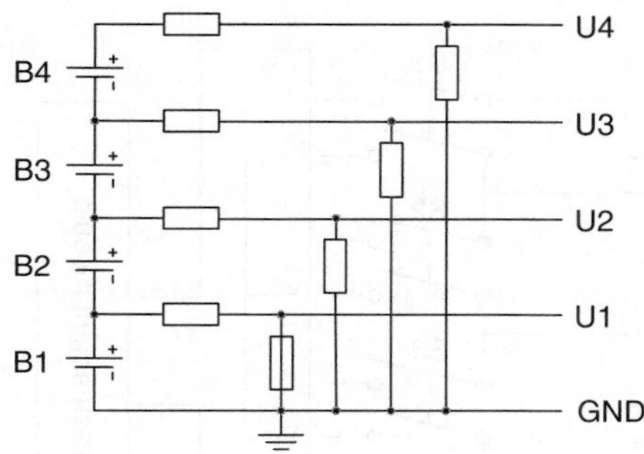

Figure 7.7 The scheme of the resistance divider.

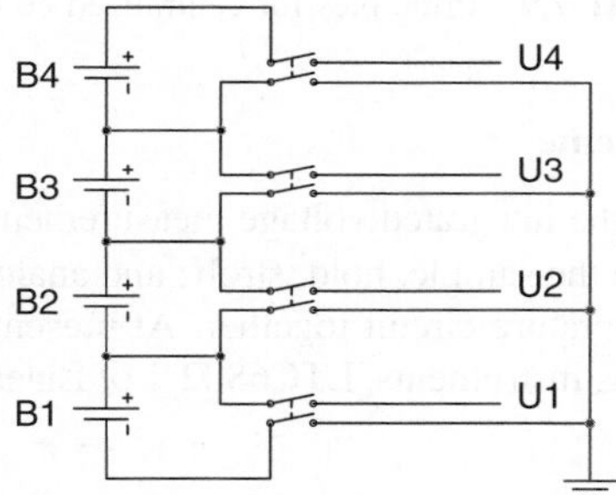

Figure 7.8 The principle of relay switching.

Relay Switching

By adopting this scheme, we could transfer the voltage of B_1 into U_1, B_2 into U_2, B_3 into U_3, and so on. Although U_1, U_2, U_3, are non-ground signals in this way, we can regard the above voltage as the common ground signals because only one relay is connected at the same time. This measurement method is simpler, as shown in Figure 7.8. Currently, with the optically triggered MOS gated solid state relay, the realization of the above circuit is easier. Moreover, the relay uses a contactless switch, which prolongs the cycle life of the circuit.

The Distributed Measurement Method and Its Control System

A structure of a measurement module is shown in Figure 7.9. In this scheme, several batteries share one measurement module. By channel switching, each cell voltage can be input to the measurement module and so there would be no common ground problem. Each measurement module is connected to the BCU by a communication bus and sends the measurement data to the host. Because the transmission signals on the communication bus are digital data (usually uses RS485 or CAN bus), it has a strong anti-interference ability. The measurement module could be installed near the batteries, thus the voltage signal acquisition line could be short and the anti-interference could also be improved.

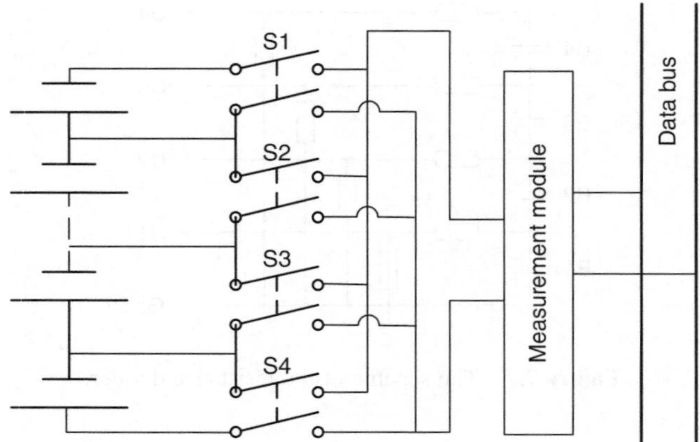

Figure 7.9 Principles for centralized control.

7.2.1.2 The Integrated Scheme

The integrated scheme adopts the integrated voltage measurement chips of LIBs series for EVs. These chips generally integrate the sample, hold, strobe and analog digital converter (ADC) circuit, balance control and temperature circuit together. At present, the mainstream chips in the market are bq76PL536 of Texas Instruments, LTC6802-1 of Linear, and MAX11068 of Maxim.

bq76PL536

The main features of bq76PL536 are:

- Supports three to six series cells;
- The high-speed SPI can be used for data communication;
- Exclusive fault signals;
- High-accuracy ADC in 14 bit resolution and 3 μs shortest switch time;
- Nine ADC inputs: six cell voltage input, two temperature input and one universal input;
- The built-in comparator (secondary protection) is used for protection from over-voltage, under-voltage and over-heating, the programmable threshold value and the delay time;
- Low power: the typical value of the static current is 12 μA and the typical value of the idle current is 48 μA;
- Integrates high precision of the 5.0 V/3 mA low dropout regulator (LDO).

The bq76PL536 is a stackable protector for a three to six series cell lithium-ion battery pack and analog front end (AFE) that incorporates a precision ADC, which has independent cell voltage and temperature protection, cell balancing and a precision of 5 V LDO to power user circuit.

The bq76PL536 provides protection for under-voltage, over-voltage and over-heating. Once the data surpasses the setting threshold value, it will automatically output the fault signals without external components configuration or start-up protection. The cell voltage and temperature protection functions are independent of the ADC system. The bq76PL536 should be used together with a main controller to improve its BMS function to the greatest extent, but its

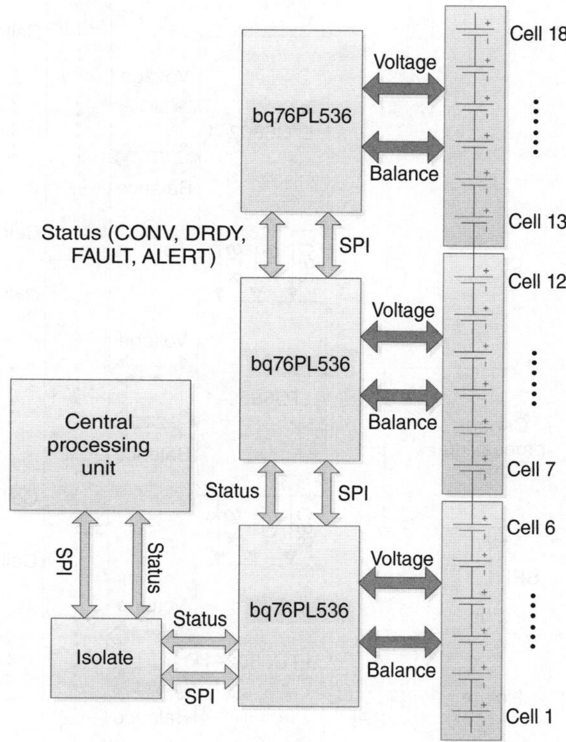

Figure 7.10 The circuit diagram for bq76PL536.

protection function is not controlled by the controller. The bq76PL536 could be stacked vertically to monitor 192 cells without isolation parts between the ICs (as shown in Figure 7.10).

LTC6802-1

The main features of the LTC6802-1 are:

1. It can be used to measure the LIB cell voltage of no more than 12 series cells (maximum total voltage is 60 V)
2. The stackable structure supports a 1000 V + system;
3. The measurement maximum error is 0.25%;
4. Battery balance: the on-chip passive battery balance switch provides the off-chip passive battery balance;
5. Two thermistor input;
6. 1 MHz serial interface for database error test;
7. High EMI immunity;
8. Built-in $\Delta\Sigma$ converter for noise filtering;
9. Fault detection for wire break connections;
10. A low power mode.

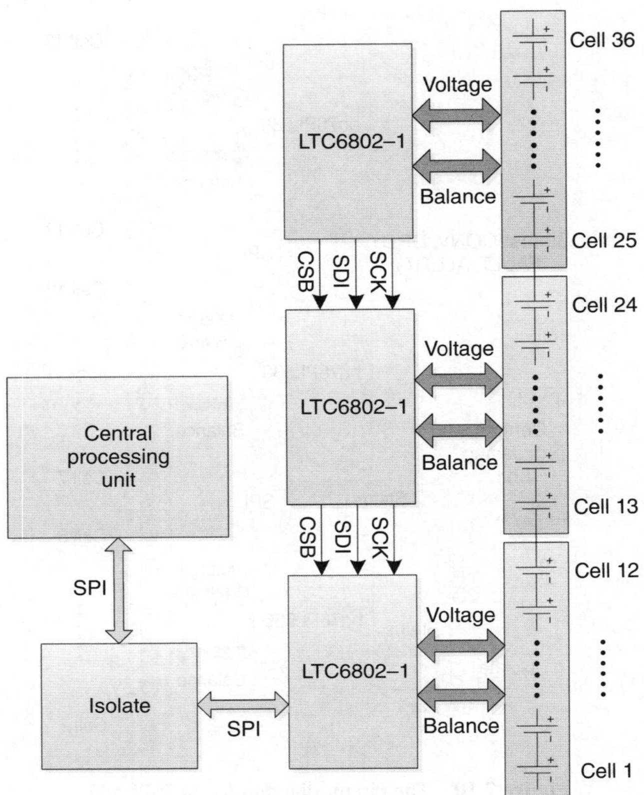

Figure 7.11 The circuit diagram for LTC6802-1.

The LTC6802-1 is the detection chip for battery voltage with a 12 bit resolution ADC, accurate voltage reference, high-voltage multi-inputs and a serial interface. In a normal mode, LTC6802-1 can mostly measure the total voltage of 12 cells in series, which is no more than 60 V. Without an optical couple or isolators, the LTC6802-1 chips can be connected with each other to monitor the voltage of each cell in a series battery pack through the special level shift interface. The circuit is shown in Figure 7.11.

When several LTC6802-1 are connected together, they could work at the same time to ensure that all the batteries can be measured in 13 ms. In addition to providing a mode of monitoring cell voltage, LTC6802-1 provides a standby mode in order to reduce power consumption.

MAX11068

The main features of MAX11068 are:

1. High-precision I/O, ±0.25% voltage measurement accuracy, ≤5 mV offset voltage.
2. Integrated 12 data acquisition system, 12-bit precision, high-speed SAR ADC, 12 cell voltage measurement in 107 μs, two additional analog inputs used for temperature measurement.

3. Battery fault detection: digital over-voltage and under-voltage threshold detection, open-circuit detection, over-heating threshold detection.
4. 12 integrated battery balancing switch supporting 200 mA current.
5. Integrated 6–70 V input linear regulator.
6. Integrated 25 ppm/°C, 2.5 V accurate reference.
7. Integrated level shifter, I^2C and SMBus.
8. Three general digital I/O.
9. Low power: 75 μA in a standby mode and 1 μA in a shut-off mode.

MAX11068 is an intelligent digital acquisition chip that is programmable, has high integration density, high voltage, and 12 channels. It can be used for the vehicle system and other equipment with series chargeable batteries. This chip combines a simple static machine and a high-speed I^2C bus, which supports a 31 chips stackable connection through an SMBus series link communication. MAX11068 AFE consists of a 12 cell voltage measurement acquisition system with a high voltage switch and an input unit. The input multiplexer/switching group allows differential measurement of each cell and all measurement values are differential data of cell voltage. Twelve cells can be measured in less than 107 μs.

MAX11068 adopts secondary scanning to measure cell voltage with error modification. The first stage of scanning is data acquisition for a 12 cell voltage. The second stage is error modification, which eliminates errors. Through the above two steps, a more accurate index can be obtained over the entire temperature range and in a noisy environment. Figure 7.12 is the circuit diagram for voltage measurement of MAX11068.

The integrated scheme is simpler in circuit design than the discrete scheme. It also solves many issues present in the discrete scheme, such as poor component matching , poor digital acquisition accuracy, more external mode, less automatic test, high test cost, low test coverage, difficult for external components power control, complex circuit wiring and large size.

7.2.2 Temperature Measurement

Temperature measurement is the real-time monitoring of the operating temperature of the battery pack. Currently, the measurement methods include use of a thermocouple, metal thermal resistance temperature detector, thermistor, analog integrated temperature sensor and intelligent temperature sensor.

1. The thermocouple is widely used in temperature measurement. Its main features include wide measurement range, stable performance, simple structure, good dynamic response, small measurement error, remote transmission 4–20 mA electrical signals, convenient automatic control and integrated control. The principles of a thermocouple are based on the thermoelectric effect. If two different conductors or semiconductors are connected into a loop and the two contact points have different temperatures, then thermoelectric power is produced in the loop.

   The output voltage of the thermocouple depends on the temperature difference between the hot and cool ends. In practice, it is necessary to add temperature compensation, use a special filter for enlarging the circuit and design a special temperature detection circuit at the cool end. When there are lots of temperature points to be measured, the connection

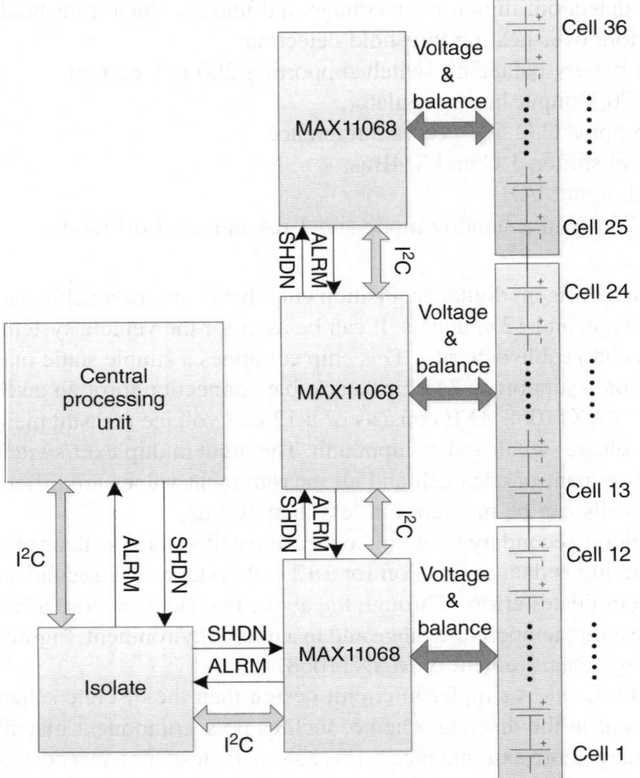

Figure 7.12 The electric circuit of MAX11068.

wiring will be complex. The large amount of wiring will decrease the reliability of the system and cause inconvenience in placing the lines.

2. A thermal resistance detector is a common temperature detector in the middle–low temperature region. Its principles are based on the fact that the resistance of a conductor or semiconductor will change as the temperature changes. Most thermal resistance detectors are made of pure metals. Currently, the widely used metals are platinum and copper, and others, such as nickel, manganese and rhodium, and so on, are beginning to be used.

   The main features of the thermal resistance detector include remote transmission signal, high sensitivity, good stability, good interchangeability and high accuracy. However, due to its large size and large thermal inertia, it cannot measure the speed of temperature changing with its low response speed. It needs external power stimulation. Its processing circuit and connection are as complex as those of the thermocouple.
3. The thermistor is a temperature measurement device comprising solid semiconductors with a high resistance temperature coefficient. According to the temperature coefficient, thermistors are classified into two types: positive temperature coefficient resistors (PTC) and negative temperature coefficient resistors (NTC).

The main features of the thermistor include high sensitivity, small size, good stability, strong overload ability, high response speed, and small delay. However, it has poor accuracy and pronounced nonlinearity. It is widely used in temperature control with less accuracy.

4. An analog integrated temperature sensor is an integrated sensor made of silicon semiconductors, it is also called a silicon sensor or monolithic integrated temperature sensor. An analog integrated temperature sensor is a special IC which integrates temperature sensors into one chip. It can complete temperature measurement and output an analog signal.

   The output signals of the analog integrated temperature sensor include current, voltage, frequency, and so on. The main features are low price, fast response speed, remote transmission distance, small size and low power. It is suitable for measuring temperature at a remote distance. Nonlinear calibration is not necessary but it needs a complex processing circuit.

5. The intelligent temperature sensor (also known as a digital temperature sensor) integrates a temperature sensor, peripheral circuit, ADC, microcontroller and interface circuit into one chip, with the ability to measure temperature and communicate with the microprocessor.

   The intelligent temperature sensor can output temperature data and relative temperature control data. It has good characteristics, such as high measurement accuracy, fast conversion time, programmability, multipoint measurement in parallel, convenient measurement and installation, and easy positioning. Table 7.1 shows a comparison of each scheme.

Take the intelligent temperature sensor DS18B20 as an example to design a BMS temperature measurement circuit.

DS18B20 is the 1-wire intelligent temperature sensor produced by DALLAS Semiconductor Inc, with a range of −55 to 125 °C and accuracy of ±0.5 °C. It can directly convert temperature into a digital signal without a peripheral ADC circuit. The time of DS18B20 conversion is related to the setting resolution. When fixed in 9-bit, the maximum conversion time is 93.75 ms, in 10-bit 187.5 ms, in 11-bit 375 ms, and in 12-bit 750 ms.

When leaving the factory, each DA18B20 must be adjusted by laser. Each DA18B20 has a unique 64-bit serial number, so several sensors can be installed on one digital transmission bus. By using the function of online programmability, we can manually set the sensor number. During the installation, it is convenient to position each sensor according to the number. It can quickly find the point and it is convenient to maintain when there is a temperature point failure.

Table 7.1 Comparison of different temperature measurement schemes.

Scheme	Range (°C)	Error (°C)	Processing circuit	Positioning
Thermocouple	0–1600	±1	Filter and amplification AD conversion	Inconvenient
Thermal resistance	−200 to 600	<1	Filter and amplification AD conversion	Inconvenient
Thermistor	−50 to 400	<1	Filter and amplification AD conversion	Inconvenient
Analog integrated temperature sensor	−40 to 150	<5	Filter and amplification AD conversion	Inconvenient
Intelligent temperature sensor	−40 to 150	<3	Not required	Convenient

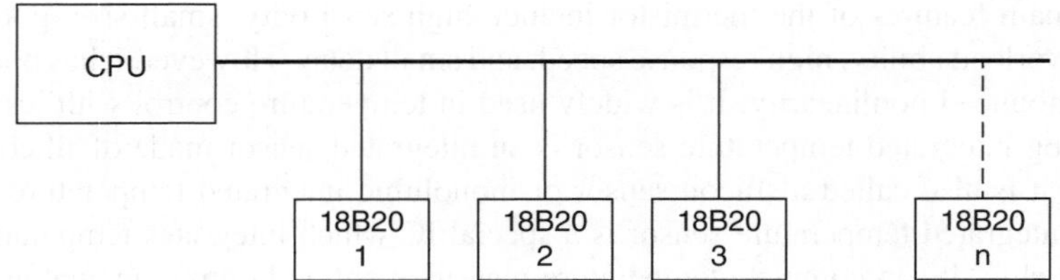

Figure 7.13 The temperature measurement module.

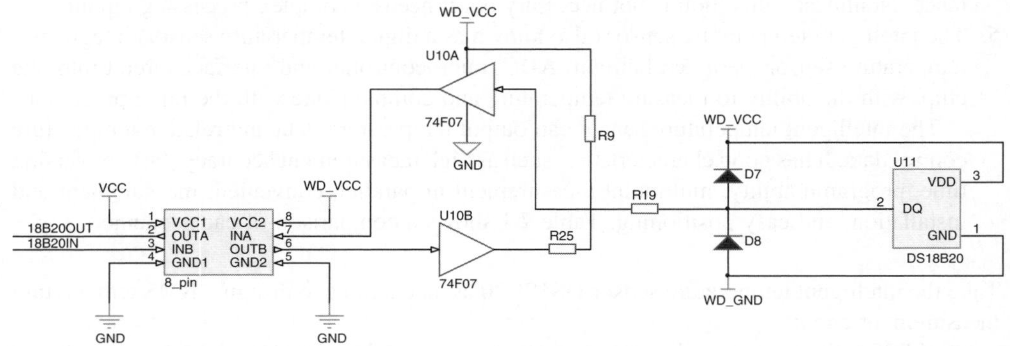

Figure 7.14 The interface circuit of the temperature measurement module.

Because of the different numbers of temperature measurements required in different battery packs, the number of a temperature measurement module needs flexible adjustment. According to the features of the 1-wire bus of DS18B20, it is easy to realize the sensitivity of the temperature measurement module that needs to adjust the number of sensors in parallel (as shown in Figure 7.13) and of software processes.

In actual application, it generally connects the temperature sensor with the tabs directly but this would connect the temperature measurement circuit directly to the high-voltage system. In order to avoid the introduction of high-voltage, it is necessary to add an isolation circuit between DS18B20 and the CPU so as to improve the anti-interference ability. Figure 7.14 is the interface circuit diagram. However, it cannot meet the requirements of drive current after the isolation. Therefore, it is essential to adopt a chip 7407 for driving.

7.2.3 *Current Measurement*

The current of charging or discharging is the external reflection of energy transfer and also the important base of the BMS energy management. The current measurement provides data for estimating SOC, and for the charging and discharging control of the BMS. It is necessary to ensure its sampling accuracy, anti-interference ability, zero drift, temperature drift and linear error. The schemes of current measurement include the Hall effect current sensor, the shunt, and so on.

7.2.3.1 Hall Effect Current Sensor

The output signal of a current sensor is a secondary current, in proportion to the input signal (primary current). The secondary current generally only has a value of 10–400 mA. If the output current goes through a measurement resistance, it will give a voltage output signal which is proportional to the primary current. Then through amplification and adjusting the circuit, the output signal is converted in the ADC.

According to the form of current, the Hall effect current sensor is classified into two types: DC and AC; by the form of the feedback, it is classified into open loop and closed loop types.

Figure 7.15 is a diagram of the open loop current sensor. Its working principle is that if semiconductor Hall components are put into the space of the magnetic loop, the Hall current (secondary current) and the Hall voltage would change as the measured conductor current I_N (primary current) changes. The change in voltage will be amplified through the amplifier that reflects the measured current.

In order to stably and accurately measure the current I_N, compensation windings are placed around the magnet ring, as shown in Figure 7.16. The compensation winding and output form

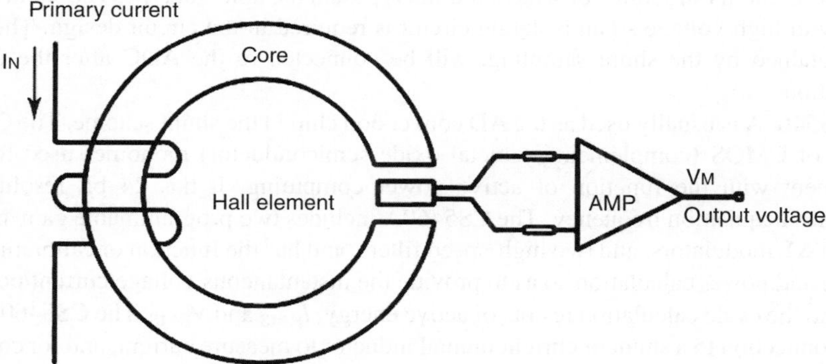

Figure 7.15 The structure diagram of the open loop current sensor.

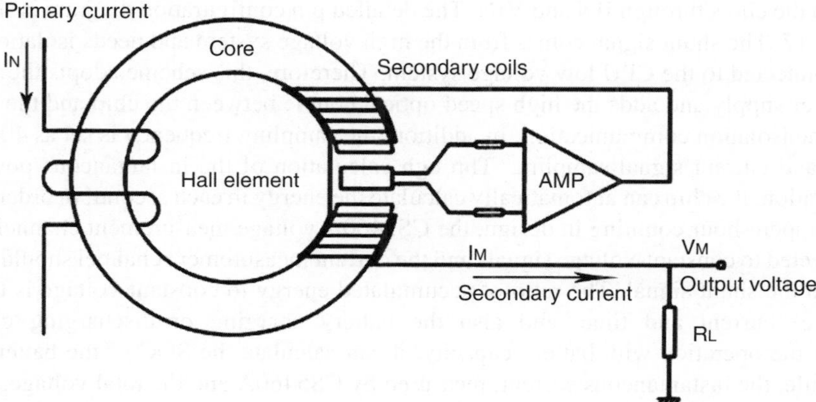

Figure 7.16 The structure diagram of the closed loop current sensor.

the closed loop of the Hall negative feedback. The Hall equipment and the ancillary circuit produce the secondary magnetic compensation current I_M, which reflects the measured current I_N. Through a sampling resistance R_L, the compensation current I_M can produce a voltage drop, which outputs in the form of voltage V. Then the voltage passes through the external filter amplification and ADC circuit, and the measured current is calculated.

The Hall effect current sensor has the characteristics of high accuracy, fast speed, wide band, better isolating function, strong overload ability and no loss of energy in the measured circuit. However, in the complex electromagnetic interference environment of vehicles, it is difficult to ensure the linearity of the measurement with some issues like the zero-drift, temperature-drift, and so on.

7.2.3.2 The Shunt

The shunt comprises a rod resistor of nickel, manganese copper alloy and copper strips with a nickel coating. When the DC current goes through the shunt, it produces a voltage at the two ends of the shunt. Its structure consists of the slot type and the non-slot type. The shunt connects directly with high voltage so an isolating circuit is required in the circuit design. The voltage signal, obtained by the shunt sampling, will be connected to the ADC after the filter and amplification.

The CS5460A is usually used as the AD conversion chip in the shunt scheme. The CS5460A is a chip of CMOS (complementary metal oxide semiconductor) monomer used for power measurement with the function of active power computing. It has 24 bit resolution and 4000 point/s acquisition frequency. The CS5460A includes two programmable gain amplifications, two $\Delta\Sigma$ modulators, and two high-speed filters, and has the function of calibrating, effective value and power calculation so as to provide the instantaneous voltage/current/power data sample and the cycle calculation results of active energy, I_{RMS} and V_{RMS}. The CS5460A is suitable for connection to a shunt or current mutual inductor to measure current, and for connection to a divider resistance or voltage transformer to measure voltage.

The CS5460A exchanges data with the CPU through an SPI series communication bus, which uses five signal lines: CS, SDI, SDO, SCLK, and INT. The current and voltage signal is sent to the chips through IIN and VIN. The detailed pin configuration diagram is shown in Figure 7.17. The shunt signal comes from the high voltage system and needs isolation when being connected to the CPU low voltage system. Therefore, this scheme adopts the isolated 5 V power supply and adds the high speed optical couple between the chip and the CPU to realize the isolation communication. In addition, the sampling frequency is set as 4000/s for voltage and current signal sampling. Through calculation of the instantaneous power and accumulation, the chip can automatically calculate the energy in each second. In order to realize the ampere-hour counting in design, the CS5460A voltage measurement channel should be connected to constant-voltage signals and the current measurement channel should be connected to the shunt signal. The ratio of accumulated energy to constant voltage is the integration of current and time, and also the battery charging or discharging capacity. Through the operation with battery capacity, it can calculate the SOC of the battery pack. Meanwhile, the instantaneous current, measured by CS5460A and the total voltage, reflects the instantaneous power of the battery.

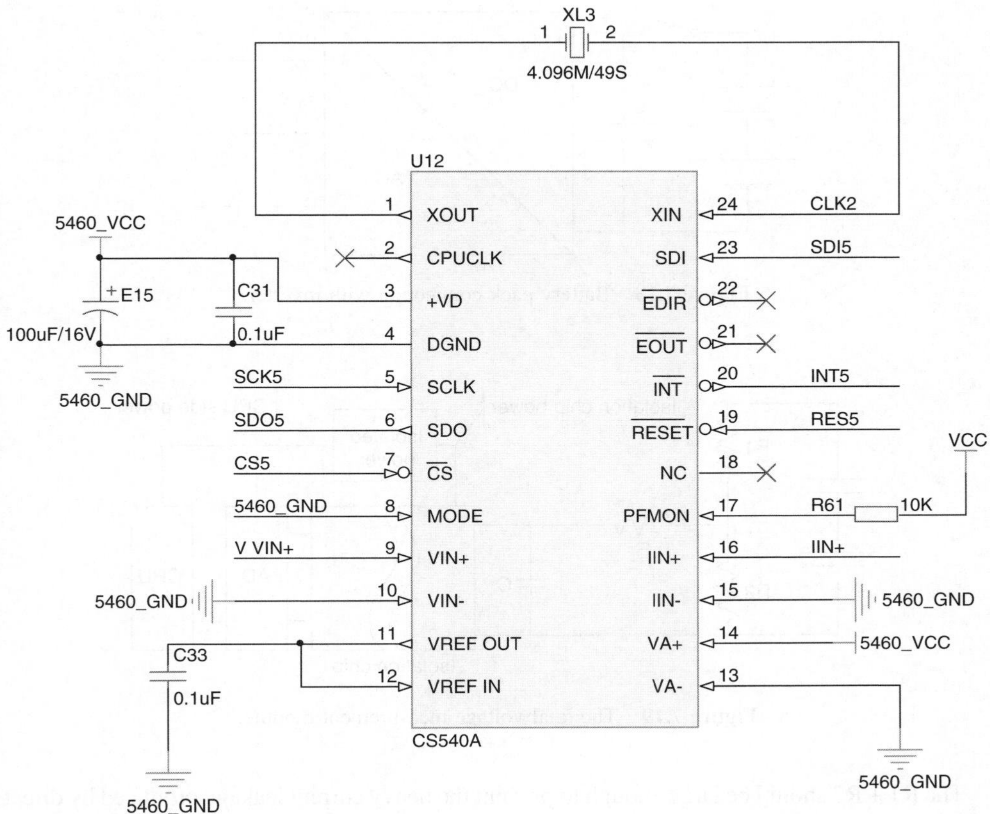

Figure 7.17 The pin configuration diagram for CS5460.

7.2.4 Total Voltage Measurement

As an important parameter for battery application and management, the total voltage can reflect the entire state of the battery. Although the total voltage could be obtained by summing the voltage of all the cells, this cannot really reflect the total battery voltage situation after connection. On the one hand, when several packs are in series, the single pack generally has over-current insurance. The packs are connected to each other with wire. When the connection is not firm, only the total voltage can reflect this. On the other hand, the pack always connects with the inverter, as shown in Figure 7.18. Before the inverter, there is a large capacity that is required in the pre-charge process to prevent a large instantaneous current. By connecting the total voltage sample line to the external end of the relay (measuring the capacity voltage), we can monitor the capacity voltage and provide data for pre-charge control during the pre-charge process.

The total voltage measurement generally uses a resistance divider, then uses the external ADC for sampling. As shown in Figure 7.19, it uses R1 and R2 for the resistance divider, then is connected through a RC filter to an ADC conversion chip.

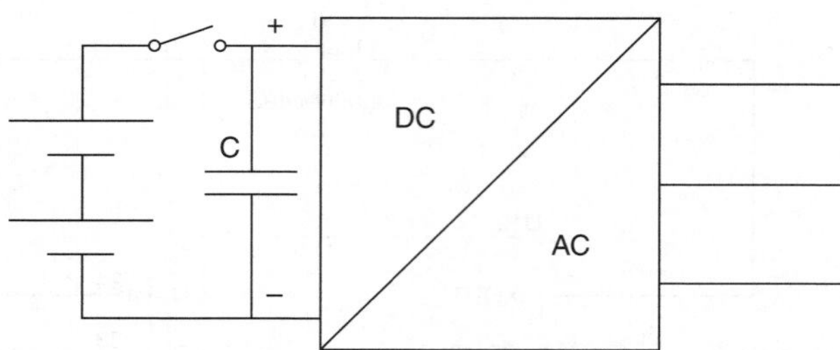

Figure 7.18 Battery pack connection with inverter.

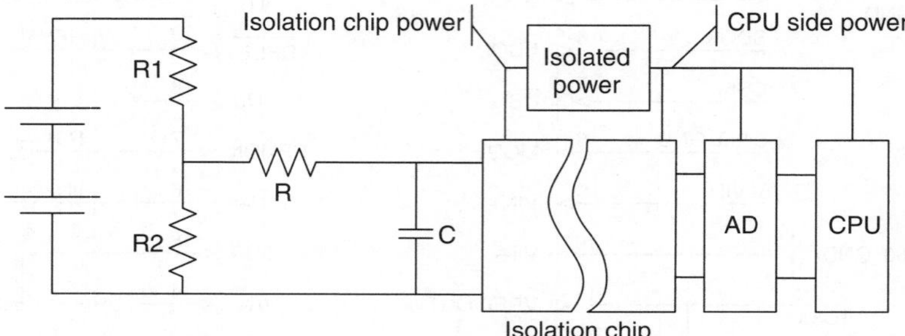

Figure 7.19 The total voltage measurement module.

The R1 + R2 should be large enough to prevent the heavy current leakage produced by direct connection with the battery pack and the wasting of energy. The voltage of the battery pack is usually high. It is necessary to add an isolator between the sample circuit and the CPU in order to avoid the high voltage breaking the measurement circuit. The isolated circuit could be put in the front of the ADC chip to make the analog isolation and it could also be put behind the ADC chip to make the digital isolation.

7.2.5 Insulation Measurement

The conventional vehicle uses natural gas or fuel oil as the power source without a high voltage insulation problem. The EV is a complex production of mechanical-electrical integration. The power supply of the EV is realized by several series battery packs, with the total voltage being in the range 100–500 V. Devices with large current and insulation are as follows: traction battery, motor, charger, energy recycle device, and auxiliary battery charge device. All these devices have high voltage components. Although their insulation issue has already been considered in the design stage, poor working conditions, such as shaking, corrosion by acid–base gas and changes in temperature and humidity, could cause fast aging of the power cable and insulation materials, or even break the isolation , which would decrease the insulation strength and endanger personnel. In order to ensure safety in use, it is essential to evaluate the insulation conditions of EVs.

According to the National Standard, 18384.3-2001, the safety requirement for electric vehicles, there are clear stipulations for the definition of insulation state, measurement method, and safety requirements. The insulation state of an EV is evaluated according to the ground insulation resistance of the DC positive and negative bus. The definition of traction battery insulation resistance in the national standard is the relative resistance to maximum leakage current (in the worst condition) when there is a short between the traction battery and ground (electric chassis). According to the requirements of EVs and personnel safety in the national standard, under the conditions that the maximum AC voltage is less than 660 V, the maximum DC voltage is less than 1000 V and the weight is less than 3500 kg, the requirements of the high voltage security are as follows: (i) Personnel's security voltage is less than 35 V, or the product of the contact current with a person and the duration time is less than 30 mA s. (ii) Insulation resistance divided by the battery rated voltage should be more than 100 Ω V^{-1}, and preferably more than 500 Ω V^{-1}. Therefore, in order to ensure the insulation security of on-board batteries, it is necessary to detect the insulation resistance and raise an alarm in time. The insulation resistance measurement is significant for the applied technology of the traction battery.

Currently, the insulation measurement methods include the AC voltage insulation measurement method and the DC voltage insulation measurement method.

The AC voltage insulation measurement method is relatively simple, detecting leakage current by detecting the resultant magnetic field of the AC voltage with a mutual inductor. However, this method requires the additional AC current test signals to be passed into the DC current system, which would certainly impact on the DC current system.

The DC current insulation measurement is generally completed by measuring the insulation resistance. There are two measuring methods: passive ground detection and active ground detection.

7.2.5.1 Passive Ground Detection

The principle for passive ground detection is shown in Figure 7.20. R_P and R_N are the ground insulation resistance of the positive and negative bus, respectively. U_P and U_N are the ground voltage of the positive and negative bus, respectively. U stands for the total voltage. R_{P1}, R_{P2}, R_{N1}, and R_{N2} are bias resistors with given known resistance. When S is switched on, the

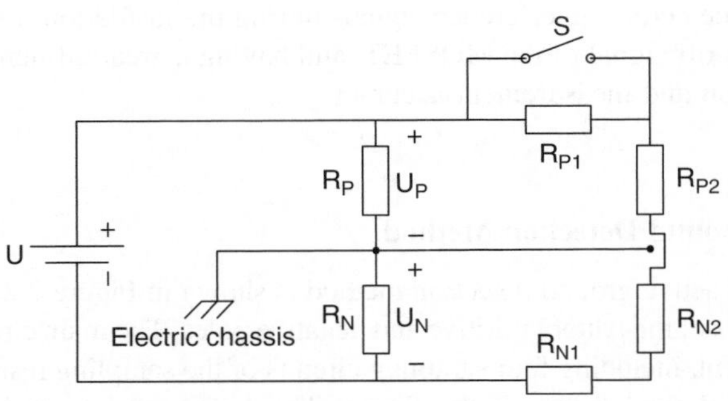

Figure 7.20 Diagram of the passive ground detection method.

measured ground voltages of the positive and negative bus are U_P and U_N. When S is switched off, the measured ground voltages of the positive and negative bus are U_P' and U_N'.

According to the circuit theory, when S is switched off:

$$\frac{U_P}{R_p} + \frac{U_P}{R_{p1} + R_{p2}} = \frac{U_N}{R_N} + \frac{U_N}{R_{N1} + R_{N2}} \tag{7.2}$$

When S is switched on and R_{P1} is bypassed,

$$\frac{U_p'}{R_p} + \frac{U_p'}{R_{p2}} = \frac{U_N'}{R_N} + \frac{U_N'}{R_{N1} + R_{N2}} \tag{7.3}$$

The solution of the simultaneous equation is:

$$R_p = \frac{R_{p^2}\left(R_{p^1} + R_{p^2}\right)\left(U_p U_N' - U_p' U_N\right)}{U_p' U_N \left(R_{p^1} + R_{p^2}\right) - U_p U_N' R_{p^2}} \tag{7.4}$$

$$R_N = \frac{\left(R_{N^1} + R_{N^2}\right)\left(R_{p^1} + R_{p^2}\right) R_{p^2} \left(U_N' U_p - U_N U_p'\right)}{U_p' U_p \left(R_{N^1} + R_{N^2}\right) - R_{p^2}\left(R_{p^1} + R_{p^2}\right)\left(U_N' U_p - U_p U_p'\right)} \tag{7.5}$$

In order to simplify let $R_{P1} = R_{P2} = R_{N1} = R_{N2}$

$$Rp = \frac{2R_{p^1}\left(U_p U_N' - U_p' U_N\right)}{2U_p' U_N - U_p U_N'} \tag{7.6}$$

$$R_N = \frac{2R_{p^1}\left(U_N' U_p - U_N U_p'\right)}{U_p' U_p - \left(U_N' U_p - U_p U_p'\right)} \tag{7.7}$$

From the equation, if we calculate the ground voltage of the positive and negative bus under the state of switch-on and switch-off, we can calculate the ground insulation resistance of the positive and negative bus. This is the most common detection method for insulation resistance.

However, switch S uses the switch tube (MOSFET); electromagnetic interference of electric vehicles will produce certain interference signals during the insulation measurement process, affecting the switch-off signal of the MOSFET, and having a great influence on the insulation resistance calculation and measurement accuracy.

7.2.5.2 Active Ground Detection Method

The principle of the active ground detection method is shown in Figure 7.21. R_+ and R_- are the insulation resistances of the battery positive and negative poles. The main circuit of the insulation monitoring equipment, including four sampling circuits of the sampling resistance (R_1', R_2', R_3', and R_4'), is in the dashed frame in the figure. The four sampling resistances can change

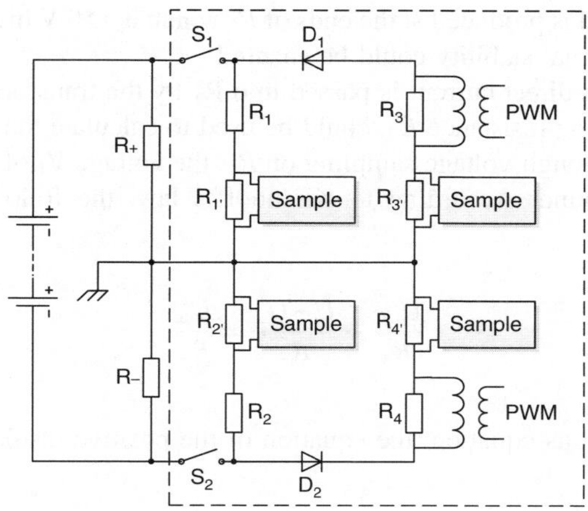

Figure 7.21 The active ground detection method.

appropriately to modify the sampling signals. According to the working principle, four sampling resistances can consist of a series of one fixed resistance and one slider-type rheostat. However, in consideration of the practical operating conditions, the resistance of the slider-type rheostat would be changed because of the interference and this will decrease the measurement accuracy. Therefore, the resistance is fixed according to the road condition after practical debugging.

The biggest feature of this method is the requisition of the active circuit which provides the DC signals produced in the transformer by the control of the CPU using the pulse width modulation (PWM) method. R_1 and R_2 are 50 MΩ divider resistances. D_1 and D_2 are unidirectional diodes. Take D_1 for example, when the PWM signals are input to the transformer, there will be DC current signals on R_3. The DC current signals will flow to R_+ by diode D_1. D_1 will prevent the power current from flowing to R_3, ensure the accuracy of the injection current and protect the transformer from interference, so as to ensure the accuracy of the calculation. D_2 has the same function as D_1. In the practical circuit connection, D_2 has the opposite direction to D_1. When using the circuit of D_2 to measure the ground insulation resistance of the battery's negative end, the ground is at high potential. The produced high potential goes through the ground insulation resistance of the battery's negative end and forms a circuit loop. So it has the opposite direction but the same function as D_1.

The insulation resistance of an electric vehicle is generally a slowly changing parameter while the process of measurement is fast. Therefore, it could be inferred that the insulation resistance is constant. Under this premise, keep relay S_1 and S_2 switched on without PWM1 and PWM2. Through the sampling for R'_1 and R'_2, we can obtain the voltages U_1 and U_2, and then obtain the total voltage.

By comparison between U_1 and U_2, we can know the relationship between the two insulation resistances R_+ and R_-. If $R_- > R_+$, we can calculate the ratio of U_1 and U_2.

Because $R_- > R_+$, in order to ensure the accuracy of monitoring, we could first calculate the R_+ value, and then according to the divider ratio we could calculate the R_- value. First, switching off relay S_2, and then permitting the CPU to generate PWM1 signals. Through the transformer a DC signal is produced at the ends of R_3 which is 150 V higher than the total voltage. In this way the signal stability could be ensured.

Through diode D_1, a direct current is passed into R_1 by the transformer. The voltage V_{CC}, obtained by the sampling resistance R'_3, could be used to calculate the direct current injected by the transformer. Through voltage sampling on R'_1, the voltage V_1 of the positive insulation resistance could be found. According to Kirchhoff's law, the following equation can be obtained:

$$\frac{U_{CC}}{R'_3} + \frac{U - U'_+}{R_-} = \frac{U'_+}{R_+} \tag{7.8}$$

Through the simultaneous equation, the equation of the positive insulation resistance can be obtained:

$$R_+ = \frac{\left(1 + \frac{1}{N}\right) U'_+ R'_3}{U_{CC}} - \frac{R'_3 U}{N U_{CC}} \tag{7.9}$$

After calculation of the R_+ value, we could calculate the negative insulation resistance $R-$ through the ratio N. The above equations are deduced when $R_- > R_+$. The insulation resistance calculation formula for $R_+ > R_-$ can also be derived based on the same principle.

All export method of equation is the same as the above situation. After calculation of the insulation resistance of the positive and negative ends, the two relays should be switched off, and the PWM1 and PWM2 signals should be shut down, preparing for the next calculation. This active ground detection method detects the insulation resistance through the method of current injection. Because the injection voltage is about 150 V higher than the total voltage, even the vehicle has a good insulation performance, ensuring that the current is not small.

The electric system of EVs is a power circuit with high voltage and current. Under normal conditions, the high voltage system is a closed system, completely insulated from the vehicle body, except for electricity leakage caused by the high voltage cable aging problem. In order to ensure the vehicle security and personnel safety, the vehicle must be equipped with an insulation detection and protection device.

7.3 Design of the Battery Equalization Management Circuit

Because of the different resistance of each cell, the different operating conditions, the inconsistency of the initial SOC and the inconsistency of self-charge, the inconsistency of each cell in a battery pack will increase after the battery pack has been used for a long time. In order to optimize the utilization of the battery capacity and energy, it is essential to equalize the battery. According to whether the equalization process consumes energy or not, the equalization control can be classified into the energy dissipative type and the energy non-dissipative type.

7.3.1 The Energy Non-Dissipative Type

The energy non-dissipative type adopts a capacitor and an electric inductance as the storage components, uses the power conversion circuit as the topology foundation and adopts the distributed or centralized structure, which achieves the scheme of unidirectional and bidirectional charging or discharging. Several energy non-dissipative types of battery equalizer are described in the following subsections.

7.3.1.1 Distributed Direct Current Conversion Module

The independent DC converter equalization method connects an independent DC converter to each cell in the battery pack. One side of the converter is connected to the two ends of each cell and the other side is connected to the two ends of the battery pack. When the voltage of a cell is too high, it will release energy by turning on the switch tube. The energy coupling to the secondary side will be absorbed by the battery pack to realize equalization.

The DC transformation can be by either a forward converter or a fly-back converter. The bidirectional DC–DC module can be used to realize the energy bidirectional flow. Figure 7.22 shows the equalization scheme for the bidirectional isolated fly-back DC–DC converter.

The distributed equalization method needs special DC–DC converters. Therefore, this method needs a number of power switch tubes, with a large number of control signals and complex control logics. The cost of the system is high.

7.3.1.2 The Centralized Equalization Converter

The centralized equalization method could reduce hardware consumption with easy control. The use of the unidirectional isolation fly-back converter and the multi-secondary side

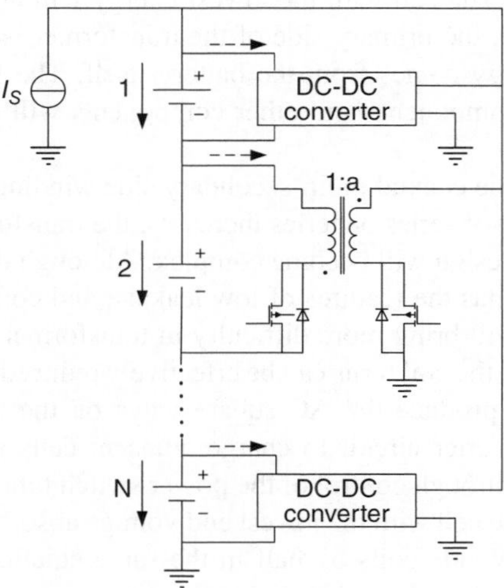

Figure 7.22 Two-way isolated fly-back DC–DC converter equalizer.

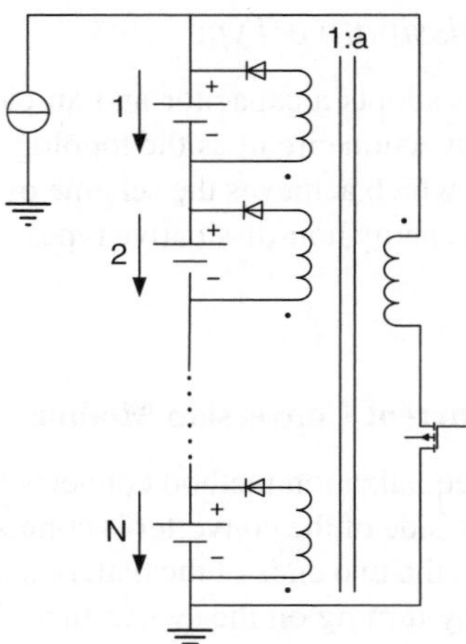

Figure 7.23 Coaxial multiple secondary winding of the transformer equalizer.

windings transformer could realize the centralized equalization control. The primary side of the transformer is connected to the battery pack ends and each secondary side is connected to each cell. An equalization method that uses a fly-back converter and a multi-secondary side windings transformer is shown in Figure 7.23.

In this way, it could send the battery energy to the low energy cell. The charging voltage of the cell can be adjusted. When the system detects the low energy cell, the transformer secondary switch will be turned on and the primary side will form the loop circuit with energy stored in the transformer coils. When it is switched off, the storage energy will be released to the battery cell and the low energy cell will absorb the energy. In the ideal situation, the voltages of the secondary sides are the same. The cell with the lowest energy will absorb the largest energy.

As shown in the scheme, the primary side of the transformer is connected to both ends of the battery pack. The energy comes from the battery itself. The loss of energy through the transformer coils, power components and other components will decrease the energy of the battery pack.

Each secondary side of the coaxial multi-secondary side windings transformer is connected to each cell. As the number of series batteries increases, the transformer secondary side windings will increase and the design will become complex. Although the coaxial multi-secondary side windings transformer has the features of low leakage and controllable leakage magnetic flux, the secondary sides will bring more difficulty in transformer design and realization.

As shown in Figure 7.24, the coil turns can be effectively reduced by half. In this method, the additional DC power will produce the AC square wave on the transformer secondary line through the half-bridge inverter circuit to charge adjacent cells in the positive or negative half-period, respectively. Through control of the power switch tube, the secondary side output voltage can be adjusted. The cell with the lowest end voltage absorbs most energy. This design could reduce the secondary side coils by half in the same equalization output situation and reduce the difficulty in transformer design and realization.

In the fly-back convertor, the system energy and the duty of the switch tube is limited strictly to preventing transformer magnetic saturation and damage to the switch tube and diode. The

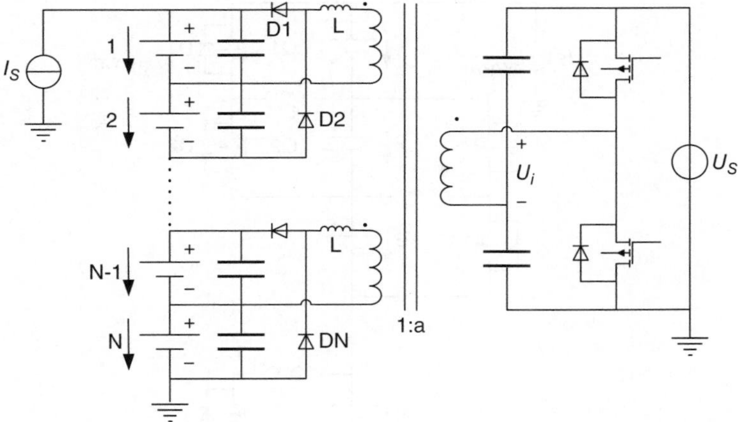

Figure 7.24 The secondary winding of the transformer of a coaxial equalizer.

main disadvantage of the forward convertor is that the transformer secondary voltage will be restricted by the minimum cell voltage. Only one cell can be changed at a time, which limits the equalization efficiency.

The centralized equalization structure is suitable for different kinds of batteries if the number of batteries is not large. Its main features are large line numbers, small total volume, light weight and high maintenance. Once the controller, switch tube or high-frequency transformer are broken, the entire equalization circuit board will have to be replaced.

7.3.1.3 The Non-Dissipative Converter

The non-dissipative converter can transform the charging current of the fully charged energy cells to an adjacent cell to avoid excessive charging. Figure 7.25 shows the scheme of a non-dissipative converter, and this structure could also be used to realize bidirectional equalization.

Each module consists of a pair of power switch tubes, storage inductance, and a pair of diodes. Each end of the module is connected to one cell from the non-isolated half-bridge converter. When one cell voltage reaches the equalization voltage, the relevant switch will be turned on and the energy will be stored in the bypass inductance. When the switch is turned off, the stored energy will flow to adjacent cells and the current direction is the same as the charging current direction of the inductance. This method can directly release the energy of high voltage cells. However, it has small energy transferring capability, complex control strategy and lots of power switches, and large switching losses.

7.3.1.4 Capacitor Equalization

Using a capacitor as the storage component could realize cell equalization and energy transmission to reduce the inconsistency of cell energy. As shown in Figure 7.26, the cell will use the capacitor as the energy storage component. Through the control of a switch, adjacent

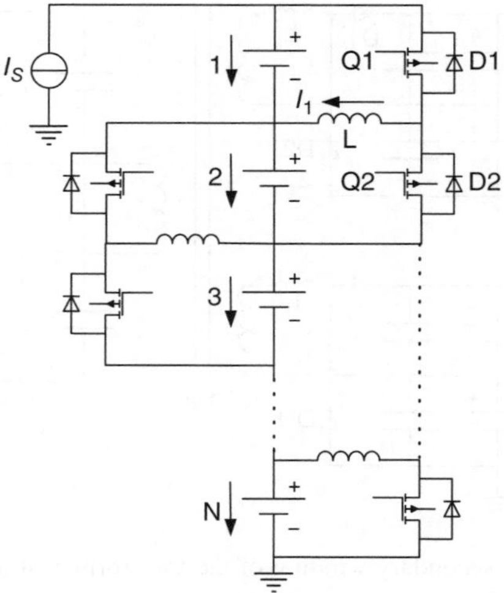

Figure 7.25 Bidirectional non-dissipative current shunt.

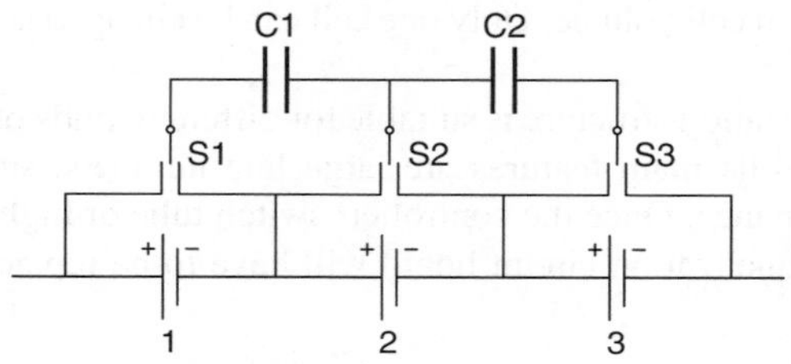

Figure 7.26 A switched capacitor network equilibrium.

cells can transfer energy from the higher voltage one to the lower voltage one to realize the voltage equalization. For *N* batteries, *N* power switches and *N*–1 energy capacitors are required to make up the switch–capacitor network.

When using this method, if high voltage cells and low voltage cells are distributed at different ends of the battery pack, the energy will be transmitted from one side to the other side, this is time-consuming and there is energy loss in each cell when transmitting the energy.

From Sections to 7.3.1.4 we see that the energy non-dissipative equalization scheme uses a unidirectional or bidirectional DC convertor transmitting energy from higher voltage cells to lower voltage cells by the controlling of switches to achieve the balanced state of each cell in a battery pack.

When adopting a voltage balance stratagem, it will avoid the over-charging and under-charging problem to some degree, which will prolong the cycle life of the battery pack.

7.3.2 The Energy Dissipative Type

The energy dissipative type equalization circuit realizes energy equalization using a bypass resistance, as shown in Figure 7.27.

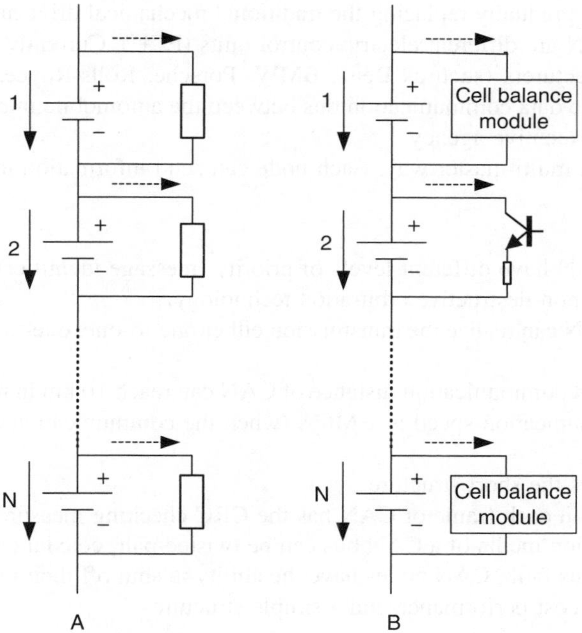

Figure 7.27 Energy equilibrium solution.

In Figure 7.27, topology A uses a battery bypass resistance to realize the self-charging balance. The energy consumption of the resistance is proportional to the cell voltage. Therefore, the highest voltage cell consumes the most amount of energy. This uncontrolled and passive equalization method has large energy loss and low efficiency. Topology B is an improved versionof A, which uses the cell equalization module to control the bypass current. If the cells that reach the equalization voltage are discharged, other bypass loops should remain off.

This method is the simplest one and has the lowest cost, but the equalization current is usually small so that it takes a long time to reach equalization.

7.4 Data Communication

Data communication is the important link for data transmission inside the BMS and data transmission between the BMS and other devices. After sampling the voltage, temperature and current, BMUs should upload these data to the BCU, then send the processed information to the VCU or display and also transmit controlled information to each sub-system. The amount of communication data is very large. The reliability, timely collection, uploading and publishing are the premises for normal performance of vehicles.

7.4.1 CAN Communication

CAN was developed by the German company Bosch and is one of the most widely used buses since the 1993 standardization (ISO 11898-1). It is the mainstream network in automobile electronics control, gradually replacing the traditional mechanical drive and control. The network nodes of CAN are different electric control units (ECU). Currently, some of the world famous car manufacturers (such as Benz, BMW, Porsche, Rolls-Royce, Jaguar, etc.) have adopted CAN as the data communication bus between the automobile internal control system and the detection executive agency.

CAN works in a multi-master way. Each node can send information to other nodes when necessary.

- The nodes of CAN have different levels of priority (message identifier).
- CAN adopts the non-destructive arbitration technology.
- The nodes of CAN can realize the transmission either one-to-one, one-to-many or by universal broadcast.
- The longest direct communication distance of CAN can reach 10 km in remote areas and the maximum communication speed is 1 Mbps (when the communication distance is no more than 40 m).
- The message uses the short structure.
- The information in each frame of CAN has the CRC checking measurement.
- The communication media of a CAN bus can be twisted pair, coaxial cable or optical fiber.
- In case of a serious fault, CAN nodes have the ability to shut off their output automatically;
- CAN has a high cost performance and a simple structure.

The basic structure diagram of a CAN bus is shown in Figure 7.28. The communication data are transmitted into the CAN through special CAN controllers.

The interface hardware circuit of a CAN bus based on CTM is illustrated in Figure 7.29. A large amount of data is required by the management system with good real-time communication. However, if all data are put on the same bus, the high load rate will lead to bus congestion, and poor real-time data communication. According to the function, structure and data type, these data could be sent to different buses.

One common method is to distribute the CAN data of BMS between three buses: the internal CAN bus, the charging CAN bus, and the vehicle CAN bus. Each bus undertakes the relevant output tasks for reliable and real-time transmission of the data.

The internal CAN bus is mainly used for collecting the BMS internal battery information and publishing control commands. The voltage of each cell, temperature, total voltage, current and relay control command are all transmitted on the internal CAN bus. Because data transmission only goes on inside the BMS, a communication protocol can be developed by BMS manufacturers to protect the product.

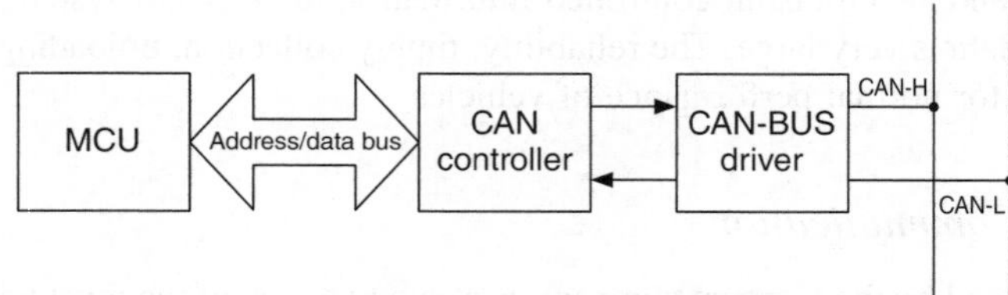

Figure 7.28 Structure diagram of the CAN node.

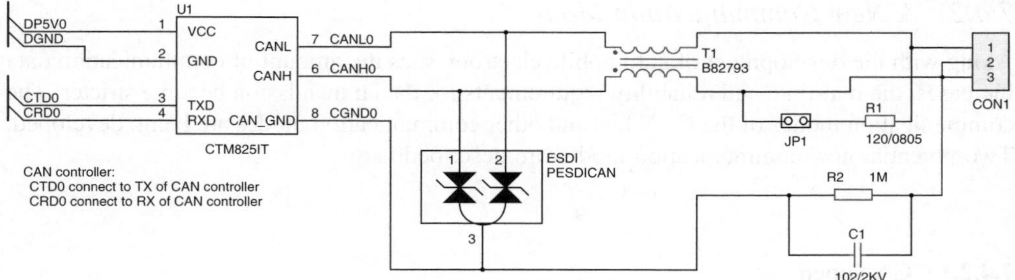

Figure 7.29 The interface hardware circuit of the CAN bus based on CTM.

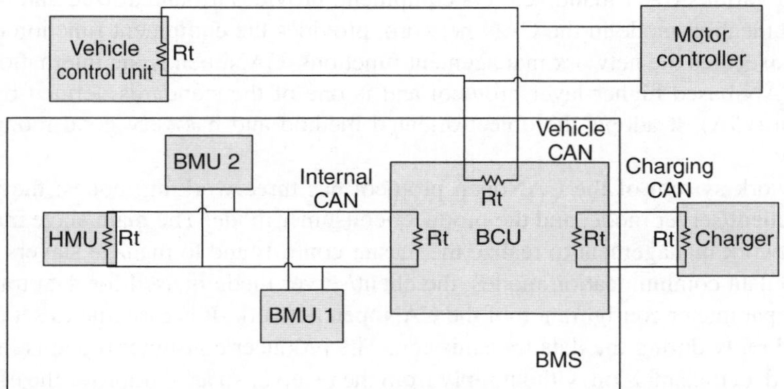

Figure 7.30 Structure diagram of the BMS CAN bus.

The charging CAN bus is mainly used for communication between the BMS and the charger. Its main information includes charger state, start or stop charging, maximum permissible charge current, and so on. In consideration of the generality of electric vehicle charging stations/piles, it is necessary to have the same regulations for the charge interface and the charge protocol. In accordance with the national standard, the communication protocol of the charger and the BMS should produce a message at the handshake stage, parameter configuration stage, charging stage, end charging stage and wrong stage.

Other points related to the charging CAN bus are that there are less data to be communicated and the vehicle cannot be charged during the operating stage. If the vehicle needs a display to show the battery state, the display will become the node of the charging CAN bus, without increasing the load rate of other CAN buses.

The vehicle CAN bus is mainly used for communication between the BMS and the VCU. Besides the BMS and VCU, there are the motor controller and other devices as nodes in this bus, which is the most important data bus. The BMS should transmit information on the SOC, total voltage, and maximum permissible power/current to the bus. Different automobile manufacturers may use different communication protocols.

The structure diagram of a BMS CAN bus is shown in Figure 7.30.

7.4.2 A New Communication Mode

Along with the development of automobile electronics, as the amount of communication data increases, the real-time and reliability requirements for data transmission become stricter. The communication modes of the CAN bus and other communication modes are being developed. Two potential new communication modes are described here.

7.4.2.1 CANopen

The CAN only defines the physical layer and data link layer, without the application layer. It is not complete and needs a high-level agreement to define the 11/29 identifier of the CAN message and 8 byte data use. In industrial automation applications based on the CAN bus, there is needed an open, standardized high-layer protocol, which supports the interoperability and compatibility of various CAN manufacturers equipment, provides a standardized and unified system communication mode in the CAN network, provides the equipment function description mode, and executes the network management functions. CANopen is the internationally standardized CAN-based higher-layer protocol and is one of the standards defined by CAN-in-Automation (CiA). It adopts the object-oriented method and has very good modularity and adaptability.

The network system of the CANopen protocol has three working modes: the main/slave mode, the client/server mode, and the producer/consumer mode. The main/slave mode is suitable for network management to realize the master controls and to manage slavers. As one of the reliable data communication modes, the client/server mode is used for data transmission, namely the parameter configuration of the CANopen network. It is essential to set up the connection and reply during the data transmission. The producer/consumer mode is mainly used for process data transmission without reply from the receiver so as to improve the efficiency of data transmission.

7.4.2.2 FlexRay

FlexRay is a recently developed data communications protocol which is designed to be faster and more reliable than the CAN bus and the LIN (local interconnect network) bus.

The onboard network standard FlexRay has become the base for products of the same type. In future, it will guide the development direction of the control structure in automobile electronics. The FlexRay Consortium has pushed the standardization of FlexRay to a higher level, which is becoming the new automobile internal network communication protocol. FlexRay focuses on meeting the core requirements of the modern automobile industry, including fast data speed, sensitive data communication, overall topology selection and fault-tolerant operation.

Therefore, FlexRay could provide the necessary speed and reliability for the next generation automobile control system. The maximum performance limit of a CAN network is 1 Mbps. The maximum data speed of each channel of FlexRay is 10 Mbps, and the total speed is 20 Mbit/s. Therefore, the band of the FlexRay bus may be 20 times as large as the CAN bus when applied in the onboard network.

FlexRay could also provide reliable features which the CAN network does not have. Especially, the redundant communication ability of FlexRay can copy the network configuration completely by hardware and monitor the process. FlexRay could also provide flexible configurations, supporting all kinds of topology, such as bus, star-type and hybrid-type. A designer could configure the distributed system according to the combination of two topologies or more.

In addition, FlexRay could carry out synchronous and asynchronous data transmission to meet all kinds of system requirements. For example, the distributed control system generally requires the synchronization of data transmission.

In order to satisfy different communication requirements, FlexRay provides static and dynamic communication segments for each communication period. Static communication segments could provide the time delay, while the dynamic communication segment could meet the different band requirements during the operation period. The fixed static segment of a FlexRay message could be transmitted by the use of fixed-time-trigger, while the dynamic segment adopts a flexible-time-trigger.

The mode of FlexRay not only works as a single-channel system like the CAN and LIN networks, but can also work as a dual channel system. A dual channel system could transmit data by the redundant network, one of the important functions for improving system reliability.

7.5 The Logic and Safety Control

When the EV is parked, it is essential to cut off the battery pack from the converter circuit to ensure battery safety and reduce static dissipation. When in operation, it is essential to connect the battery with the circuit for power supply. Meanwhile, when charging, it is necessary to connect the battery pack with the charger. When the temperature reaches a fixed limit, fans or heaters are required to be opened or closed to control the battery temperature. It usually uses the relay to control the switch of fans or heater strips. A typical BMS should detect and control the total voltage positive relay, total voltage negative relay, pre-charge relay, charge relay, fan relay and heater strip relay. Some relays (such as fans relays) do not need to consider the sequence with other relays, and only use the independent, simple method for control but other relays (such as the total voltage positive relay) should consider the sequence with other relays and the control sequence is relatively complex.

During the operating or charging process, in order to ensure safety, it is essential to judge the battery safety according to the battery status. If there are safety issues with the battery pack, the BMS should send the alarm information in time to the controller and display, reminding the VCU and the driver to adopt measures to ensure safe driving.

7.5.1 The Power-Up Control

The BMS is supplied by a low voltage battery which is of limited power. However, the total negative relay, total positive relay, pre-charging relay and charge relay need high power to operate. In order to drive the relay stably, a middle relay is used to drive the target relay. It can not only ensure the isolation of the BMS from high voltage to make the BMS work more stably, but also increases the power of these relays to realize the reliable turn-on. Figure 7.31 is the high voltage relay connection diagram of the EV.

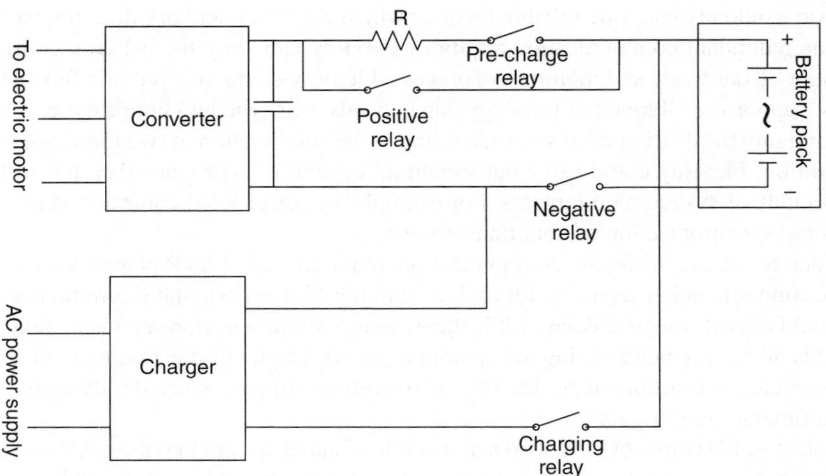

Figure 7.31 Structure diagram of a high voltage relay.

The input end of the converter has a filter capacitor, so it cannot directly connect the battery with the converter loop circuit, otherwise, on the instant of connection, it would produce a large current which would harm the battery. Therefore, it is necessary to connect a suitable resistance in series in the loop during the power-up procedure. When the voltage of the capacitor reaches a fixed value, it can cut off the resistance, which can prevent the battery from being harmed when connected to the circuit loop.

The typical control flow is shown in Figure 7.32.

The smallest pre-charging voltage can be calculated by the following equation:

$$U_1 = U - Ir \tag{7.10}$$

where r is resistance, I the maximum permissible instantaneous current,and U the total voltage.

The time from the start of pre-charge to the end of pre-charge is:

$$T = RC\ln\frac{U_1}{U} \tag{7.11}$$

where R is the pre-charge resistance and C the input capacity of the converter.

7.5.2 Charge Control

According to the regulations of the national standard, the complete charge flow is as shown in Figure 7.33. After completion of handshaking between the BMS and the charger, and the configuration of the battery parameters, it can begin the charging process. The charge flow is as shown in Figures 7.34 and 7.35.

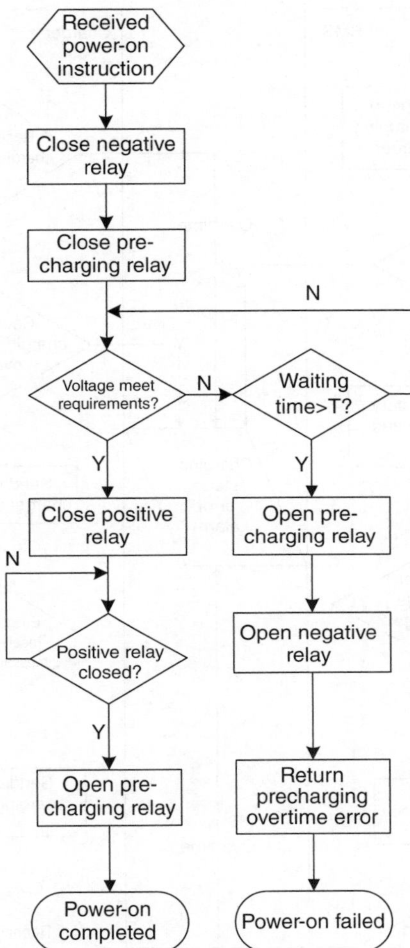

Figure 7.32 The high voltage power-up flow.

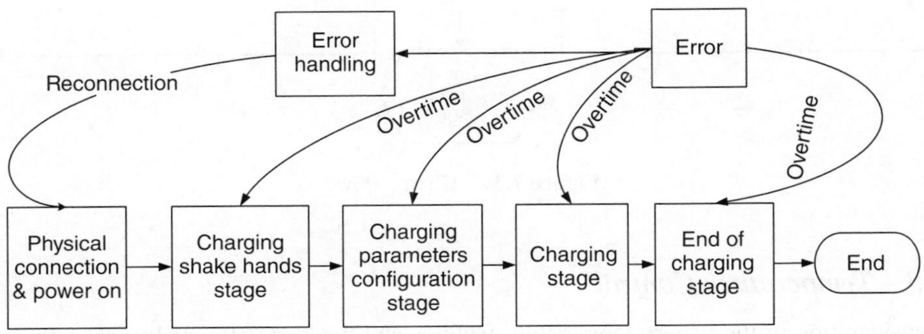

Figure 7.33 Charge flow.

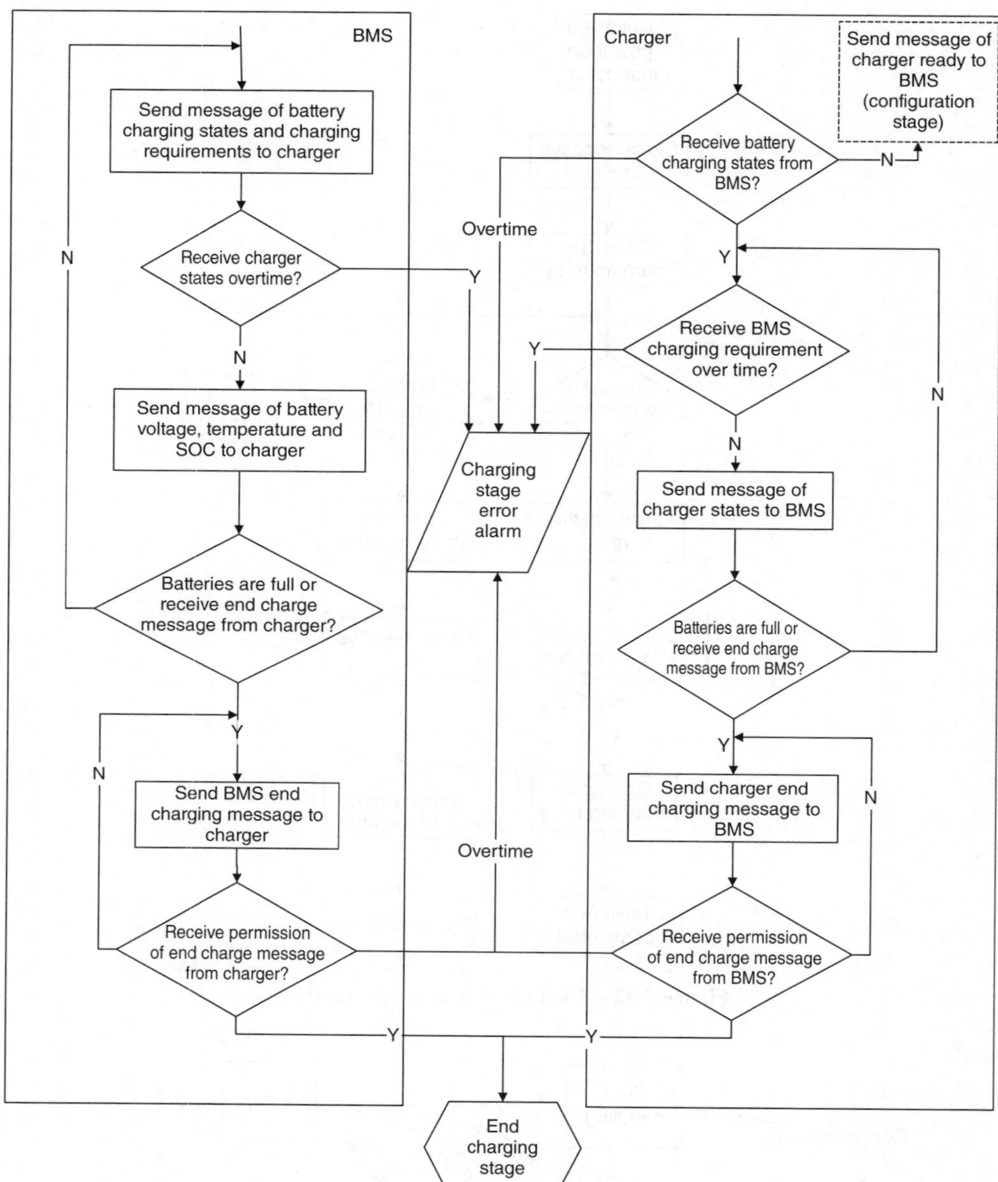

Figure 7.34 Charge flow.

7.5.3 *Temperature Control*

In consideration of the battery temperature features and the hysteretic quality of temperature control, it is essential to set up the control threshold value. Different structures of battery

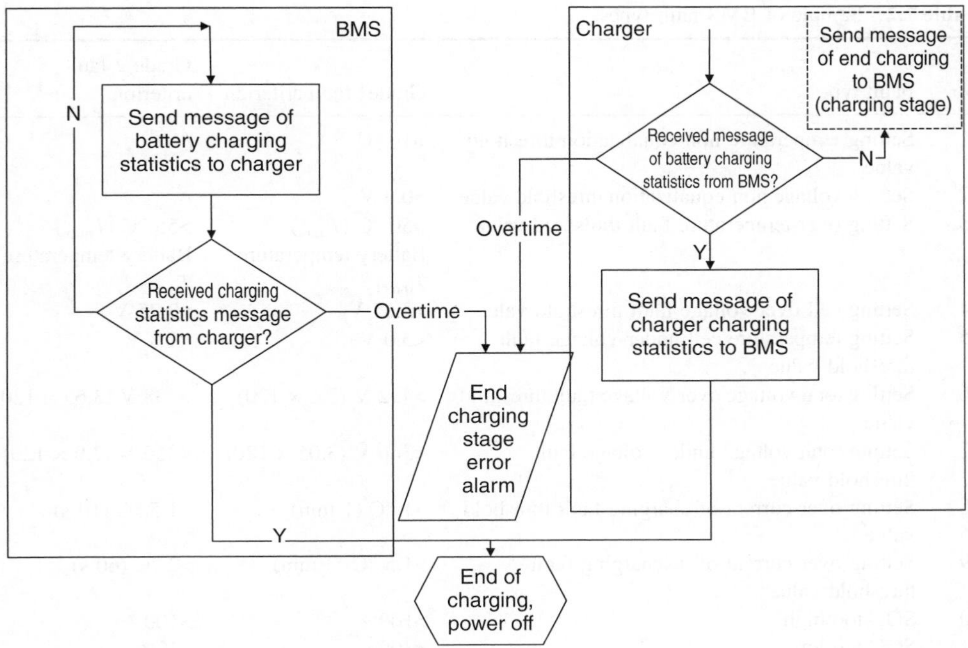

Figure 7.35 End of charge flow.

box have great influence, so a large number of tests or correction analyses are required to define these threshold values.

When the temperature or temperature difference is too high, it is essential to start the fans to perform thermal dissipation and balance.

Generally, there are four threshold values: level 1 and 2 threshold values of high temperature, level 1 and 2 threshold values of low temperature. If the threshold values surpass the level 1 of high temperature, a cooling process for the battery will be executed and the battery output power decreased. If the threshold values surpass the level 2 of high temperature which shows that the battery working temperature reaches the limit value, it is required to stop battery output to avoid thermal runaway which will lead to dangerous accidents. In practical application, the restarting temperature is different from the shut-down temperature. The restarting temperature is lower than the shut-down temperature, which will produce the hysteresis, avoiding the frequent switching of battery relays.

7.5.4 Fault Alarm and Control

Table 7.2 shows two levels of alarm situations. The threshold values of the safety alarm are related to the battery performance, the packing method and the vehicle type. The table lists the alarm threshold values for a ferrous phosphate lithium battery pack, comprising 120 series batteries, with 210 Ah capacity and 384 V rated voltage.

Table 7.2 Sample of BMS fault types.

No.	Fault type	Grade1 fault criterion	Grade 2 fault criterion
1	Setting temperature non-equalization threshold value	>10 °C	/
2	Setting voltage non-equalization threshold value	>0.3 V	/
3	Setting over-temperature fault threshold value	>50 °C (T_{max}) Battery temperature T_{max1}	>55 °C (T_{max}) Battery temperature T_{max2}
4	Setting cell over-voltage fault threshold value	>3.65 V	>3.75 V
5	Setting temperature cell under-voltage fault threshold value	<3.0 V	<2.8 V
6	Setting total voltage over-voltage fault threshold value	>432 V (3.6 × 120)	>438 V (3.65 × 120)
7	Setting total voltage under-voltage fault threshold value	<366 V (3.05 × 120)	<360 V (3.0 × 120)
8	Setting over-current of charging fault threshold value	>1 °C (1 min)	>1.5 °C (10 s)
9	Setting over-current of discharging fault threshold value	>1.5 °C (3 min)	>2 °C (60 s)
10	SOC too high	>100%	>100%
11	SOC too low	<30%	<10%
12	Insulation fault	<500 Ω/V, but > 100 Ω/V	≤100 Ω/V

When the BMS raises the alarm, it should be dealt with according to the different grades of the alarm and the alarm type. For example, when the SOC is too low, the driver could begin to consider charging the battery at a nearby charging station. When the SOC is too low and reaches the grade 2 alarm, the driver should immediately stop the car. The common fault type and safety measurements of the BMS are shown in Table 7.3.

7.6 Testing the Stability of the BMS

The conditions for electric vehicles are changeable, and the working condition for the BMS is poor. So the test on the stability of the BMS is necessary and includes:

7.6.1 Dielectric Resistance

Apply 500 V DC between the live part and the shell in the BMS, and test the dielectric resistance. Generally, the insulation resistance should be no less than 2 MΩ.

Table 7.3 The BMS common safety measurements.

No.	Fault type	Fault grade	Criterion	Measurement	Note
1	Cell battery under-voltage	Grade 1	$V_{Max} > V_{H1}$	Alarm + buzzer	Derating current charging
2		Grade 2	$V_{Max} > V_{H2}$	Cut-off main return circuit	Stop charging
3	Cell battery under-voltage	Grade 1	$V_{Min} < V_{L1}$	Alarm + buzzer	Derating current discharging
4		Grade 2	$V_{Min} < V_{L12}$	Cut-down main return circuit	Stop dischar-ging
5	Battery over-temperature	Grade 1	$T_{max} > T_{H1}$	Alarm + buzzer	Derating current use
6		Grade 2	$T_{max} > T_{H2}$	Cut-down main return circuit	Stop use
7	Over-current	Grade 1	$I > I_{H1}$	Alarm + buzzer	Derating current use
8		Grade 2	$I > I_{H2}$	Cut-down main return circuit	Stop use
9	SOC too high	Grade 1	$SOC > SOC_{H1}$	Alarm + buzzer	Derating current charging
10		Grade 2	$SOC > SOC_{H2}$	Cut-down main return circuit	Stop charging
11	SOC too low	Grade 1	$SOC < SOC_{L1}$	Alarm + buzzer	Derating current discharging
12		Grade 2	$SOC < SOC_{L2}$	Cut-down main return circuit	Stop charging
13	Insulation fault	Grade 1	$R_{min} < R_{L1}$	Alarm + buzzer	Quick detection
14		Grade 2	$R_{min} < R_{L2}$	Cut-down main return circuit	Prohibit use
15	Contactor fault	Grade 2	Non conducting: $I_{c+} < I_{L+}$	Cut-down main return circuit	Prohibit use
16		Grade 2	Over-current: $I_{c+} > I_{H+}$	Cut-down main return circuit	Prohibit use
17	Pre-charging contactor fault	Grade 2	Non-conducting: $I_p + < I_P$	Cut-down main return circuit	Prohibit use
18		Grade 2	Over-current: $I_p + > I_P -$	Cut-down main return circuit	Prohibit use
19	Contactor fault	Grade 2	Non-conducting: $I_{c-} < I_{L-}$	Cut-down main return circuit	Prohibit use
20		Grade 2	Over-current: $I_c - > I_H -$	Cut-down main return circuit	Prohibit use
21	Air-machine fault	Grade 2	Non-conducting: $I_{Fan} < I_{LF}$	Alarm + buzzer	Quick detection
22		Grade 2	Over-current: $I_{Fan} > I_{HF}$	Alarm + buzzer	Quick detection
23	CAN communi-cation fault	Grade 2	Overtime: $CAN > t_H$	Alarm + buzzer	Stop car nearby

(*continued overleaf*)

Table 7.3 (*continued*)

No.	Fault type	Fault grade	Criterion	Measurement	Note
24	Pre-charging fault	Grade 2	Overtime: $t_{Pre} > t_{Ph}$	Cut-down main return circuit	Prohibit use
25	BMS fault	Grade 2	Voltage detection abnormal	Alarm + buzzer	Quick detection
26		Grade 2	Temperature detection abnormal	Alarm + buzzer	Quick detection
27		Grade 2	Current detection abnormal	Alarm + buzzer	Quick detection
28		Grade 2	Total voltage detection abnormal	Alarm + buzzer	Quick detection
29		Grade 2	Insulation monitoring abnormal	Alarm + buzzer	Quick detection
30		Grade2	Storage abnormal	Alarm + buzzer	Quick detection
31		Grade 2	Internal communication abnormal	Alarm + buzzer	Quick detection
32		Grade 2	Real-time clock abnormal	Alarm + buzzer	Quick detection
33	Battery inconsistency fault	Grade 1	Voltage inconsistency: $U_{max} - U_{min} > U_{B1}$	Alarm	Quick maintenance
34		Grade 2	Votage non-match: $U_{max} - U_{min} > U_{B2}$	Cut-down main circuit	Stop nearby
35		Grade 1	Temperature inconsistency: $T_{max} - T_{min} > T_B$	Alarm	Quick maintenance
36	Too full storage	Grade 1	Storage full	Alarm	Quick read
37	Battery magnetic fault	Grade 2	Battery anti-connection	Cut-down main circuit	Prohibit use

7.6.2 Insulation Withstand Voltage Performance

Apply 50–60 Hz sine wave AC voltage to the voltage sampling circuit loop, the test voltage is $(2U + 1000)$ V, U being the nominal voltage, and the duration is 1 min. No electric discharge phenomena such as breaking down or flashover should occur during the test.

7.6.3 Test on Monitoring Functions of BMS

1. Install or connect the BMS according to the normal working requirement, or provide the BMS with a suitable electrical and temperature environment to be monitored through the simulation system. Install the voltage, current and temperature sensor correctly and turn on the BMS.
2. Compare the data from the BMS (cell or module voltage acquisition channels should be no less than 5 points, current collection points should be no less than 2 points, temperature acquisition channel should be no less than 2 points, and they should be distributed reasonably) with the data obtained from the facilities testing, and ascertain the error.

Generally, the parameters required in monitoring the BMS are:

1. Total voltage value ≤ ±1% FSR;
2. Current value ≤ ±0.3 A (≤30 A),≤ ±1% (>30 A);
3. Temperature value ≤ ±2 °C
4. Module voltage value ≤ ±0.5% FSR

7.6.4 SOC Estimation

For the BMS of a BEV or PHEV (plug-in hybrid electric vehicle), the test should include testing that SOC ≥ 80%. For other types of EVs, whether the test should include the condition of SOC ≥ 80% needs to be in accordance with the actual condition. The SOC estimation accuracy should be:

1. When SOC ≥ 80%, error should be ≤6%
2. When 80% > SOC > 30%, error should be ≤ 10%
3. When SOC ≤ 30%, error should be ≤ 6%

7.6.5 Battery Fault Diagnosis

Change input signals such as voltage, current or temperature through the simulation system to meet the toggling conditions of a fault. Monitor the data reported by the communication interface in the BMS. Keep a record of relevant fault items and toggle conditions.

7.6.6 Security and Protection

Change the input signals, such as voltage, current or temperature, through the simulation system to meet the toggle conditions in Table 7.3, and monitor the software and hardware response of the BMS.

7.6.7 Operating at High Temperatures

Put the BMS into the high-temperature cabinet (at first the cabinet is at a normal temperature), and get it into the working state. Keep it working for 2 h when the temperature reaches 65 ± 2 °C. Record the parameters of the battery system measured by the BMS during the test, and analyze the error. The BMS should work normally during and after the test, and should meet the requirements.

7.6.8 Operating at Low Temperatures

Put the BMS into the low-temperature cabinet (at first the cabinet is at a normal temperature), and get it into the working state. Keep it working for 2 h when the temperature reaches −25 ± 2 °C. Record the parameters of the battery system measured by BMS during the test, and analyze the error. The BMS should work normally during and after the test, and should meet the requirements.

7.6.9 High-Temperature Resistance

Put the BMS into the high-temperature cabinet (at first the cabinet is at a normal temperature), and keep it there for 4 h when the temperature reaches 85 ± 2 °C. Record the parameters of the battery system measured by the BMS after the temperature gets to the normal state, and analyze the error. The BMS should work normally after the test, and should meet the requirements.

7.6.10 Low-Temperature Resistance

Put the BMS into the low-temperature cabinet (at first the cabinet is at a normal temperature), and keep it there for 4 h when the temperature reaches –40 ± 2 °C. Record the parameters of the battery system measured by BMS after the temperature gets to the normal state, and analyze the error. The BMS should work normally after the test, and should meet the requirements.

7.6.11 Salt Spray Resistance

The test should be done according to the provisions of *Basic Environmental Testing Procedures for Electric and Electronic Products, Salt Spray Test Approach (GB/T 2423.17)*. The BMS should be installed in the test chamber in accordance with the actual installation state or similar conditions, and the connector should be in the normal state. The test duration is 16 h. After the test, put the BMS at the normal temperature for 1–2 h until it reaches this temperature, and analyze the error. The BMS should work normally after the test, and should meet the requirements.

7.6.12 Wet-Hot Resistance

The test should be done according to the provisions of *Basic Environmental Testing Procedures for Electric and Electronic Products, Alternating Wet and Heat Test Method (GB/T2423.4)*. The test duration is two cycles (48 h). After the test, put the BMS at normal temperature for 1–2 h until it reaches this temperature, and analyze the error of the parameters of the battery system measured by the BMS. The BMS should work normally after the test, and should meet the requirements.

7.6.13 Vibration Resistance

The test should be done according to the provisions of *Basic Technical Conditions for Automobile Electrical Equipment (QC/T 413-2002)*. The BMS should be tested when moving from up to down, left to right and front to back, with each test having a duration of 8 h. The BMS should be confined to the vibration test-bed in a normal installation state and tested in the off state. The vibration of the vibratory testing machine should be sine wave. The distortion of the acceleration wave should be less than 25%. The testing instrumentation should be installed on the BMS or on its jig.

Conditions of sweeping frequency test:

Sweeping frequency range: 10–500 Hz
Amplitude or acceleration: when 10–25 Hz, amplitude 0.35 mm; when acceleration 25–500 Hz, amplitude 30 m s^{-1}
Sweeping frequency rate: 1 oct/min

After the test, analyze the error of the battery system parameters as measured by the BMS. The BMS should work normally after the test, and should meet the requirements.

7.6.14 Resistance to Power Polarity Reverse Connection Performance

Reverse the input voltage connection and connect the BMS to the power supply for 1 min. After the test, keep the BMS power supply in a normal state and check whether it works normally. If it does, analyze the error of the battery system parameters measured by the BMS. The BMS should work normally after the test, and should meet the requirements.

7.6.15 Electromagnetic Radiation Immunity

The test should be done according to the provisions of *Limits and Methods of Testing for Immunity of Electrical/Electronic Sub-assemblies in Vehicles to Electromagnetic Radiation (GB/T 17619-1998)*. The test frequency is 400–1000 MHz. Analyze the error of the battery system parameters measured by the BMS. The BMS should work normally after the test, and should meet the requirements.

7.7 Practical Examples of BMS

Electric vehicles include pure electric vehicles, hybrid vehicles and fuel-cell vehicles. Here we mainly discuss the examples of BMS on lithium-ion battery powered EVs and HEVs, including a pure electric bus, a hybrid electric bus, a pure electric passenger car vehicle and a hybrid passenger car vehicle.

7.7.1 Pure Electric Bus (Pure Electric Bus for the Beijing Olympic Games)

The pure electric buses used for the Beijing Olympic Games (Figure 7.36), running on the three roads of the Olympic village inner ring with low noise and no exhaust gas, were "zero emission" clean vehicles with a long cycle life. Every charging ensured that the bus could run for 130 km on the bus line with air conditioning on and at a speed of 80 km h^{-1} with 80 passengers on board. The bus adopts the dispersion rapid replacement method. The integrated electric drive system of an integrated AC motor and the automatic transmission system improve the efficiency of the motor and prolong the cycle life of the batteries.

The pure electric bus for the Beijing Olympic Games used LMO batteries. The normal voltage of the cell is 3.6 V. The capacity of the battery pack is 360 Ah. The whole battery system consists of 104 series batteries distributed in different locations of the vehicle in 8 packs.

Figure 7.36 Pure electric bus for Olympic.

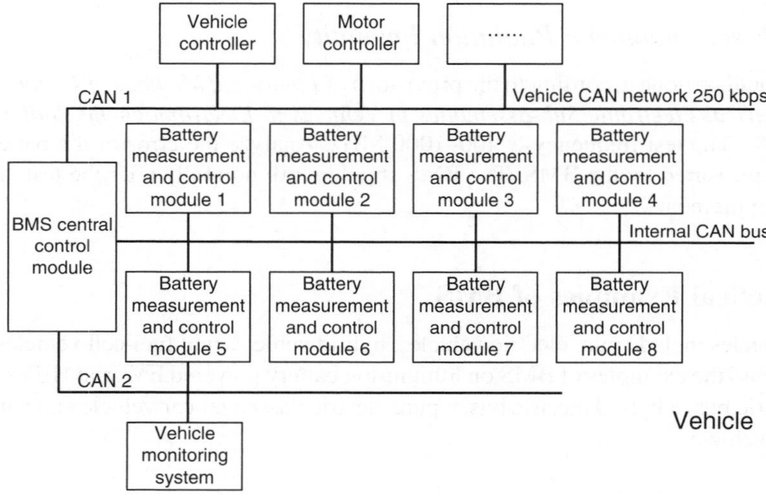

Figure 7.37 The structure of BMS in the vehicle mode.

Because it uses the combination charge model of rapid replacement, charging in different packs and charging for emergency use, the BMS operating mode can be classified into modes such as the onboard mode, the rapid replacement mode and the emergency charging mode.

7.7.1.1 The Onboard Mode

The structure of the BMS in the onboard mode is shown in Figure 7.37. The BCU delivers real-time information on the battery state to the VCU and the motor control unit (MCU) through a CAN1 high speed bus so the bus can adopt a reasonable strategy to finish its operation task and prolong the cycle life of the batteries. Meanwhile, the BCU delivers detailed information on the battery pack to the vehicle monitoring system through a CAN2 high speed bus to

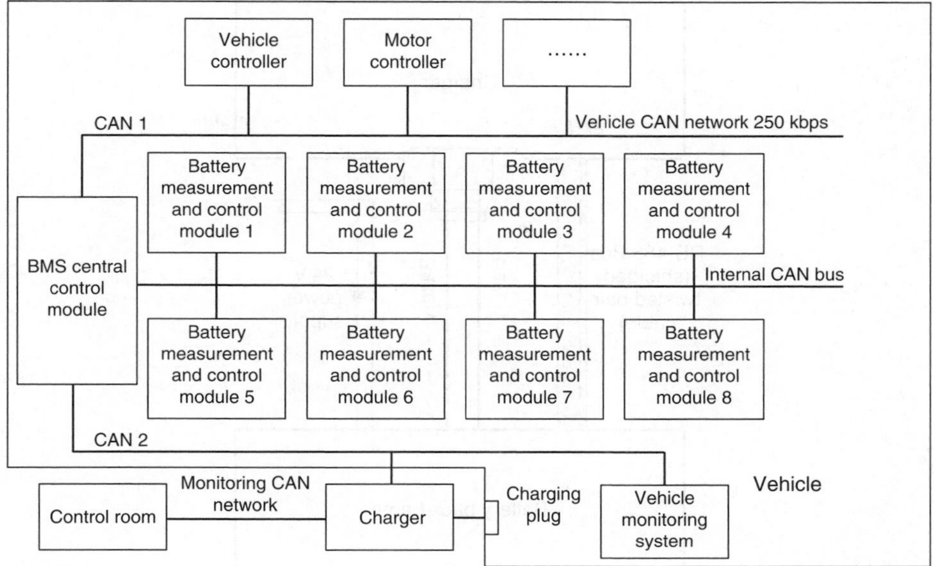

Figure 7.38 The structural diagram of BMS in emergency charging.

realize the functions of showing states and failure warnings, and providing evidence for maintenance and replacement.

7.7.1.2 The Emergency Charging Mode

The topological structure of the BMS in the emergency charging mode is shown in Figure 7.38. In this mode, the connector of the charger is inserted directly into the socket of the bus. During charging, the vehicle CAN2 high speed bus adds a charging equipment node. The charger gets into the real-time state of the battery through the CAN2 to adjust the charging strategy to ensure safety during the charging process.

7.7.1.3 The Rapid Replacement Mode

Charging different packs is adopted in the rapid replacement mode and the structure of the BMS is shown in Figure 7.39. The unloaded parts and the BMU are all removed from the bus, while the BCU is still on the bus in the rapid replacement charging mode. So the battery pack no longer has the functions of current detection, SOC estimation and communication ability with the charger through the CAN bus. The BMU communicates with the charger by the RS-485 bus. The BMU transfers information, such as cell voltage, temperature and faults, to the charger in real-time to ensure safety when charging.

During the rapid replacement charging mode, the battery packs that are unloaded from the bus are charged on different charging platforms. If the battery pack in the bus needs replacing, not all the battery packs with the same size can be put in. Because each BMU in the battery packs has a unique address, when the BMUs with the same address are put in one bus, a

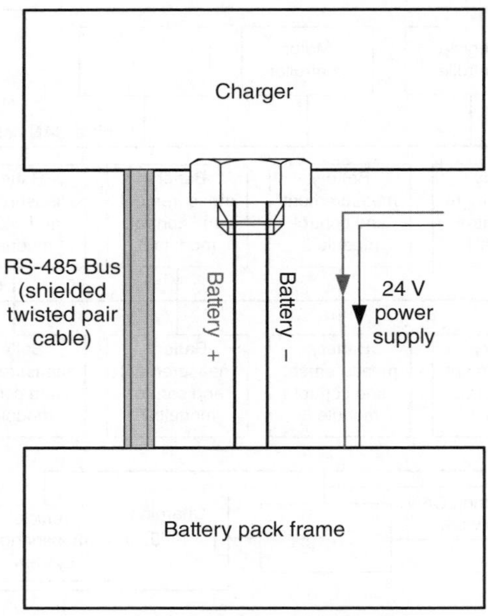

Figure 7.39 The structure of BMS in the rapid replacement mode.

collision will happen so the communication cannot be performed. So the battery packs should be labeled with their address and the addresses should be in line with the corresponding BMU. When the battery packs are put back into the bus, they are grouped based on the addresses to ensure that each group has a unique BMU address from one to eight. In order to find where the failure happens most frequently, the location of the packs should be recorded and labeled. When putting in the packs, it is necessary to ensure the address on the pack corresponds to the location number. In an emergency situation, the packs can be temporarily addressed in a simple way, but when labeling the BMU, make sure the address corresponds with the pack number and the location number.

The insulation of the BMS is completed in the BCU, so when charging different packs insulation detection cannot be performed. In this situation, such measures as connecting to the ground, laying rubber insulation, and the wearing of insulating gloves by charging workers are adopted to solve the insulation issues.

The BMS of the pure electric bus for the Beijing Olympic Games can realize the following functions: the measuring of cell voltage, battery temperature, the working current of the battery pack and insulation resistance; the controlling of the cooling fan; recording the time of charging and discharging; estimation of the battery pack SOC; fault analysis of batteries and online alarm; evaluation of each individual pack and on-board communication to provide the necessary battery data for the whole bus (through CAN-bus 1); communicating with on-board charging equipment to send battery information to the display panel (through CAN-bus 2); communicating with the charger to guarantee charging safety (through RS-485); the initialized function of the BMS can be realized with simple equipment, meeting the requirements of charging the battery quickly and regrouping the battery packs. It can operate stably within −25 to

75 °C and ensure that the voltage measuring error of the BMS is less than 0.5% within 3–6 V, the temperature measuring error is less than 1 °C, the measurement error of the current is less than 0.5% within the range –300 A to +300 A, and the estimation error of the SOC is less than 8%.

7.7.2 Pure Electric Vehicles (JAC Tongyue)

The maximum speed of the second generation of the Tongyue BEV (Figure 7.40) is 100 km h^{-1} and the maximum distance travelled per charge is 100 km. As for the charging time, the Tongyue electric car can be fully charged within 2 h via the special charging pile, and within 6 h with the household 220 V power supply. It has been equipped with a battery thermal management system to further strengthen the safety.

The car uses LFP batteries. The capacity of one cell is 10 Ah. The battery is packed into 2 packs, using 5 parallel and 95 serial connections. The rated voltage of the battery pack is 304 V and the capacity is 50 Ah. The BMS structure of the car is seen in Figure 7.41.

The main functions of the BMS include cell voltage detection, battery temperature detection, the operating current detection, insulation testing, thermal management, battery pack SOC estimation, battery failure analysis and online alarm, communication with the vehicle equipment, contactor control, and charging management.

7.7.3 Hybrid Electric Bus (FOTON Plug-In Range Extended Electric bus)

The FOTON AUV plug-in extended-range electric city bus (Figure 7.42) uses the series hybrid drive system. The drive system includes the engine and the AC induction motor, and it is suitable for low speed in urban areas and frequent starting and stopping conditions.

Figure 7.40 Pure electric vehicles (JAC Tongyue).

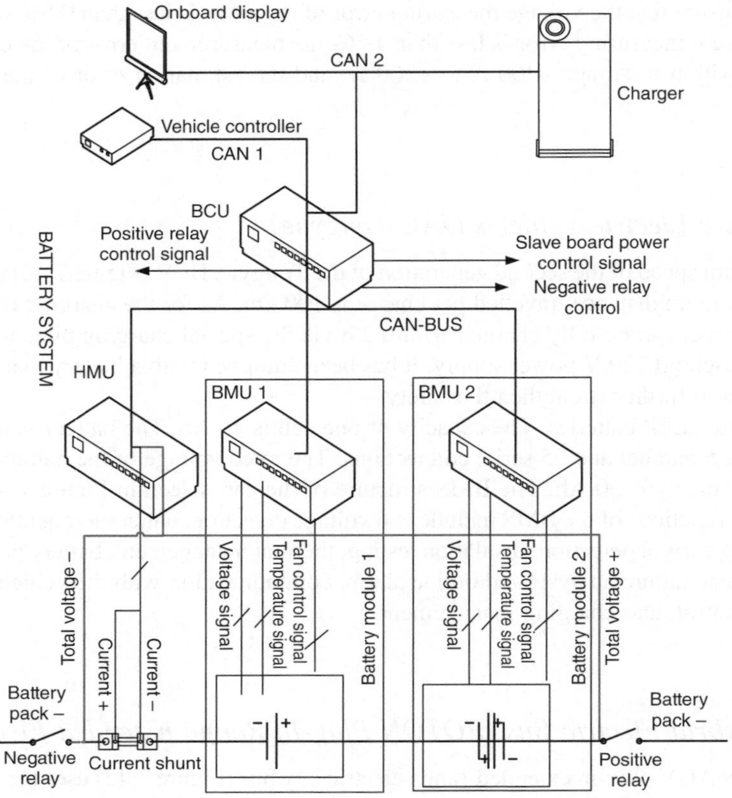

Figure 7.41 The BMS structure of the Tongyue BEV.

Figure 7.42 The FOTON hybrid bus.

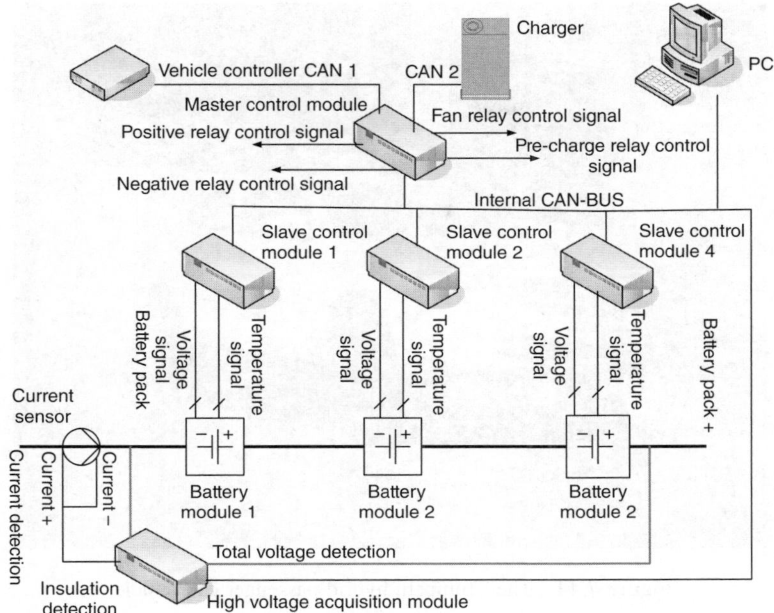

Figure 7.43 The structural diagram of the BMS of the AUV hybrid bus.

The capacity of the LMO cell used in the bus is 35 Ah. The battery system has a total voltage of 651 V and a capacity of 70 Ah. The batteries are divided into 4 packs with 2 parallel connections and 95 serial connections. The batteries are located in different areas of the bus. The structure of the BMS is shown in Figure 7.43.

The functions of the BMS include: cell voltage detection, battery temperature detection, the operating current detection, insulation testing, thermal management, the estimation of the battery pack SOC, battery discharging capacity counting, battery failure analysis and online alarm, communication with the vehicle equipment, the battery high voltage control, charging and discharging management, and recording the latest battery operation data.

7.7.4 Hybrid Passenger Car Vehicle (Trumpchi)

The Trumpchi (Figure 7.44) has an ICE as well as an integrated starter generator (ISG). At the starting stage, the ISG can replace the ICE temporarily to start the vehicle and reduce idle loss and pollution of the ICE; in normal driving, the ICE can drive the vehicle and the ISG motor has the function of power balance, which can keep the ICE in the best working condition; in the braking process, the ISG can regenerate electric energy to charge the batteries, thus recycling the brake energy and improving energy efficiency.

The battery system of the car consists of 90 LFP cells, each cell has a capacity of 6.5 Ah; all the batteries are grouped in one pack, and the system has a nominal voltage of 288 V. The structural diagram of the BMS is shown in Figure 7.45.

Figure 7.44 The Trumpchi hybrid passenger car vehicle.

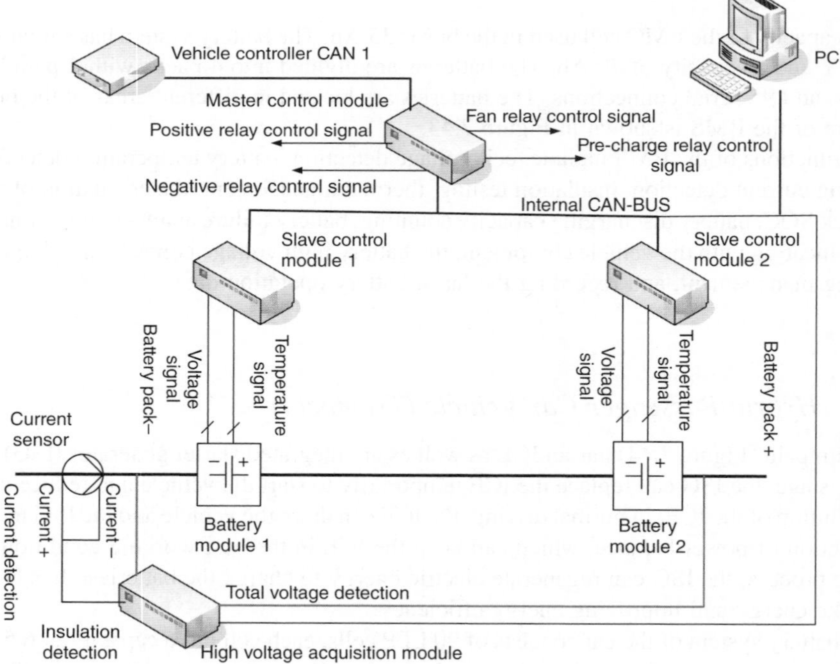

Figure 7.45 Structural diagram of the BMS of the Trumpchi hybrid passenger car vehicle.

The BMS realizes the traditional battery state measurement and the estimation of parameters, as well as the balancing functions of batteries that can reduce maintenance and prolong the service life of the batteries.

7.7.5 The Trolley Bus with Two Kinds of Power

The trolley bus with two kinds of power takes trolley wires as the source of power, making use of an onboard charger for battery pack charging. The power is supplied by onboard batteries when crossing districts without trolley wires. One kind of trolley bus with two kinds of power is shown in Figure 7.46.

The configuration of the traction power supply system is shown in Figure 7.47.

Figure 7.46 Trolley bus with two kinds of power.

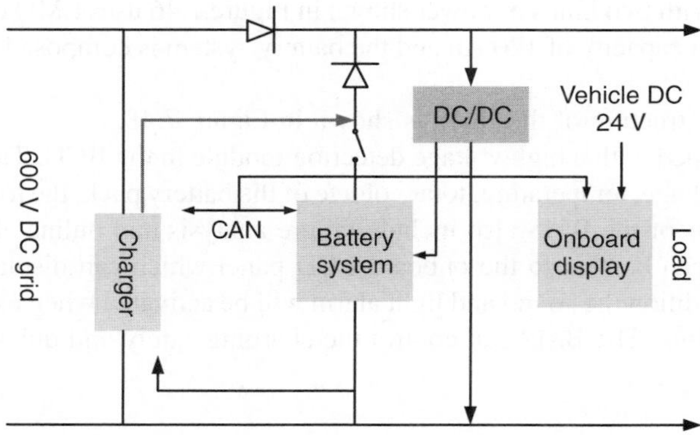

Figure 7.47 The system configuration of a trolley bus with two kinds of power.

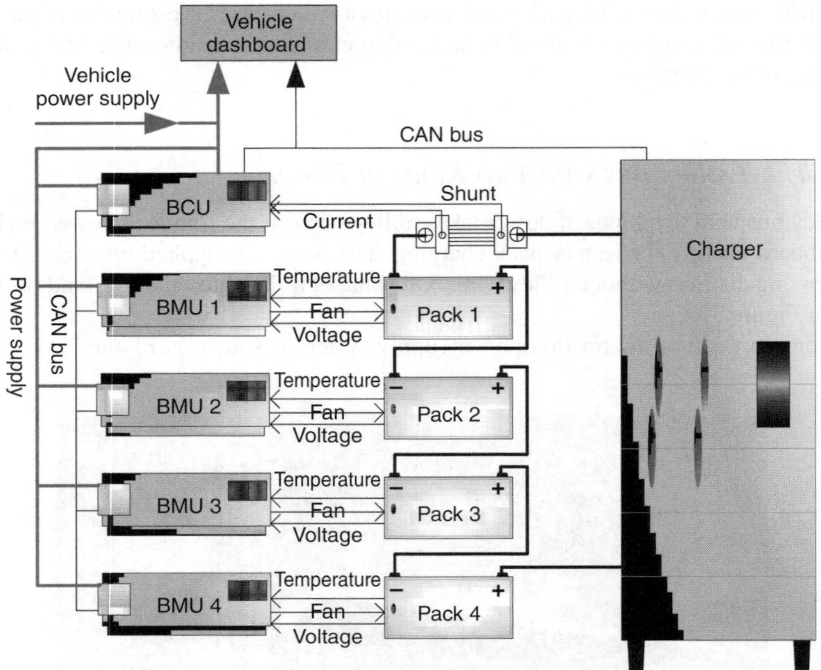

Figure 7.48 The structural diagram of the BMS of a trolley bus with two kinds of power.

The development of traction battery technology and charging online technology makes the further development and application of the double power source trolley bus possible. The new traction battery, which is of high specific energy, fully sealed and maintenance-free, can provide highly efficient traction power for electric buses, thus trolley buses can operate without power lines set along their routes. Meanwhile, the technical restriction on electric buses is removed because of the establishment of charging stations. Electric buses with two kinds of power that can be charged online are becoming more widely-used green traffic tools.

The trolley bus with two kinds of power shown in Figure 7.46 uses LMO cells as the traction battery, which has a capacity of 120 Ah and the battery system is composed of 4 packs, that is, 104 batteries.

The topological structure of the BMS is shown in Figure 7.48.

The bus is equipped with a high voltage detection module in the BCU. The whole BMS can measure the cell voltage, temperature, total voltage of the battery pack, the total current, and the SOC. The functions of the BMS also include failure analysis and online alarm. Major information and faults can be sent to the onboard LCD panel which can display the information correctly, and in addition the sound and light alarm will be activated when exceeding threshold values for a CAN bus. The BMS can control the charging safely and quickly.

Index

Fundamentals and Applications of Lithium-ion Batteries in Electric Drive Vehicles, First Edition.
Jiuchun Jiang and Caiping Zhang.